CAMBRIDGE TRACTS IN MATHEMATICS
General Editors
B. BOLLOBAS, H. HALBERSTAM, C.T.C. WALL

96: Dependence with complete connections and its applications

MARIUS IOSIFESCU
SERBAN GRIGORESCU
Centre of Mathematical Statistics, Bucharest

Dependence with complete connections and its applications

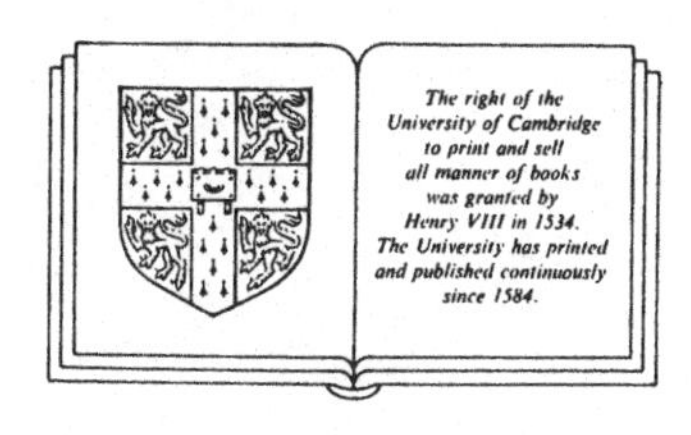
The right of the University of Cambridge to print and sell all manner of books was granted by Henry VIII in 1534. The University has printed and published continuously since 1584.

CAMBRIDGE UNIVERSITY PRESS
Cambridge
New York   Port Chester
Melbourne   Sydney

CAMBRIDGE UNIVERSITY PRESS
Cambridge, New York, Melbourne, Madrid, Cape Town, Singapore, São Paulo, Delhi

Cambridge University Press
The Edinburgh Building, Cambridge CB2 8RU, UK

Published in the United States of America by Cambridge University Press, New York

www.cambridge.org
Information on this title: www.cambridge.org/9780521101806

First published 1990
This digitally printed version 2009

A catalogue record for this publication is available from the British Library

ISBN 978-0-521-33331-3 hardback
ISBN 978-0-521-10180-6 paperback

The original hardback edition of 1990 was dedicated to the memory of Octav Onicescu (1892-1983) and Gheorghe Mihoc (1906-1981), co-founders of the subject. This new paperback edition is published in memory of my co-author and friend Serban Grigorescu (1946-1997) and of our former colleague Adriana Berechet (1939-2006) who took a great interest in our book and suggested several corrections included in the present edition.

Contents

Preface

Le moment actuel d'un corps vivant ne trouve pas sa raison d'être dans le moment immédiatement anterieur ... il faut y joindre tout le passé de l'organisme, son hérédité, enfin l'ensemble d'une très longue histoire.

H. Bergson *L'évolution créatrice*, p. 21. Alcan, Paris, 1907.

The theory of dependence with complete connections was initiated by a note published in 1935 in *Comptes Rendus de l'Académie des Sciences de Paris* by the Romanian mathematicians Octav Onicescu and Gheorghe Mihoc. Intended to be a non-trivial extension of Markovian dependence, in the spirit of the Bergsonian quotation above, during its development this theory was condescendingly accompanied by the view that it does not go much beyond its celebrated Markovian predecessor or even reduces to the latter. We hope that the present work will help to rule out this misunderstanding, showing, in particular, that even the associated Markov chain is of a special type and cannot be brought under established theories for Markov chains.

The Romanian version of this work, published in 1982, was the third monograph devoted to dependence with complete connections following *Processes with Complete Connections* by G. Ciucu and R. Theodorescu (1960) and *Random Processes and Learning* by M. Iosifescu and R. Theodorescu (1969). As a consequence, our 1982 book aimed at being a state-of-the-art survey building on the results obtained during the period 1969–1982. Within that period the applicability of the concept was largely extended, a lot of work was done on the associated Markov chain, the classical limit theorems got a functional form, and the rôle played by dependence with complete connections in the metric theory of continued fractions was entirely clarified. All these results appeared in the five chapters of the 1982 book.

The present book is a thoroughly revised, updated and somewhat expanded version of our 1982 work. Scarcely a page has escaped either amendment, deletion or addition. Moreover, parts of Chapters 3,4, and 5 have been completely rewritten, and a new section on piecewise monotonic transformations has been included in Chapter 5. This topic, which has undergone an explosive development in recent years, is the main newcomer to the field of dependence with complete connections.

As with the Romanian version, to make the book self-contained, we have included three appendices containing those notions and results from measure theory, functional analysis, and dependent random variables that are basic to the theory of dependence with complete connections.

The concluding notes and comments aimed at tracing the historical developments of the subject have also been revised and updated to suit the new contents.

The bibliography mainly covers the period 1969–1988. For a more complete bibliographic picture the reader should also consult the reference lists of the 1960 and 1969 volumes mentioned above.

The authors hope that the present work will testify both to the importance and vitality of the concept of dependence with complete connections in its sixth decade of existence and to the fact that interesting work is being done and has yet to be done in the field.

Acknowledgements

Much of our original work included in this book was done in the framework of our association with the Centre of Mathematical Statistics of Bucharest under generous financial support extended over years from the National Committee for Science and Technology of Romania.

Thanks are due to our colleagues and friends Joel E. Cohen (New York), Harry Cohn (Melbourne), Allan Gut (Uppsala), Sofia Kalpazidou (Salonika), Bo Henry Lindqvist (Trondheim), Magda Peligrad (Cincinnati), Helmut Pruscha (Munich), and Petre Tăutu (Heidelberg) for invaluable help with bibliographic materials.

We wish to acknowledge the technical help we have received from Sergiu Celac, Rodica Culcer, Hélène Ely, and Adriana Grădinaru.

We wish to express our deep gratitude to Joel E. Cohen without whose perseverence and interest this English version might not have existed.

Our colleague and friend Gheorghe Popescu read a first version of the manuscript and with his crystal-clear mind detected some inaccuracies and slips. Expressing our great indebtedness to him, we wish to make it clear that any remaining errors are ours.

Finally, we are especially grateful to our wives Ştefania Iosifescu and Monica Grigorescu for their constant support and patience.

Bucharest Marius Iosifescu
Serban Grigorescu

Abbreviations and notation

a.e.	almost everywhere
a.s.	almost surely
GRSCC	generalized random system with complete connections
iff	if and only if
MC	Markov chain
pw.m.t.	piecewise monotonic transformation
RSCC	random system with complete connections
r.v.	random variable
t.p.f.	transition probability function
w.r.t.	with respect to
$\mathbb{N}^*$	$\{1,2,\ldots,n,\ldots\}$
$\mathbb{N}$	$\{0,1,2,\ldots,n,\ldots\}$
$-\mathbb{N}$	$\{\ldots,-n,\ldots,-2,-1,0\}$
$\mathbb{Z}$	$(-\mathbb{N})\cup\mathbb{N}^*$
$\mathbb{R}$	the collection of all real numbers
$\mathbb{R}_+$	$\{a\in\mathbb{R}: a\geqslant 0\}$
$\mathbb{R}_-$	$\{a: a<0\}$
$\mathbb{C}$	the collection of all complex numbers
$[a]$	the integer part of $a\in\mathbb{R}_+$
$\{a\}$	the fractionary part of $a\in\mathbb{R}_+$
Φ	standard normal distribution function
χ_A	the indicator function of the set A

$\mathcal{P}(X)$, 241
$\mathcal{B}, \mathcal{B}_M (M\subset\mathbb{R})$, 241
$\mathcal{B}_X$, 251
$\sigma((f_i)_{i\in\Lambda})$, 243
$\|\mu\|$, μ a measure, 244
$\mathbf{E}(f)$, 248
$\mathbf{P}\equiv\mathbf{Q}, \mathbf{P}\ll\mathbf{Q}$, 248
$\mathbf{B}$, 254
$B(W), B(W,\mathcal{W})$, 259
$\mathrm{ba}(W,\mathcal{W})$, $\mathrm{ca}(W,\mathcal{W})$, 259
$\mathrm{ca}(W)$, $\mathrm{rca}(W)$, 259
$C(W), C_r(W)$, 259
$\|\cdot\|_L$, 77, 264
$|\cdot|$, 259
$\|\cdot\|_v$, 265
$\|\cdot\|_1$, 265
BV_λ, L^1_λ, 198
$\{(W,\mathcal{W}),(X,\mathcal{X}),u,P\}$, 5
P^n_j, 6
$\mathbf{P}^\infty_j$, 40, 42, 46
Q^n, 8, 69
Q_n, 71
$\mathbf{Q}^\infty$, 86
$C=C_r([0,1])$, 254

Numbering Section 1.2 means Section 2 of Chapter 1; Subsection 1.2.3 means Subsection 3 of Section 1.2; (1.2.3) means equation 3 in Section 1.2; Problem 1 means Problem 1 of the current chapter; Section A1.2 means Section 2 of Appendix 1. Smith (1980, p. 1) means page 1 of the item Smith (1980) of the reference list.

Note Positive means >0, non-negative means ≥ 0, denumerable means either finite or countably infinite, countable means countably infinite, log denotes the natural logarithm.

Introduction

Let us consider a finite set $X = \{1, \ldots, m\}$, a stochastic matrix

$$\mathbf{P} = \begin{pmatrix} p_{11} & p_{12} & \cdots & p_{1m} \\ p_{21} & p_{22} & \cdots & p_{2m} \\ \vdots & \vdots & & \vdots \\ p_{m1} & p_{m2} & \cdots & p_{mm} \end{pmatrix},$$

and a probability distribution $\mathbf{p}_0 = (p_1, \ldots, p_m)$ on X. These elements completely define a Markov chain (an MC for short) $(\xi_n)_{n \in \mathbb{N}^*}$ with state space X, transition matrix $\mathbf{P}$ and initial probability distribution $\mathbf{p}_0$. More precisely, it is possible to construct a probability space $(\Omega, \mathscr{K}, \mathsf{P}_{\mathbf{p}_0})$, and a sequence of X-valued random variables $(\xi_n)_{n \in \mathbb{N}^*}$ defined on Ω, such that

$$\begin{aligned} &\mathsf{P}_{\mathbf{p}_0}(\xi_1 = i) = p_i, \\ &\mathsf{P}_{\mathbf{p}_0}(\xi_{n+1} = i_{n+1} \mid \xi_1 = i_1, \ldots, \xi_n = i_n) \\ &\quad = \mathsf{P}_{\mathbf{p}_0}(\xi_{n+1} = i_{n+1} \mid \xi_n = i_n) = p_{i_n i_{n+1}} \end{aligned}$$

for all $n \in \mathbb{N}^*$, $i, i_1, \ldots, i_{n+1} \in X$; the last equation above indicates the Markovian nature of the sequence $(\xi_n)_{n \in \mathbb{N}^*}$.

It is, however, possible to take a different look at an MC (cf. Iosifescu (1980, 2.1.3)). Let us consider the set

$$W = \left\{ \mathbf{p} = (p_1, \ldots, p_m) : p_i \geqslant 0, 1 \leqslant i \leqslant m, \sum_{i=1}^{m} p_i = 1 \right\}$$

and the elements $\mathbf{p}_i \in W, 1 \leqslant i \leqslant m$, where $\mathbf{p}_i = (p_{i1}, \ldots, p_{im})$ is the vector made up of the elements of the ith row of the matrix $\mathbf{P}$. Let us remark that, if the chain is in state i_n at time n, its next state will be chosen according to the probability distribution $\mathbf{p}_{i_n} \in W$. In other words, we can consider another sequence of W-valued random variables $(\zeta_n)_{n \in \mathbb{N}}$ on Ω where $\zeta_0 = \mathbf{p}_0$, and $\zeta_n =$ the probability distribution according to which the state of the chain at time $n+1$ is chosen. To understand the situation better, let us depict the paths of the two chains

$$\begin{array}{l} \zeta_0 \equiv \mathbf{p}_0 \rightsquigarrow \xi_1 = i_1 \\ \zeta_1 = \mathbf{p}_{i_1} \rightsquigarrow \xi_2 = i_2 \\ \zeta_2 = \mathbf{p}_{i_2} \rightsquigarrow \xi_3 = i_3 \\ \vdots \qquad\qquad \vdots \end{array}$$

In the scheme above, the broken arrows indicate that ξ_{n+1} depends on ζ_n in a stochastic way, while the straight ones indicate that ζ_{n+1} depends on ξ_{n+1} in a deterministic way, $n \in \mathbb{N}$. In fact, the stochastic dependence is described by the transition probability function (the t.p.f. for short) P from W to X, defined by the equation

$$P(\mathbf{p}_i, j) = p_{ij} = \text{the } j\text{th component of } \mathbf{p}_i, i, j \in X,$$

while the deterministic dependence is described by a function $u: X \to W$, defined by the equation

$$u(i) = \mathbf{p}_i, i \in X.$$

So, in fact, we may consider that an MC is completely determined by the quadruple $\{W, X, u, P\}$, the elements of which were defined above, and by an initial probability distribution $\mathbf{p}_0 \in W$. Next, given these two things, we can construct not only the MC $(\xi_n)_{n \in \mathbb{N}^*}$, but also the sequence of random variables $(\zeta_n)_{n \in \mathbb{N}}$ defined as

$$\zeta_0 = \mathbf{p}_0,$$
$$\zeta_n = u(\xi_n), n \in \mathbb{N}^*.$$

Hence ζ_n is a function of ξ_n, which is not at all exciting. A possible generalization would be to make ζ_n depend also on ζ_{n-1}. This can be done by substituting the function u above by a function $u: W \times X \to W$. Moreover, it is possible to take a different function at each step, i.e. to consider a sequence $(u_n)_{n \in \mathbb{N}}$ of functions from $W \times X$ into W, such that we have $\zeta_n = u_{n-1}(\zeta_{n-1}, \xi_n)$ for all $n \in \mathbb{N}^*$. Is this a generalization for generalization's sake or does it correspond to interesting real situations? The example below gives an answer to this question and motivates the definitions in the next chapter.

Let us consider an initial urn U_0 containing $a_j^{(0)} = a_j$ balls of colour j, $1 \leqslant j \leqslant m$, and denote by $a_j^{(n)}, 1 \leqslant j \leqslant m$, the structure of the urn $U_n, n \in \mathbb{N}^*$, given by the following rule: if the structure of the urn U_{n-1} was $a_j^{(n-1)}, 1 \leqslant j \leqslant m$, and on trial n (which is a drawing from U_{n-1}) a ball of colour i was drawn, then the structure of U_n is specified by

$$a_j^{(n)} = a_j^{(n-1)} + \delta_{ij} d_j,$$

where the $d_j, 1 \leqslant j \leqslant m$, are non-negative integers. This amounts to the fact that, if on trial n a ball of colour i is drawn, then this ball is replaced, together with d_i balls of the same colour.

This scheme was devised and studied by Onicescu & Mihoc (1936b).

The special case $m = 2, d_1 = d_2 = d$ is known as the Pólya urn (see Pólya (1930)), and is, in turn, a generalization of Markov's urn, for which $d = 1$.

Define

$$\xi_n = \text{the colour of the ball drawn on trial } n$$

and set

$$p_j^{(n)} = \mathbf{P}(\xi_{n+1} = j \mid \xi_1, \ldots, \xi_n), \quad n \in \mathbb{N}^*,$$

$$p_j^{(0)} = \mathbf{P}(\xi_1 = j) = \frac{a_j}{M}, \qquad 1 \leqslant j \leqslant m,$$

where $M = \sum_{j=1}^m a_j$.

We want to find the relationship between $\mathbf{p}_n = (p_j^{(n)})_{1 \leqslant j \leqslant m}$ and $\mathbf{p}_{n-1} = (p_j^{(n-1)})_{1 \leqslant j \leqslant m}$, given that $\xi_n = i$. Clearly,

$$p_j^{(n-1)} = \frac{a_j^{(n-1)}}{M^{(n-1)}}, \quad p_j^{(n)} = \frac{a_j^{(n-1)} + \delta_{ij} d_j}{M^{(n-1)} + d_i}, \quad 1 \leqslant j \leqslant m, \tag{0.1}$$

where $M^{(n-1)} = \sum_{j=1}^m a_j^{(n-1)} (M^{(0)} = M)$. Now, denoting by $v_j^{(n)}$ the number of balls of colour j that occurred on the first n drawings ($v_j^{(0)} = 0$), we can write

$$a_j^{(n)} = a_j + v_j^{(n)} d_j, \quad 1 \leqslant j \leqslant m, \quad n \in \mathbb{N}. \tag{0.2}$$

Assuming, for the sake of simplicity, that all the $d_j > 0$, equations (0.1) and (0.2) and the obvious equation $\sum_{j=1}^m v_j^{(n)} = n$ yield

$$M^{(n-1)} = \frac{\sum_{k=1}^m \dfrac{a_k}{d_k} + n - 1}{\sum_{k=1}^m \dfrac{p_k^{(n-1)}}{d_k}},$$

$$a_j^{(n-1)} = M^{(n-1)} p_j^{(n-1)}.$$

Hence, by the second equation of (0.1), given that $\xi_n = i$, the probability $p_j^{(n)}$ is a function $\psi_{i,j}^{n-1}$ of $\mathbf{p}_{n-1} = (p_j^{(n-1)})_{1 \leqslant j \leqslant m}$, i.e.

$$p_j^{(n)} = \psi_{i,j}^{n-1}(p_1^{(n-1)}, \ldots, p_m^{(n-1)}), \quad 1 \leqslant i, j \leqslant m, \quad n \in \mathbb{N}^*.$$

Therefore the evolution of the probabilistic structure of the urn scheme we have considered can be described as follows. Starting with a probability vector $\mathbf{p}_0 = (p_1^{(0)}, \ldots, p_m^{(0)})$ we select a colour $\xi_1 = i$ according to $\mathbf{p}_0$, and then construct the new probability vector

$$\zeta_1 = \mathbf{p}_1 = (\psi_{i,j}^0(p_1^{(0)}, \ldots, p_m^{(0)}))_{1 \leqslant j \leqslant m}.$$

Next, a colour $\xi_2 = k$ is selected according to $\mathbf{p}_1$, then the new probability vector

$$\zeta_2 = \mathbf{p}_2 = (\psi_{k,j}^1(p_1^{(1)}, \ldots, p_m^{(1)}))_{1 \leqslant j \leqslant m}$$

is constructed and so on.

In this example, we have

$$W=\left\{\mathbf{p}=(p_1,\ldots,p_m):p_i\geqslant 0, 1\leqslant i\leqslant m, \sum_{i=1}^{m} p_i=1\right\},$$

$$X=\{1,\ldots,m\},$$

the t.p.f. from W to X is given by

$$P(\mathbf{p},\{i\})=p_i, \mathbf{p}\in W, i\in X,$$

if $\mathbf{p}=(p_1,\ldots,p_m)$, and the functions $u_n: W\times X\to W, n\in\mathbb{N}$, are specified by the formula

$$u_n(\mathbf{p},i)=(\psi^n_{i,j}(\mathbf{p}))_{1\leqslant j\leqslant m}.$$

Two sequences of random variables $(\zeta_n)_{n\in\mathbb{N}}$ and $(\xi_n)_{n\in\mathbb{N}^*}$ are associated with the quadruple $\{W, X, (u_n)_{n\in\mathbb{N}}, P\}$. The variable ζ_{n-1} is the (random) probability vector according to which we select a ball on trial $n\in\mathbb{N}^*$, and ξ_n is the colour of this ball. It is now clear that

$$\zeta_n=u_{n-1}(\zeta_{n-1},\xi_n), \quad n\in\mathbb{N}^*,$$

and

$$\mathsf{P}(\xi_1=i\,|\,\zeta_0)=P(\zeta_0,i),$$

$$\mathsf{P}(\xi_{n+1}=i\,|\,\xi_n,\zeta_n,\ldots,\xi_1,\zeta_1,\zeta_0)=P(\zeta_n,i)$$

for all $i\in X$ and $n\in\mathbb{N}^*$. Onicescu & Mihoc (1935b) say that the random variables $\xi_1,\xi_2,\ldots$ (which are no longer Markovian) are connected into a *chain with complete connections*.

1

Fundamental notions

1.1 The concept of a random system with complete connections

1.1.1 The homogeneous case

We start with a formal definition.

Definition 1.1.1. *A* random system with complete connections (*an* RSCC *for short) is a quadruple* $\{(W,\mathscr{W}),(X,\mathscr{X}),u,P\}$, *where*

(i) $(W,\mathscr{W})$ *and* $(X,\mathscr{X})$ *are arbitrary measurable spaces;*

(ii) $u: W\times X\to W$ *is a* $(\mathscr{W}\otimes\mathscr{X},\mathscr{W})$*-measurable map;*

(iii) P *is a t.p.f. from* $(W,\mathscr{W})$ *to* $(X,\mathscr{X})$.

Throughout this book we shall denote by $x^{(n)}$ the element $(x_1,\ldots,x_n)\in X^n$. If in the same formula there appear $x^{(m)}$ and $x^{(n)}, m<n$, then the first m coordinates of $x^{(n)}$ are precisely the coordinates of $x^{(m)}$, i.e. $x^{(n)}=(x^{(m)}, x_{m+1},\ldots,x_n)$.

For any $n\in\mathbb{N}^*$, let us define recursively the maps $u^{(n)}: W\times X^n\to W$ by the equation

$$u^{(n+1)}(w,x^{(n+1)})=\begin{cases}u(w,x_1), & \text{if } n=0\\ u(u^{(n)}(w,x^{(n)}),x_{n+1}), & \text{if } n\in\mathbb{N}^*.\end{cases} \tag{1.1.1}$$

From now on, we shall simply write wx^n for $u^{(n)}(w,x^{(n)})$, whenever no confusion is possible. Using this convention, the above equation becomes

$$wx^{(n+1)}=\begin{cases}wx_1, & \text{if } n=0\\ (wx^n)x_{n+1}, & \text{if } n\in\mathbb{N}^*.\end{cases}$$

Condition (ii) in Definition 1.1.1 yields the $(\mathscr{W}\otimes\mathscr{X}^n,\mathscr{W})$-measurability of the map $u^{(n)}$ (see Problem 1). Hence all the integrals in the definitions below make sense. For every $w\in W, r\in\mathbb{N}^*$, and $A\in\mathscr{X}^r$, let us define

$$P_r(w,A)=\begin{cases}P(w,A), & \text{if } r=1\\ \displaystyle\int_X P(w,dx_1)\int_X P(wx_1,dx_2)\cdots\int_X P(wx^{(r-1)},dx_r)\chi_A(x^{(r)}), & \text{if } r>1.\end{cases} \tag{1.1.2}$$

(When confusion can arise, we shall write $A^{(r)}$ for an arbitrary set in $\mathscr{X}^r$.)

It is obvious that, for $r \in \mathbb{N}^*$ fixed, P_r is a t.p.f. from $(W, \mathscr{W})$ to $(X^r, \mathscr{X}^r)$; moreover, by the definition of P_r, we have

$$P_{n+r}(w, A^{(n)} \times A^{(r)}) = \int_{A^{(n)}} P_n(w, dx^{(n)}) P_r(wx^n, A^r) \tag{1.1.3}$$

for all $n, r \in \mathbb{N}^*$, $A^{(n)} \in \mathscr{X}^n$ and $A^{(r)} \in \mathscr{X}^r$.

For every $w \in W, n, r \in \mathbb{N}^*$, and $A \in \mathscr{X}^r$, let us define

$$P_r^n(w, A) = P_{n+r-1}(w, X^{n-1} \times A), \tag{1.1.4}$$

with the convention $X^0 \times A = A$. It is obvious that P_r^n is a t.p.f. from $(W, \mathscr{W})$ to $(X^r, \mathscr{X}^r)$ for any fixed $n, r \in \mathbb{N}^*$.

1.1.2 The existence theorem

The probabilistic meaning of the quantities defined by (1.1.2) and (1.1.4) is to be revealed in the existence theorem below (see also Theorem 5.5.1).

Theorem 1.1.2. (i) *For a given RSCC $\{(W, \mathscr{W}), (X, \mathscr{X}), u, P\}$ and an arbitrarily fixed $w_0 \in W$, there exist a probability space $(\Omega, \mathscr{K}, \mathsf{P}_{w_0})$ and a sequence $(\xi_n)_{n \in \mathbb{N}^*}$ of X-valued random variables defined on Ω, such that, for all m, n, $r \in \mathbb{N}^*$ and $A \in \mathscr{X}^r$, we have*

$$\mathsf{P}_{w_0}([\xi_n, \ldots, \xi_{n+r-1}] \in A) = P_r^n(w_0, A), \tag{1.1.5}$$

$$\mathsf{P}_{w_0}([\xi_{n+m}, \ldots, \xi_{n+m+r-1}] \in A \mid \xi^{(n)}) = P_r^m(w_0 \xi^{(n)}, A), \ \mathsf{P}_{w_0}\text{-a.s.}, \tag{1.1.6}$$

$$\mathsf{P}_{w_0}([\xi_{n+m}, \ldots, \xi_{n+m+r-1}] \in A \mid \xi^{(n)}, \zeta^{(n)}) = P_r^m(\zeta_n, A), \ \mathsf{P}_{w_0}\text{-a.s.}, \tag{1.1.7}$$

where $\xi^{(n)} = (\xi_1, \ldots, \xi_n)$, $\zeta_n = w_0\, \xi^{(n)}$, and $\zeta^{(n)} = (\zeta_1, \ldots, \zeta_n)$.

(ii) *The sequence $(\zeta_n)_{n \in \mathbb{N}}$, with $\zeta_0 = w_0$, is a W-valued homogeneous MC, whose initial distribution is concentrated at w_0 and whose transition operator is given by the equation*

$$Uf(w) = \int_X P(w, dx) f(wx), \quad f \in B(W, \mathscr{W}). \tag{1.1.8}$$

(Here $B(W, \mathscr{W})$ is the Banach space of all bounded $\mathscr{W}$-measurable complex-valued functions defined on W – see Section A2.4.)

Proof. (i) In a manner similar to the proof of the existence theorem for MCs, we define $(\Omega, \mathscr{K}) = (X^{\mathbb{N}^*}, \mathscr{X}^{\mathbb{N}^*})$. In other words, Ω is the set of all sequences $(x_1, x_2, \ldots)$ of elements from X, and $\mathscr{K}$ is the smallest σ-algebra containing all the rectangles $A_1 \times \cdots \times A_r \times X \times \cdots$ in Ω, $A_i \in \mathscr{X}, 1 \leqslant i \leqslant r$, $r \in \mathbb{N}^*$ (see Section A1.2). By Ionescu Tulcea's theorem (see Section A1.7), there exists a probability $\mathbf{P}_{w_0}$ on $(\Omega, \mathscr{K})$ such that

$$\mathbf{P}_{w_0}\mathrm{pr}_{[1,r]}^{-1}(\cdot) = P_r(w_0, \cdot) \tag{1.1.9}$$

for all $r \in \mathbb{N}^*$, where $\mathrm{pr}_{[1,r]}: X^{\mathbb{N}^*} \to X^r$ is the projection map on the first r coordinates of $X^{\mathbb{N}^*}$. For $\omega = (x_1, x_2, \ldots) \in \Omega$, put $\xi_n(\omega) = x_n$, and, as already indicated,

$$\zeta_0 = w_0, \zeta_n = w_0 \xi^{(n)}, \quad n \in \mathbb{N}^*. \tag{1.1.10}$$

Use (1.1.4) and (1.1.9) to get

$$\begin{aligned}
&\mathbf{P}_{w_0}([\xi_n, \ldots, \xi_{n+r-1}] \in A) \\
&\quad = \mathbf{P}_{w_0}([\xi_1, \ldots, \xi_n, \ldots, \xi_{n+r-1}] \in X^{n-1} \times A) \\
&\quad = (\mathbf{P}_{w_0}\mathrm{pr}_{[1,n+r-1]}^{-1})(X^{n-1} \times A) = P_{n+r-1}(w_0, X^{n-1} \times A) = P_r^n(w_0, A),
\end{aligned}$$

for all $n, r \in \mathbb{N}^*$ and $A \in \mathscr{X}^r$.

Denote by $\mathscr{K}_{[1,n]}$ and $\mathscr{L}_{[1,n]}$ the σ-algebras generated by the random variables $\xi_1, \ldots, \xi_n$, respectively $\xi_1, \zeta_1, \ldots, \xi_n, \zeta_n$. Since $P_r^m(w_0\xi^{(n)}, A)$ is $\mathscr{K}_{[1,n]}$-measurable, to prove (1.1.6) we only have to show that

$$\int_{\mathrm{pr}_{[1,n]}^{-1}(A')} P_r^m(w_0\xi^{(n)}, A)\,\mathrm{d}\mathbf{P}_{w_0} = \mathbf{P}_{w_0}(\mathrm{pr}_{[1,n]}^{-1}(A') \cap \mathrm{pr}_{[n+m,n+m+r-1]}^{-1}(A))$$

for any $m, n, r \in \mathbb{N}^*$, $A' \in \mathscr{X}^n$, and $A \in \mathscr{X}^r$. Using (1.1.9), (1.1.4), (1.1.3) and Proposition A1.5, we can indeed write

$$\begin{aligned}
&\int_{\mathrm{pr}_{[1,n]}^{-1}(A')} P_r^m(w_0\xi^{(n)}, A)\,\mathrm{d}\mathbf{P}_{w_0} \\
&\quad = \int_{A'} P_r^m(w_0 x^{(n)}, A) P_n(w_0, \mathrm{d}x^{(n)}) \\
&\quad = \int_{A'} P_{m+r-1}(w_0 x^{(n)}, X^{m-1} \times A) P_n(w_0, \mathrm{d}x^{(n)}) \\
&\quad = P_{n+m+r-1}(w_0, A' \times X^{m-1} \times A) = (\mathbf{P}_{w_0}\mathrm{pr}_{[1,n+m+r-1]}^{-1})(A' \times X^{m-1} \times A) \\
&\quad = \mathbf{P}_{w_0}(\mathrm{pr}_{[1,n]}^{-1}(A') \cap \mathrm{pr}_{[n+m,n+m+r-1]}^{-1}(A)).
\end{aligned}$$

Finally, by the definition (1.1.10) of the ζ_n, $n \in \mathbb{N}^*$ it turns out that $\mathscr{L}_{[1,n]} = \mathscr{K}_{[1,n]}$, from which we get

$$\begin{aligned}
&\mathbf{P}_{w_0}([\xi_{n+m}, \ldots, \xi_{n+m+r-1}] \in A \mid \xi^{(n)}, \zeta^{(n)}) \\
&\quad = \mathbf{P}_{w_0}([\xi_{n+m}, \ldots, \xi_{n+m+r-1}] \in A \mid \xi^{(n)}) \\
&\quad = P_r^m(w_0\xi^{(n)}, A) = P_r^m(\zeta_n, A), \; \mathbf{P}_{w_0}\text{-a.s.},
\end{aligned}$$

which concludes the proof of (i).

(ii) Let us remark that $\zeta_{n+1} = \zeta_n \xi_{n+1}, n \in \mathbb{N}^*$, by (1.1.10). Taking $r = m = 1$ in (1.1.7) and using the properties of the conditional expectation (see Section A1.11), we can then write

$$\begin{aligned}
&\mathbf{E}_{w_0}(f(\zeta_{n+1})|\zeta^{(n)}) \\
&= \mathbf{E}_{w_0}(\mathbf{E}_{w_0}(f(\zeta_n\xi_{n+1})|\zeta^{(n)}, \xi^{(n+1)})|\zeta^{(n)}) \\
&= \mathbf{E}_{w_0}\left(\int_X P(\zeta_n, \mathrm{d}x) f(\zeta_n x)\middle|\zeta^{(n)}\right) = \int_X P(\zeta_n, \mathrm{d}x) f(\zeta_n x), \ \mathbf{P}_{w_0}\text{-a.s.},
\end{aligned}$$

for any $f \in B(W, \mathscr{W})$. (Here $\mathbf{E}_{w_0}$ is the expectation operator w.r.t. $\mathbf{P}_{w_0}$.) Hence (1.1.8) should hold (see Section A.3.2). □

Remarks. 1. Taking $m = r = 1$ in (1.1.6) yields

$$\mathbf{P}_{w_0}(\xi_{n+1} \in A | \xi^{(n)}) = P(w_0\xi^{(n)}, A), \qquad \mathbf{P}_{w_0}\text{-a.s.}$$

This shows that the distribution of ξ_{n+1}, given the past $\xi^{(n)}$, does depend on it, by means of the map $u^{(n)}$, which appears in the first argument of the t.p.f. P. This justifies the nomenclature 'chain of infinite order' or, alternatively, 'chain with complete connections', used to designate $(\xi_n)_{n \in \mathbb{N}^*}$.

2. The MC $(\zeta_n)_{n \in \mathbb{N}}$ is called the *associated* Markov chain and has a special structure, simply because of the special form of the operator U, in whose definition the map u plays an essential role. Taking $f(w) = \chi_B(w)$, $B \in \mathscr{W}$, in (1.1.8), we get the formula for the t.p.f. of the associated MC, namely

$$Q(w, B) = \int_X P(w, \mathrm{d}x)\chi_B(wx) = P(w, B_w), \quad w \in W, \quad B \in \mathscr{W}, \quad (1.1.11)$$

where $B_w = \{x \in X : wx \in B\}$. Since, as is easily seen, the iterates of the operator U are given by

$$U^n f(w) = \int_{X^n} P_n(w, \mathrm{d}x^{(n)}) f(wx^{(n)}), \quad f \in B(W, \mathscr{W}),$$

for all $n \in \mathbb{N}^*$, we deduce immediately that the n step t.p.f. is given by

$$Q^n(w, B) = P_n(w, B_w^{(n)}), \quad w \in W, \quad B \in \mathscr{W}, \quad n \in \mathbb{N}^*,$$

where $B_w^{(n)} = \{x^{(n)} : wx^{(n)} \in B\}$.

3. We may imagine the following diagram representing both sequences associated with an RSCC given an arbitrary fixed point $w_0 \in W$:

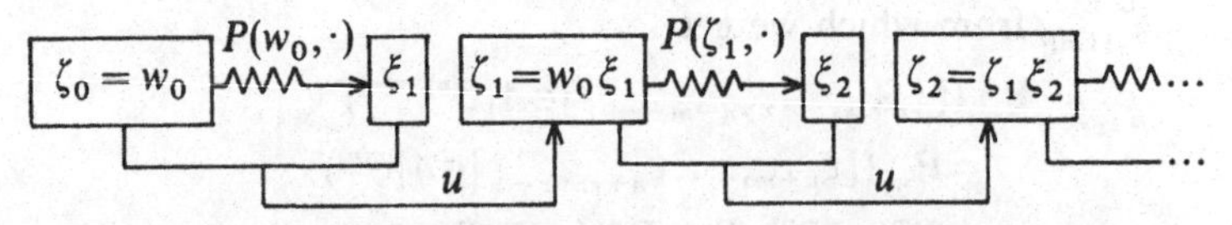

In this diagram, the broken arrows indicate a stochastic dependence, while the straight ones indicate a deterministic dependence.

4. Equation (1.1.5) reveals the probabilistic meaning of the quantities P_r^n, namely $P_r^n(w_0, A)$ is the probability that the r-dimensional random

vector $[\xi_n, \ldots, \xi_{n+r-1}]$ is in the r-dimensional set $A \in \mathscr{X}^r$, given that the process $(\zeta_n)_{n\in\mathbb{N}}$ started from w_0. The probabilistic meaning of the P_r is obvious from the equality $P_r = P_r^1$. Since the concept of an RSCC is a generalization of that of an MC, one would expect the P_r^n and P_r to satisfy a Chapman–Kolmogorov type equation and that is the case as put forward in Corollary 1.1.3.

Corollary 1.1.3. *We have*

$$P_r^n(w, A) = \int_{X^s} P_s(w, dx^{(s)}) P_r^{n-s}(wx^{(s)}, A), \quad 1 \leqslant s < n, \tag{1.1.12}$$

for all $n, r \in \mathbb{N}^*$, $w \in W$ and $A \in \mathscr{X}^r$.

Proof. We have successively

$$\begin{aligned}
P_r^n(w, A) &= \mathbf{P}_w([\xi_{s+1}, \ldots, \xi_n, \ldots, \xi_{n+r-1}] \in X^{n-s-1} \times A) \\
&= \mathbf{E}_w(\mathbf{P}_w([\xi_{s+1}, \ldots, \xi_n, \ldots, \xi_{n+r-1}] \in X^{n-s-1} \times A \mid \xi^{(s)})) \\
&= \mathbf{E}_w(P_{n-s+r-1}(w\xi^{(s)}, X^{n-s-1} \times A)) = \mathbf{E}_w(P_r^{n-s}(w\xi^{(s)}, A)) \\
&= \int_{X^s} P_s(w, dx^{(s)}) P_r^{n-s}(wx^{(s)}, A).
\end{aligned}$$

□

Remark. Equation (1.1.12) shows that $P_r^n(\cdot, A)$, is obtained as a function on W, by applying the operator U to the function $P_r(\cdot, A)$ $(n-1)$ times, i.e.

$$P_r^n(\cdot, A) = U^{n-1}(P_r(\cdot, A)), \quad n \in \mathbb{N}^*,$$

with the convention $U^0 f = f$ for all $f \in B(W, \mathscr{W})$.

Let us remark that the constructive proof of Theorem 1.1.2 makes it possible to point out two other MCs associated with an RSCC; this is done in Corollary 1.1.4.

Corollary 1.1.4. *The sequences $(\zeta_n')_{n\in\mathbb{N}^*} = (\zeta_n, \xi_n)_{n\in\mathbb{N}^*}$ and $(\zeta_n'')_{n\in\mathbb{N}} = (\zeta_n, \xi_{n+1})_{n\in\mathbb{N}}$ are Markovian, their transition operators being given by*

$$U'f(w, x) = \int_X P(w, dx') f(wx', x')$$

and

$$U''f(w, x) = \int_X P(wx, dx') f(wx, x'),$$

respectively, for all $f \in B(W \times X, \mathscr{W} \otimes \mathscr{X})$.

It is clear, from the first equation above, that the function $U'f(w, x)$ does not depend on the second argument, so that we may write $U'f(w, x) = f'(w)$

with $f' \in B(W, \mathscr{W})$. Analogously we have $U''f(w,x) = f''(wx)$ with $f'' \in B(W, \mathscr{W})$.

1.1.3 The non-homogeneous case

Until now we have confined ourselves to the case when the map u and the t.p.f. P do not depend on the time parameter. Now we shall focus on the case where both u and P do depend on the time, which leads to introducing the concept of a non-homogeneous random system with complete connections (a non-homogeneous RSCC for short).

Definition 1.1.5. *A* non-homogeneous RSCC *is a quadruple* $\{(W, \mathscr{W}), (X, \mathscr{X}), (u_t)_{t \in \mathfrak{T}}, ({}^tP)_{t \in \mathfrak{T}}\}$ *(here* $\mathfrak{T}$ *is either the set* $\mathbb{N}$ *of natural numbers or the set* $\mathbb{Z}$ *of integers), where*

(i) $(W, \mathscr{W})$ *and* $(X, \mathscr{X})$ *are arbitrary measurable spaces;*
(ii) *for any* $t \in \mathfrak{T}, u_t: W \times X \to W$ *is a* $(\mathscr{W} \otimes \mathscr{X}, \mathscr{W})$*-measurable map;*
(iii) *for any* $t \in \mathfrak{T}, {}^tP$ *is a t.p.f. from* $(W, \mathscr{W})$ *to* $(X, \mathscr{X})$.

For any fixed $t \in \mathfrak{T}$ and any $n \in \mathbb{N}^*$, let us define iteratively the maps $u_t^{(n)}: W \times X^n \to W$ by the equation

$$u_t^{(n+1)}(w, x^{(n+1)}) = \begin{cases} u_t(w, x_1), & \text{if } n = 0 \\ u_{t+n}(u_t^{(n)}(w, x^{(n)}), x_{n+1}), & \text{if } n \in \mathbb{N}^*. \end{cases} \tag{1.1.13}$$

Similarly to the homogeneous case, we shall write $(wx^{(n)})_t$ instead of $u_t^{(n)}(w, x^{(n)})$ whenever no confusion is possible; with this convention, (1.1.13) becomes

$$(wx^{(n+1)})_t = \begin{cases} (wx_1)_t, & \text{if } n = 0 \\ ((wx^{(n)})_t x_{n+1})_{t+n}, & \text{if } n \in \mathbb{N}^*. \end{cases}$$

Condition (ii) in Definition 1.1.5 yields the $(\mathscr{W} \otimes \mathscr{X}^n, \mathscr{W})$-measurability of the map $u_t^{(n)}$ for all $t \in \mathfrak{T}$; hence all the integrals in the definitions below make sense.

For any $t \in \mathfrak{T}, w \in W, r \in \mathbb{N}^*$, and $A \in \mathscr{X}^r$, let us define

$$\begin{aligned} &{}^tP_r(w, A) \\ &= \begin{cases} {}^tP(w, A), & \text{if } r = 1 \\ \displaystyle\int_X {}^tP(w, dx_1) \int_X {}^{t+1}P((wx_1)_t, dx_2) \cdots \int_X {}^{t+r-1}P((wx^{(r-1)})_t, dx_r)\chi_A(x^{(r)}), & \text{if } r > 1. \end{cases} \end{aligned} \tag{1.1.14}$$

It is clear that, for any fixed $t \in \mathfrak{T}$ and $r \in \mathbb{N}^*$, ${}^tP_r$ is a t.p.f. from $(W, \mathscr{W})$ to $(X^r, \mathscr{X}^r)$; moreover, by the definition of ${}^tP_r$, we get

$$ {}^tP_{n+r}(w, A^{(n)} \times A^{(r)}) = \int_{A^{(n)}} {}^tP_n(w, dx^{(n)})^{t+n}P_r((wx^{(n)})_t, A^{(r)}) \quad (1.1.15)$$

for all $n, r \in \mathbb{N}^*$, $A^{(n)} \in \mathscr{X}^n$, and $A^{(r)} \in \mathscr{X}^r$.

For any $t \in \mathfrak{T}, w \in W, r \in \mathbb{N}^*$, and $A \in \mathscr{X}^r$, let us also define

$$ {}^tP_r^n(w, A) = {}^tP_{n+r-1}(w, X^{n-1} \times A)$$

with the convention $X^0 \times A = A$. It is clear that ${}^tP_r^n$ is a t.p.f. from $(W, \mathscr{W})$ to $(X^r, \mathscr{X}^r)$ for any fixed $t \in \mathfrak{T}$ and $n, r \in \mathbb{N}^*$. Now it is possible to give an existence theorem in the case $\mathfrak{T} = \mathbb{N}$.

Theorem 1.1.6. (i) *For a given non-homogeneous RSCC* $\{(W, \mathscr{W}), (X, \mathscr{X}), (u_n)_{n \in \mathbb{N}}, ({}^nP)_{n \in \mathbb{N}}\}$ *and an arbitrarily fixed* $w_0 \in W$, *there exist a probability space* $(\Omega, \mathscr{K}, \mathsf{P}_{w_0})$ *and a sequence* $(\xi_n)_{n \in \mathbb{N}^*}$ *of X-valued random variables defined on* Ω, *such that for all* $m, n, r \in \mathbb{N}^*$ *and* $A \in \mathscr{X}^r$ *we have*

$$\mathsf{P}_{w_0}([\xi_n, \ldots, \xi_{n+r-1}] \in A) = {}^0P_r^n(w_0, A), \quad (1.1.17)$$

$$\mathsf{P}_{w_0}([\xi_{n+m}, \ldots, \xi_{n+m+r-1}] \in A \mid \xi^{(n)}) = {}^nP_r^m((w_0\xi^{(n)})_0, A), \ \mathsf{P}_{w_0}\text{-a.s.}$$

$$\mathsf{P}_{w_0}([\xi_{n+m}, \ldots, \xi_{n+m+r-1}] \in A \mid \xi^{(n)}, \zeta^{(n)}) = {}^nP_r^m(\zeta_n, A), \ \mathsf{P}_{w_0}\text{-a.s.} \quad (1.1.18)$$

where $\xi^{(n)} = (\xi_1, \ldots, \xi_n)$, $\zeta_n = (w_0\xi^{(n)})_0$, *and* $\zeta^{(n)} = (\zeta_1, \ldots, \zeta_n)$.

(ii) *The sequence* $(\zeta_n)_{n \in \mathbb{N}}$, *with* $\zeta_0 = w_0$, *is a W-valued non-homogeneous MC, whose initial distribution is concentrated at* w_0 *and whose transition operators* ${}^nU, n \in \mathbb{N}$, *are given by the equation*

$$ {}^nUf(w) = \int_X {}^nP(w, dx) f((wx)_n), \quad f \in B(W, \mathscr{W}). \quad (1.1.19)$$

We shall also state the analogue of Corollary 1.1.3 in this case in Corollary 1.1.7.

Corollary 1.1.7. *We have*

$$ {}^nP_r^m(w, A) = \int_{X^s} {}^nP_s(w, dx^{(s)})^{n+s}P_r^{m-s}((wx^{(s)})_n, A), \quad 1 \leqslant s < m, \quad (1.1.20)$$

for all $m, r \in \mathbb{N}^*, n \in \mathbb{N}, w \in W$, *and* $A \in \mathscr{X}^r$.

We do not give the details, since the proofs of both Theorem 1.1.6 and Corollary 1.1.7 are completely similar to those of Theorem 1.1.2 and Corollary 1.1.3.

Finally, let us mention a slight generalization of the notion of a non-homogeneous RSCC, due to Le Calvé & Theodorescu (1967). Instead of the pair $(W, \mathscr{W})$ and $(X, \mathscr{X})$ they consider two families $\{(W_t, \mathscr{W}_t)\}_{t\in\mathfrak{T}}$ and $\{(X_{t+1}, \mathscr{X}_{t+1})\}_{t\in\mathfrak{T}}$ of measurable spaces, $u_t: W_t \times X_{t+1} \to W_{t+1}$ being a $(\mathscr{W}_t \otimes \mathscr{X}_{t+1}, \mathscr{W}_{t+1})$-measurable map and ${}^tP$ a t.p.f. from $(W_t, \mathscr{W}_t)$ to $(X_{t+1}, \mathscr{X}_{t+1})$ for all $t\in\mathfrak{T}$. All the considerations and results above carry over to this more general framework.

1.1.4 Generalized systems

In this subsection we shall present a generalization due to Le Calvé & Theodorescu (1971), which essentially consists in substituting the deterministic map u by a t.p.f. Π. For the sake of simplicity, we shall only consider the homogeneous case.

Definition 1.1.8. *A homogeneous* generalized *random system with complete connections (a homogeneous GRSCC for short) is a quadruple* $\{(W, \mathscr{W}), (X, \mathscr{X}), \Pi, P\}$, *where*

(i) $(W, \mathscr{W})$ *and* $(X, \mathscr{X})$ *are arbitrary measurable spaces;*
(ii) Π *is a t.p.f. from* $(W \times X, \mathscr{W} \otimes \mathscr{X})$ *to* $(W, \mathscr{W})$;
(iii) P *is a t.p.f. from* $(W, \mathscr{W})$ *to* $(X, \mathscr{X})$.

For any $x\in X$, and $f\in B(W, \mathscr{W})$ let us define Γ_x by the equation

$$\Gamma_x f(w) = \int_W \Pi(w, x, dw') f(w'), \quad w\in W.$$

It is clear that $\Gamma_x(B(W, \mathscr{W})) \subseteq B(W, \mathscr{W})$ and Γ_x is a linear operator of norm 1 on $B(W, \mathscr{W})$. Note also that $\Gamma_{.} f(\cdot)$ is $\mathscr{W} \otimes \mathscr{X}$-measurable.

For $x^{(n)}\in X^n$ let us define

$$\Gamma_{x^{(n)}} = \begin{cases} \Gamma_{x_1}, & \text{if } n = 1 \\ \Gamma_{x_1} \circ \Gamma_{x_2} \circ \cdots \circ \Gamma_{x_n}, & \text{if } n > 1. \end{cases}$$

It is possible now to state the following existence theorem, whose proof is completely similar to that of Theorem 1.1.2.

Theorem 1.1.9. *For a given homogeneous GRSCC and an arbitrarily fixed* $w_0\in W$, *there exist a probability space* $(\Omega, \mathscr{K}, \mathbf{P}_{w_0})$, *a sequence* $(\xi_n)_{n\in\mathbb{N}^*}$ *of* X*-valued random variables and a sequence* $(\zeta_n)_{n\in\mathbb{N}}$ *of* W*-valued random variables, both defined on* Ω, *such that:*

(i) *for any* $A\in\mathscr{X}$, $B\in\mathscr{W}$ *and* $n\in\mathbb{N}^*$ *we have*

$$\mathbf{P}_{w_0}(\zeta_0\in B) = \delta(w_0, B), \mathbf{P}_{w_0}(\xi_1\in A) = P(w_0, A),$$

where $\delta(w_0, \cdot)$ is the probability distribution on $\mathscr{W}$ concentrated at w_0, and

$$\mathbf{P}_{w_0}(\xi_{n+1} \in A \mid \xi^{(n)}, \zeta^{(n)}) = P(\zeta_n, A), \mathbf{P}_{w_0}\text{-a.s.}$$

$$\mathbf{P}_{w_0}(\zeta_{n+1} \in B \mid \xi^{(n+1)}, \zeta^{(n)}) = \Pi(\zeta_n, \xi_{n+1}, B), \mathbf{P}_{w_0}\text{-a.s.}$$

(ii) *the sequence $(\zeta_n)_{n\in\mathbb{N}}$ is a homogeneous MC, whose t.p.f. Q and transition operator U are given by the equations*

$$Q(w, B) = \int_X P(w, \mathrm{d}x)\Pi(w, x, B) = \int_X P(w, \mathrm{d}x)\Gamma_x\chi_B(w),$$

for all $w \in W$, $B \in \mathscr{W}$, and

$$Uf(w) = \int_X P(w, \mathrm{d}x)\Gamma_x f(w) = \int_{W \times X} P(w, \mathrm{d}x)\Pi(w, x, \mathrm{d}w')f(w')$$

for all $f \in B(W, \mathscr{W})$.

The sequences $(\xi_n)_{n\in\mathbb{N}^*}$ and $(\zeta_n)_{n\in\mathbb{N}}$ above will be called the generalized chain with complete connections and the Markov chain associated with the given GRSCC, respectively.

Now we shall introduce the quantities P_r and P_r^n which are analogous to those defined by equations (1.1.2) and (1.1.3).

To do that, let us first define

$$S_r(w_0, C) = \int_X P(w_0, \mathrm{d}x_1) \int_W \Pi(w_0, x_1, \mathrm{d}w_1) \cdots \int_X P(w_{r-1}, \mathrm{d}x_r)$$
$$\times \int_W \Pi(w_{r-1}, x_r, \mathrm{d}w_r)\chi_C(x^{(r)}, w^{(r)})$$

for all $r \in \mathbb{N}^*$, $w_0 \in W$, and $C \in (\mathscr{X} \otimes \mathscr{W})^r$. Then, for any $r \in \mathbb{N}^*$, $w \in W$, and $A \in \mathscr{X}^r$, put

$$P_r(w, A) = \begin{cases} P(w, A), & \text{if } r = 1 \\ \displaystyle\int_{(X \times W)^{r-1}} S_{r-1}(w, \mathrm{d}x^{(r-1)} \times \mathrm{d}w^{(r-1)}) \int_X P(w_{r-1}, \mathrm{d}x_r)\chi_A(x^{(r)}), & \text{if } r > 1. \end{cases}$$

For every $w \in W$, $n, r \in \mathbb{N}^*$ and $A \in \mathscr{X}^r$, let us define

$$P_r^n(w, A) = P_{n+r-1}(w, X^{n-1} \times A)$$

with the convention $X^0 \times A = A$.

By standard arguments and computations, we get

$$P_r^n(w, A) = \begin{cases} P_r(w, A), & \text{if } n = 1 \\ \displaystyle\int_X P(w, \mathrm{d}x)(\Gamma_x P_r^{n-1}(\cdot, A))(w), & \text{if } n > 1. \end{cases} \tag{1.1.21}$$

and

$$P_r^n(w, A) = \mathbf{P}_w([\xi_n, \ldots, \xi_{n+r-1}] \in A). \tag{1.1.22}$$

It is also easy to prove an analogue of the Chapman–Kolmogorov equation, namely

$$P_r^n(w, A) = \int_{(X \times W)^s} S_s(w, \mathrm{d}x^{(s)} \times \mathrm{d}w^{(s)}) P_r^{n-s}(w_s, A) \qquad (1.1.23)$$

for all $n \geqslant 2$ and $1 \leqslant s < n$.

Remark. In a manner similar to the case of an RSCC, equation (1.1.23) enables us to deduce that $P_r^n(\cdot, A)$, as a function on W, is obtained by applying the operator U to $P_r(\cdot, A)$ $(n-1)$ times, i.e.

$$P_r^n(\cdot, A) = U^{n-1}(P_r(\cdot, A)), \quad n \in \mathbb{N}^*,$$

with the convention $U^0 f = f$ for all $f \in B(W, \mathscr{W})$.

The result below makes it clear why the concept of a GRSCC is actually a generalization of that of an RSCC.

Assume that there is a map $u: W \times X \to W$ such that

$$\Pi(w, x, B) = \delta(wx, B) \qquad (1.1.24)$$

for all $w \in W, x \in X$ and $B \in \mathscr{W}$. Here $\delta(y, B) = 1$ if $y \in B$ and $= 0$ otherwise. It is easy to see that such a u is $(\mathscr{W} \otimes \mathscr{X}, \mathscr{W})$-measurable.

Proposition 1.1.10. *Given a homogeneous GRSCC* $\{(W, \mathscr{W}), (X, \mathscr{X}), \Pi, P\}$ *for which* (1.1.24) *holds, let* $(\xi_n)_{n \in \mathbb{N}^*}, (\zeta_n)_{n \in \mathbb{N}}$ *be the sequences of random variables associated with it by Theorem* 1.1.9. *Let* $(\xi'_n)_{n \in \mathbb{N}^*}$ *and* $(\zeta'_n)_{n \in \mathbb{N}}$ *be the sequences of random variables associated by Theorem* 1.1.2 *with the RSCC* $\{(W, \mathscr{W}), (X, \mathscr{X}), u, P\}$. *Then the sequences* $(\xi_n)_{n \in \mathbb{N}^*}$ *and* $(\xi'_n)_{n \in \mathbb{N}^*}$, *and* $(\zeta_n)_{n \in \mathbb{N}}$ *and* $(\zeta'_n)_{n \in \mathbb{N}}$, *respectively, are equivalent (see Section* A1.5).

The proof is obtained by straightforward computations, using (1.1.24).

Remarks. 1. In the case of an RSCC we have $\mathsf{P}_{w_0}(\xi_{n+1} \in A \mid \xi^{(n)}) = P(w_0 \xi^{(n)}, A)$, P_{w_0}-a.s. Since, in the case of a GRSCC for which (1.1.24) holds, we have $\Gamma_{x^{(n)}} f = f(u(\cdot, x^{(n)}))$, $f \in B(W, \mathscr{W})$, we would expect the equation

$$\mathsf{P}_{w_0}(\xi_{n+1} \in A \mid \xi^{(n)}) = (\Gamma_{\xi^{(n)}} P(\cdot, A))(w_0), \ \mathsf{P}_{w_0}\text{-a.s.},$$

to hold true for any GRSCC. In fact, this equation is given in Le Calvé & Theodorescu (1971), but, as remarked by G. Popescu (personal communication), it fails to be true in the general case.

2. It is natural to investigate whether it is possible to reduce the study of a GRSCC, for which (1.1.24) does not hold, to that of an RSCC by a suitable modification of the spaces W and X. This problem is answered

in the affirmative by Le Calvé & Theodorescu (1971, Theorem 6.2), but the proof given is not correct by virtue of the previous remark.

1.2 Examples

In this section we shall give a few examples of RSCCs and GRSCCs, which occur either in various chapters of probability theory or as a result of modelling phenomena in various fields.

Conceptually, there are two extreme special cases of RSCCs:

(i) the case where the maps u_t of $W \times X$ into W are actually maps of X into W, i.e. $u_t(w, \cdot)$ does not depend on $w \in W$ for all $t \in \mathfrak{T}$;
(ii) the case where the t.p.f.s ${}^tP$ from $(W, \mathscr{W})$ to $(X, \mathscr{X})$ are actually probability distributions on $\mathscr{X}$, i.e. ${}^tP(w, \cdot)$ does not depend on $w \in W$ for all $t \in \mathfrak{T}$.

The first case leads to the most general discrete parameter Markovian dependence (see Example 1 below), while the second one leads to seemingly special MCs, whose variables are compositions of random maps with an underlying random mechanism consisting of independent trials (see Examples 4, 7 and 8 below). Actually, the second instance is as general as the first (see Problem 4).

Example 1 *Markov chains.* Markovian dependence can be viewed from various angles as a special case of dependence with complete connections. One possible approach has already been described in the Introduction. The most natural course leading from an RSCC $\{(W, \mathscr{W}), (X, \mathscr{X}), u, P\}$ to a homogeneous MC is to take $(W, \mathscr{W}) = (X, \mathscr{X}), u(w, x) = x, w \in W, x \in X$, so that P is simply a t.p.f. on $(X, \mathscr{X})$. Analogously, a non-homogeneous MC is obtained from a non-homogeneous RSCC $\{(W, \mathscr{W}), (X, \mathscr{X}), (u_n)_{n \in \mathbb{N}^*}, ({}^nP)_{n \in \mathbb{N}}\}$ by taking $(W, \mathscr{W}) = (X, \mathscr{X}), u_n(w, x) = x, w \in W, x \in X, n \in \mathbb{N}^*$, so that ${}^nP, n \in \mathbb{N}$, is a t.p.f. on $(X, \mathscr{X})$. It is obvious that, in both cases, the two sequences of random variables associated with these special RSCCs reduce to just one sequence (more precisely, $\zeta_n \equiv \xi_n, n \in \mathbb{N}^*$), namely the MC under consideration.

A multiple MC of order $l > 1$ can be viewed as the infinite order chain associated with the RSCC $\{(X^l, \mathscr{X}^l), (X, \mathscr{X}), u, ({}^nP)_{n \in \mathbb{N}}\}$, where $u(x^{(l)}, x) = (x_2, \ldots, x_l, x)$ for $x^{(l)} = (x_1, \ldots, x_l) \in X^l, x \in X$. It is noteworthy that the random variables of the MC associated with this RSCC are l-dimensional vectors, whose components are l successive variables in the multiple MC under consideration. Note that this simple MC is a well-known tool in the study of multiple MCs.

Example 2 *Stochastic learning models.* The concept of learning is one of the most important concepts of the theory of behaviour. Learning may be defined as an adaptive modification of behaviour in the course of repeated trials. By mathematical learning theory we mean the body of research methods and results concerned with the conceptual representation of learning phenomena, the mathematical formulation of hypotheses about learning, and the derivation of testable theorems. The purpose of mathematical learning theory is to provide simple, quantitative descriptions of processes which are basic to behavioural modifications. Learning models are in fact devices for providing simple descriptions of basic learning processes. Prior to 1950, learning models were concerned with predicting, at most, the mean learning curves† obtained from experiments. Although a few of the earlier models involved probability measures and processes, no properties other than the mean learning curve were deduced.

The main feature of the modern formulations initiated since 1950 by Estes (1950) and Bush & Mosteller (1951; 1955) is that they imply random processes to which behaviour in simple learning experiments conforms. Another important feature is the step-by-step nature of the learning process.

All stochastic models for learning studied so far fit the following general theoretical scheme. The behaviour of the subject on trial n is determined by its state S_n (an indicator of the subject's response tendencies) at the beginning of the trial; S_n is a random variable taking on values in a measurable state space $(S, \mathscr{S})$. On trial n an event E_{n+1} occurs that results in a change of state; E_{n+1} is a random variable taking on values in a measurable event space $(E, \mathscr{E})$, and it specifies those occurrences on trial n that affect subsequent behaviour. Typically, E_{n+1} includes a specification of the subject's response and its observable outcome or pay-off. To represent the fact that the occurrence of an event affects a change of state, we consider a $(\mathscr{S} \otimes \mathscr{E}, \mathscr{S})$-measurable map v from $S \times E$ into S and postulate that $S_{n+1} = v(S_n, E_{n+1}), n \in \mathbb{N}$. Finally, we assume that the probability distribution of E_{n+1}, given $S_n, E_n, \ldots, S_1, E_1, S_0$, depends only on the state S_n, and we denote it by $Q(S_n, \cdot)$. By a general learning model we mean the collection $\{(S, \mathscr{S}), (E, \mathscr{E}), v, Q\}$, which is trivially an RSCC. (Notice that in fact we only changed the notation!) Various special learning models are obtained by simply particularizing S, E, v and Q. The interested reader may consult Iosifescu (1983), Iosifescu & Theodorescu (1969), and Norman (1972).

† The notion of mean learning curve measures the 'performances' of an individual as a function of training times or trials. It was proposed in 1919 by L. L. Thurstone.

Example 3 *Stochastic approximation (Robbins & Monro* (1951)). Let $F_a, a \in \mathbb{R}$, be a family of distribution functions on the real line. Assume that

$$M(a) = \int_{\mathbb{R}} y \, \mathrm{d}F_a(y)$$

exists, and the equation $M(a) = 0$ has a unique root θ. Robbins & Monro (1951) gave a statistical method for estimating θ. They chose an arbitrary $\theta_0 \in \mathbb{R}$ and defined recursively the random variables θ_n by

$$\theta_{n+1} = \theta_n - \frac{c}{n+1} \eta_{n+1},$$

where c is a positive constant and η_{n+1} is a random variable whose conditional distribution function, given $\theta_0, \ldots, \theta_n$, is F_{θ_n}. Under suitable conditions the convergence (either in probability or almost sure) of θ_n to θ as $n \to \infty$ can be proved.

It is clear that the sequence $(\theta_n)_{n \in \mathbb{N}}$ is equivalent to the MC associated with the RSCC $\{(W, \mathscr{W}), (X, \mathscr{X}), (u_n)_{n \in \mathbb{N}}, P\}$, where

$$(W, \mathscr{W}) = (X, \mathscr{X}) = (\mathbb{R}, \mathscr{B}),$$

$$u_n(w, x) = w - \frac{c}{n+1} x, n \in \mathbb{N}, \quad w, x \in \mathbb{R},$$

$$P(w, A) = \int_A \mathrm{d}F_w(y), \quad w \in \mathbb{R}, \quad A \in \mathscr{B}.$$

Various alterations and multivariate versions of the Robbins–Monro procedure (see, e.g. Wasan (1969)) can also be shown to fit the framework of RSCCs. In this respect see Macchi (1976).

For the use of stochastic approximation in the Monte Carlo method see Ermakov (1975) and Pugh (1966). For the application of stochastic approximation to pattern recognition see Perennou (1976) and the references therein. For a discussion of stochastic approximation in terms of probabilistic tools and relationships with other parts of probability theory, as well as convex analysis and stochastic optimization theory, see Schmetterer (1979).

Example 4 *Galton–Watson–Bienaymé processes with random environments (Smith & Wilkinson (1969))*. Let $(\varepsilon_n)_{n \in \mathbb{N}^*}$ be a random sequence of 'environmental variables' taking on values in some measurable space $(E, \mathscr{E})$. We suppose that, for each $e \in E$, a probability generating function $\varphi_e(s) = \Sigma_{j \in \mathbb{N}} p_j(e) s^j$, $s \in [0, 1]$ is given, for which the $p_j(\cdot)$, $j \in \mathbb{N}$, are $\mathscr{E}$-measurable. For each realization of the environmental sequence $\varepsilon = (\varepsilon_1, \varepsilon_2, \ldots)$ and the associated sequence $(\varphi_{\varepsilon_n})_{n \in \mathbb{N}^*}$ of probability generating functions, there evolves a population governed by the laws of the standard Galton–Watson–Bienaymé process. Specifically, denote by Z_n the size of the nth generation, and assume $Z_0 = a$, where a is an arbitrarily given natural integer. The size Z_{n+1} of the $(n+1)$th generation is the cumulative progeny of the Z_n individuals of the nth generation, each creating independently according to the probability generating function $\varphi_{\varepsilon_{n+1}}$. (If $Z_n = 0$, then $Z_{n+1} = 0$.) In other words, Z_{n+1} is the sum of Z_n independent random variables, each with probability generating function $\varphi_{\varepsilon_{n+1}}$. The process $(Z_n)_{n\in\mathbb{N}}$ is said to be the Galton–Watson–Bienaymé process conditioned on the environment ε and starting with a individuals. (The case of the classical Galton–Watson–Bienaymé process is obtained for a space E consisting of a single point.)

Let $\theta_n, n\in\mathbb{N}^*$, denote the (random) probability generating function of Z_n given $\varepsilon_1,\ldots,\varepsilon_n$ and assuming that $a=1$. Then it is easily seen that

$$\theta_{n+1} = \varphi_{\varepsilon_{n+1}}(\theta_n), \quad n\in\mathbb{N},$$

with $\theta_0(s) = s$, $s\in[0,1]$. Assuming for the sake of simplicity the ε_n, $n\in\mathbb{N}^*$, to be independent and identically distributed with common probability distribution $\mathbf{p}$ on $\mathscr{E}$, and keeping s fixed, the sequence $(\theta_n(s))_{n\in\mathbb{N}}$ is clearly equivalent to the MC $(\zeta_n)_{n\in\mathbb{N}}$, $\zeta_0 = s$, associated with the RSCC $\{(W,\mathscr{W}),(X,\mathscr{X}),u,P\}$, where

$$W = [0,1], \quad \mathscr{W} = \mathscr{B}_{[0,1]}, \quad (X,\mathscr{X}) = (E,\mathscr{E}),$$
$$u(w,e) = \varphi_e(w), \quad P(w,A) = \mathbf{p}(A), \quad w\in W, \quad e\in E, \quad A\in\mathscr{E}.$$

Example 5 *Hawkins Sieve* (*Williams* (1974)). The Hawkins Sieve is a probabilistic analogue of the well-known Eratosthenes Sieve. It yields a random sequence $(\alpha_n)_{n\in\mathbb{N}}$ by the following inductive sieving procedure. Let $A_0 = \{2,3,4,\ldots\}$. At step 0 put $\alpha_0 = \min A_0 = 2$; from the set $A_0 - \{\alpha_0\}$ each number in turn (and each independently of the others) is deleted with probability α_0^{-1} and the set of remaining elements of $A_0 - \{\alpha_0\}$ is denoted $A_1 \ldots$. At step n put $\alpha_n = \min A_n$; from the set $A_n - \{\alpha_n\}$ each number in turn (and each independently of the others) is deleted with probability α_n^{-1} and the set of remaining elements of $A_n - \{\alpha_n\}$ is denoted $A_{n+1}\ldots$.

Let us define $\beta_n = \prod_{k=0}^{n}(1-\alpha_k^{-1})^{-1}, n\in\mathbb{N}$. It is obvious that $\beta_{n+1} = \beta_n(1-\alpha_{n+1}^{-1})^{-1}, n\in\mathbb{N}$, and, by the definition of the procedure, we get

$$\mathbf{P}(\alpha_{n+1} - \alpha_n = j \mid \alpha_n, \beta_n, \ldots, \alpha_0, \beta_0) = \beta_n^{-1}(1-\beta_n^{-1})^{j-1}, \quad j\in\mathbb{N}^*.$$

It comes out that $(\beta_n, \alpha_n)_{n\in\mathbb{N}}$ is equivalent to the MC $(\zeta_n)_{n\in\mathbb{N}}$, $\zeta_0 = (2,2)$, associated with the RSCC $\{(W,\mathscr{W}),(X,\mathscr{X}),u,P\}$, where

$$W = \{r:r \text{ rational} \geqslant 2\} \times \{2,3,4,\ldots\}, \mathscr{W} = \mathscr{P}(W),$$
$$X = \{2,3,4,\ldots\}, \quad \mathscr{X} = \mathscr{P}(X),$$
$$u((r,m),x) = (r(1-x^{-1})^{-1}, x),$$
$$P((r,m),x) = \begin{cases} (1-r^{-1})^{x-m-1}r^{-1}, & \text{if } x \geqslant m+1 \\ 0, & \text{if } x \leqslant m \end{cases}$$

for all $(r,m) \in W$ and $x \in X$.

It is also clear that $(\alpha_n)_{n \in \mathbb{N}^*}$ is equivalent to the chain of infinite order associated with this RSCC.

Example 6 *Continued fractions* (*Iosifescu* (1974)). Any irrational number y in the unit interval $[0,1]$ has a unique infinite continued fraction expansion of the form

$$y = \cfrac{1}{a_1(y) + \cfrac{1}{a_2(y) + \cdots}},$$

where the $a_n(y)$, $n \in \mathbb{N}^*$, are natural integers determined as follows. Put $\tau(y) = 1/y \pmod 1$. Then $a_1(y) = 1/y - \tau(y)$, and $a_{n+1}(y) = a_n(\tau(y)) = a_1(\tau^n(y))$, $n \in \mathbb{N}^*$. It is obvious that, when the unit interval is endowed with the σ-algebra $\mathscr{B}_{[0,1]}$ of its Borel subsets, the a_n are random variables defined almost everywhere w.r.t. any probability measure on $\mathscr{B}_{[0,1]}$ assigning probability 0 to the set of rational numbers (thus, in particular, w.r.t. Lebesgue measure λ).

The metric theory of continued fractions is concerned with the study of the random sequence $(a_n)_{n \in \mathbb{N}^*}$ and related sequences. The first problem of this theory was raised in 1812 by Gauss who, in a letter to Laplace,† stated that

$$\lim_{n \to \infty} \lambda(y:r_n^{-1}(y) < v) = \frac{\log(1+v)}{\log 2}, \quad 0 \leqslant v \leqslant 1,$$

and asked for an estimate of the error

$$E_n(v) = \lambda(y:r_n^{-1}(y) < v) - \frac{\log(1+v)}{\log 2}.$$

Here $r_n(y) = 1/\tau^{n-1}(y), n \in \mathbb{N}^*$, i.e.

$$r_n = a_n + \cfrac{1}{a_{n+1} + \cfrac{1}{a_{n+2} + \cdots}}.$$

† The letter has been published on pages 371–372 of Gauss' *Werke*, 1917, Vol. 10, Section 1, Teubner, Leipzig. Almost the whole letter is reproduced on pages 396–397 of J. V. Uspensky's *Introduction to Mathematical Probability*, 1937, McGraw–Hill, New York. See also Gray (1984, p. 123) for other historical details on Gauss' problem.

It is not difficult to show (see Subsection 5.2.1) that for any real number $t \geqslant 1$ we have

$$\lambda(r_1 > t) = 1/t,$$

$$\lambda(r_{n+1} > t \mid a_1, \ldots, a_n) = \frac{s_n + 1}{s_n + t}, \quad n \in \mathbb{N}^*,$$

where

$$s_1 = \frac{1}{a_1}, \quad s_{n+1} = \frac{1}{s_n + a_{n+1}}, \quad n \in \mathbb{N}^*.$$

This is known as the Borel–Lévy formula.† It implies at once

$$\lambda(a_1 = i) = \frac{1}{i(i+1)},$$

$$\lambda(a_{n+1} = i \mid a_1, \ldots, a_n) = \frac{s_n + 1}{(s_n + i)(s_n + i + 1)}, \quad n \in \mathbb{N}^*,$$

for any natural integer i.

Let us consider the RSCC $\{(W, \mathscr{W}), (X, \mathscr{X}), u, P\}$, where

$$W = [0, 1], \quad \mathscr{W} = \mathscr{B}_{[0,1]},$$

$$X = \mathbb{N}^*, \quad \mathscr{X} = \mathscr{P}(\mathbb{N}^*),$$

$$u(w, x) = \frac{1}{w + x}, \quad P(w, x) = \frac{w + 1}{(w + x)(w + x + 1)}, \quad w \in [0, 1], \quad x \in \mathbb{N}^*.$$

Clearly, the sequences $(a_n)_{n \in \mathbb{N}^*}$ and $(s_n)_{n \in \mathbb{N}}$, $s_0 = 0$, under λ, are equivalent to the chain of infinite order $(\xi_n)_{n \in \mathbb{N}^*}$ and the MC $(\zeta_n)_{n \in \mathbb{N}}$ associated with the above RSCC, under $\mathbf{P}_0$ (i.e. starting with $\zeta_0 = 0$, see Theorem 1.1.2). More precisely, defining the one-to-one map θ from $\Omega = (\mathbb{N}^*)^{\mathbb{N}^*}$ into $[0, 1]$ by

$$\theta(a_1, a_2, \ldots) = \cfrac{1}{a_1 + \cfrac{1}{a_2 + \cdots}}, \quad a_i \in \mathbb{N}^*, i \in \mathbb{N}^*,$$

we have $\mathbf{P}_0 \theta^{-1} = \lambda$ and $\xi_n(\omega) = a_n(\theta(\omega))$, $\zeta_n(\omega) = s_n(\theta(\omega))$, $n \in \mathbb{N}^*$, $\omega \in \Omega$.

Example 7 *Simple models for turbulence in dynamical systems with internal friction.* Fluid turbulence is one of the great unsolved problems of theoretical physics. Its nature remains rather mysterious and controversial. A simple instance of turbulence can be easily produced by turning on the tap over the kitchen sink. The study of turbulent time evolutions in natural

† For $t \in \mathbb{N}^*$ this formula has been obtained by the Swedish mathematician T. Brodén as early as 1900 (see Brodén (1900 p. 246)), nine years before Borel. This is why we think it should be referred to as the Brodén–Borel–Lévy formula.

phenomena has progressed very slowly owing to the insufficient development of the theory on the one hand and to experimental difficulties on the other. The absence of a satisfactory mathematical theory has meant that computers have played an important role in the interpretation of data. In the last 25 years it has become apparent that the very simplest non-linear difference equations can possess a surprisingly rich spectrum of dynamical behaviour. This ranges from stable points, through cascades of stable cycles, to a regime in which the behaviour (though fully deterministic) is in many respects 'chaotic' or 'turbulent', i.e. indistinguishable from the sample paths of a random process. For details the reader is referred to the review papers by Eckmann (1981), May (1976), Ruelle (1980), and Whitley (1982), as well as the Collet & Eckmann (1980) book and the collections of papers edited by Barnsley & Demko (1986), Holden (1986), and Ikegami (1986).

In the light of the above remarks the models proposed for the study of turbulence are of the form

$$\mathbf{x}_{n+1} = f_\mu(\mathbf{x}_n) + \xi_{n+1}, \quad n \in \mathbb{N},$$

with arbitrarily given $\mathbf{x}_0 \in \mathbb{R}^m$. Here $f_\mu: \mathbb{R}^m \to \mathbb{R}^m$ is a continuous map depending on a parameter μ, and $(\xi_n)_{n\in\mathbb{N}^*}$ is a sequence of independent identically distributed m-dimensional random vectors with common probability distribution p. Clearly, the sequence $(\mathbf{x}_n)_{n\in\mathbb{N}}$ is equivalent to the MC $(\zeta_n)_{n\in\mathbb{N}}, \zeta_0 = \mathbf{x}_0$, associated with the RSCC $\{(W, \mathscr{W}), (X, \mathscr{X}), u, P\}$, where

$$(W, \mathscr{W}) = (X, \mathscr{X}) = (\mathbb{R}^m, \mathscr{B}^m),$$

$$u(\mathbf{w}, \mathbf{x}) = f_\mu(\mathbf{w}) + \mathbf{x}, \quad \mathbf{w}, \mathbf{x} \in \mathbb{R}^m,$$

$$P(\mathbf{w}, A) = \mathsf{p}(A), \quad \mathbf{w} \in \mathbb{R}^m, \quad A \in \mathscr{B}^m.$$

Actually, the greatest attention has been paid to the special case $\xi_n \equiv 0$, $n \in \mathbb{N}^*$, i.e. to the non-stochastic iteration scheme

$$\mathbf{x}_{n+1} = f_\mu(\mathbf{x}_n), \quad n \in \mathbb{N},$$

primarily in the one- and two-dimensional cases $m = 1$ and $m = 2$ with f_μ mapping some one- or two-dimensional interval into itself. Here (genuine) randomness of the iterates can only arise out of the randomness of $\mathbf{x}_0$ (if postulated as such). Notice that the case where $f_\mu = \tau$ (see Example 6), connected with the continued fraction expansion, is a classical one. It is worth mentioning that iterations of deterministic maps were extensively studied by G. Julia and P. Fatou just after the First World War and that at present there is a great revival of interest in this topic. See e.g. Liedl, Reich & Targonski (Eds.) (1985), Preston (1983), Peitgen & Richter (1986), and Thompson & Stewart (1986).

Example 8 *A class of stochastic difference equations* (*Vervaat* (1979)). Let us consider the difference equation

$$Y_n = A_n Y_{n-1} + B_n, \quad n \in \mathbb{N}^*,$$

where (A_n, B_n), $n \in \mathbb{N}^*$, are two-dimensional independent and identically distributed random vectors; denote by $\mathbf{p}$ their common probability distribution. This equation arises in various fields, e.g. economics, physics, nuclear technology, biology, and sociology. In applications, Y_n is a stock of certain objects at time n, B_n is the quantity which is added to (if $B_n \geqslant 0$) or taken away (if $B_n < 0$) from the stock just before time n, and the coefficient A_n indicates the intrinsic decrease or increase of the stock within the time interval $(n-1, n]$, $n \in \mathbb{N}^*$. The applications pertain to savings accounts, radioactive materials, brightness of the Milky Way, environmental pollution, investments, atomic cascade models, random walks in random environments, etc. The appropriate references can be found in Vervaat (1979).

It is obvious that the sequence $(Y_n)_{n \in \mathbb{N}}$ is equivalent to the MC associated with the RSCC $\{(W, \mathscr{W}), (X, \mathscr{X}), u, P\}$, where

$$(W, \mathscr{W}) = (\mathbb{R}, \mathscr{B}), \quad (X, \mathscr{X}) = (\mathbb{R}^2, \mathscr{B}^2),$$
$$u(w, (a, b)) = aw + b, \quad w \in \mathbb{R}, \quad (a, b) \in \mathbb{R}^2,$$
$$P(w, A) = \mathbf{p}(A), \quad w \in \mathbb{R}, \quad A \in \mathscr{B}^2.$$

For further consideration of the above stochastic difference equation see Grincevičius (1981). A recent paper (Brandt, 1986) takes up the case where $(A_n, B_n)_{n \in \mathbb{N}^*}$ is a strictly stationary sequence. See also Lisek (1982).

Example 9 *The motion of a particle in a random environment* (*Iordache & Iosifescu* (*1981*)). Some phenomena in the chemical industries motivate the following model for the motion of a particle subject to several random walks. Assume that a particle may be, at any time $n \in \mathbb{N}$, in one of the points $i = 0, \pm 1, \pm 2, \ldots$ of the one-dimensional integer lattice $\mathbb{Z}$. At any time n, the particle performs a jump of (random) amplitude $a \in \mathbb{Z}$ with probability $p_k(a)$ conditional on some environmental condition $k \in K$ (K is a denumerable set). If at time n the particle is in state $i \in \mathbb{Z}$, and the environmental condition is $k \in K$, then the probability of changing, within the time interval $(n, n+1]$, to a new condition $l \in K$ is $p_{kl}(i)$. Obviously

$$\sum_{a \in \mathbb{Z}} p_k(a) = 1, \quad \sum_{l \in K} p_{kl}(i) = 1, \quad k \in K, \quad i \in \mathbb{Z}.$$

Therefore, if in $i \in \mathbb{Z}$ under condition $k \in K$, the particle performs a jump of amplitude $a \in \mathbb{Z}$ and the environment moves to condition $l \in K$ with probability $p_k(a)p_{kl}(i)$. Next, starting at $i+a$[†] under condition $l \in K$, the particle performs a jump of amplitude $a' \in \mathbb{Z}$ and comes under condition $l' \in K$ with probability $p_l(a')p_{ll'}(i+a)$ and so on.

Let us consider the two-dimensional random variables

$$\lambda_{n-1} = \begin{pmatrix} \text{location of the particle at time } n-1 \\ \text{environmental condition at time } n-1 \end{pmatrix},$$

$$\alpha_n = \begin{pmatrix} \text{amplitude of the jump at time } n-1 \\ \text{environmental condition at time } n \end{pmatrix}, \quad n \in \mathbb{N}^*.$$

Then the sequences $(\lambda_n)_{n \in \mathbb{N}}$ and $(\alpha_n)_{n \in \mathbb{N}^*}$ are equivalent to the MC and the chain of infinite order associated with the RSCC $\{(W, \mathscr{W}), (X, \mathscr{X}), u, P\}$, where

$$W = X = \mathbb{Z} \times K, \quad \mathscr{W} = \mathscr{X} = \mathscr{P}(\mathbb{Z} \times K),$$

$$u((i,k),(a,l)) = (i+a, l), \quad P((i,k),(a,l)) = p_k(a)p_{kl}(i)$$

for all $(i,k) \in W$, $(a,l) \in X$.

Example 10 *A formalization of debate dynamics in small groups* (*Chevrolet & Le Calvé* (1973)). Let us consider a set $\mathfrak{J}$ of h individuals who have a task to fulfil or a problem to solve (to study a given situation, to make a decision, etc.).

From the moment of acquiring the available information, each individual is characterized, in terms of social psychology, by an 'attitude', i.e. an aprioristic state of mind resulting from a specific way of information processing. (This state of mind is subject to change, distortion, interference, etc.) Such a subjective structure of information related to the task in hand is epitomized by what we shall call a predisposition. Let us denote by Z the (finite) set of all possible states of mind; then $Z^{\mathfrak{J}}$ is the set of all possible predispositions. Let us denote by $z_n \in Z^{\mathfrak{J}}$ the predisposition at time $n \in \mathbb{N}$.

From an ethological point of view, at the time the h individuals meet, a comprehensive dynamic system of interrelations and exchanges comes into being and starts functioning, relying on a series of observable behaviours. At a verbal level it is expressed by compatibilities and incompatibilities, agreements and disagreements, acceptance or rejection of other perceptions. In other words, emotional links are immediately and spontaneously established between any two individuals in the group; such links may oscillate in time within a bipolar continuum ranging from 'sympathy' to 'antipathy'. The state of these links will be represented by a real $h \times h$ matrix $\theta = (\theta(i,j))_{i,j \in \mathfrak{J}}$, which is called the sympathy matrix, the

[†] Here $+$ stands either for usual addition or, more generally, for an arbitrary operation on $\mathbb{Z}$.

elements of which belong to the unit interval. Thus for instance, $\theta(i, j) = 1$ stands for the complete agreement of i with the statement made by j. It is obvious that, generally, $\theta(i, j) \neq \theta(j, i)$, $i \neq j$. With respect to $\theta(i, i)$, i.e. the relation of individual i to what he expresses, we shall assume that it is a measure of the authenticity of his expression. The equation $\theta(i, i) = 1$ stands for the complete authenticity of the expression of i. We shall denote by $\theta_n = (\theta_n(i, j))_{i,j \in \mathfrak{J}}$ the sympathy matrix at time $n \in \mathbb{N}$.

A group can exist and function only to the extent that it shows a certain cohesiveness. This cohesiveness – which can be perceived objectively at the level of the group functioning as such (generation of rules, values, establishment of ceremonial procedures, emergence of moral standards for the group as a whole, etc.) – appears as the end result of many concurrent factors, such as the impact of the group prestige upon the individuals, the appeal of the goals pursued by the group, the needs it seeks to meet, etc. For this reason every individual can be assigned a certain wish to participate in the group's work, a certain willingness to see that it makes progress, to wit a coefficient describing the degree of an individual's integration with the group, a measure of the level of his adhesion and solidarity. Similarly to the case of sympathy, we shall assume that the integration coefficient ranges between 0 and 1. Denoting by $\alpha_n(i)$ the integration coefficient of individual i at time $n \in \mathbb{N}$, the vector $\alpha_n = (\alpha_n(i))_{i \in \mathfrak{J}} \in [0, 1]^{\mathfrak{J}}$ can be looked at as the group's cohesiveness at time $n \in \mathbb{N}$, the way it can be perceived objectively, e.g. at the level of group 'morale'.

Finally, we shall denote by i_n the individual who intervenes in the debate at time n and by $a_n \in A$ his intervention, $n \in \mathbb{N}^*$. (Here A is the finite set of all possible interventions about the issue.)

The variables $z_n, \theta_n, \alpha_n, i_{n+1}, a_{n+1}, n \in \mathbb{N}$, will be assumed to be random. Concerning their dynamics, we shall assume that the triple predisposition–sympathy–morale $(z_{n+1}, \theta_{n+1}, \alpha_{n+1})$ is a function of the same triple $(z_n, \theta_n, \alpha_n)$ at the preceding time as well as of i_{n+1} and a_{n+1}, i.e. of the intervening individual and his intervention at the actual time. We shall write

$$(z_{n+1}, \theta_{n+1}, \alpha_{n+1}) = F[(z_n, \theta_n, \alpha_n), i_{n+1}, a_{n+1}], \quad n \in \mathbb{N}.$$

At any time there is a certain probability for a certain individual to intervene. Obviously, this probability depends on the triple predisposition–sympathy–morale. Then the probability of a certain intervention depends on both the intervening individual and the triple predisposition–sympathy–morale. More precisely we shall denote

$$\mathsf{P}(i_{n+1} = i \mid z_n, \theta_n, \alpha_n) = Q(z_n, \theta_n, \alpha_n; i),$$
$$\mathsf{P}(a_{n+1} = a \mid z_n, \theta_n, \alpha_n, i_{n+1}) = m(z_n, \theta_n, \alpha_n, i_{n+1}; a), \quad n \in \mathbb{N}.$$

It is obvious that the sequences $(z_n, \theta_n, \alpha_n)_{n\in\mathbb{N}}$ and $(i_n, a_n)_{n\in\mathbb{N}^*}$ are equivalent to the MC and the chain of infinite order associated with the RSCC $\{(W, \mathscr{W}), (X, \mathscr{X}), u, P\}$, where

$$W = Z^{\mathfrak{J}} \times [0,1]^{\mathfrak{J}\times\mathfrak{J}} \times [0,1]^{\mathfrak{J}}, \quad \mathscr{W} = \mathscr{P}(Z^{\mathfrak{J}}) \otimes \mathscr{B}^{\mathfrak{J}\times\mathfrak{J}}_{[0,1]} \otimes \mathscr{B}^{\mathfrak{J}}_{[0,1]},$$

$$X = \mathfrak{J} \times A, \mathscr{X} = \mathscr{P}(\mathfrak{J} \times A),$$

$$u = F,$$

$$P(w, x) = m(z, \theta, \alpha, i; a)Q(z, \theta, \alpha; i), \quad w = (z, \theta, \alpha) \in W, \quad x = (i, a) \in X.$$

An application of this model in diplomatic talks can be found in Grigorescu (1977b). For Markovian models of group behaviour see e.g. Kelly (1981) and the references therein, and Nowakowska (1981).

Example 11 *Partially observed random chains (Dynkin (1965), Širjaev (1967)).* The basic mathematical object dealt with by mathematical statistics is a random pair (θ, x) on a probability space $(\Omega, \mathscr{K}, \mathbf{P})$, where $\theta \in \Theta$ is a parameter (either random or non-random) whose observation is inaccessible to the statisticians, while $x \in \mathfrak{X}$ is an observed random element whose probability distribution depends on θ. In what follows, to avoid technicalities, we shall assume that the sets Θ and $\mathfrak{X}$ are denumerable. The consideration of general spaces Θ and $\mathfrak{X}$ does not involve new ideas. Most of mathematical statistics problems amount to indicating ways to get, on account of the observed values of x, the best estimation (in a specified meaning) of the unobserved component θ.

In a dynamical approach to the problem, let us consider a sequence of trials: the trial of rank n has the unobserved outcome θ_n and the observed outcome x_n, $n \in \mathbb{N}^*$. Starting from an 'initial' value θ_0 let us denote

$$x^{(n)} = (x_1, \ldots, x_n), \quad n \in \mathbb{N}^*, \quad \theta^{(n)} = (\theta_0, \theta_1, \ldots, \theta_n), \quad n \in \mathbb{N}.$$

We shall assume that for any time n the conditional distribution of (θ_{n+1}, x_{n+1}), given $\theta^{(n)}$ and $x^{(n)}$, is known, and denote

$$\mathbf{P}(\theta_{n+1} = \theta, x_{n+1} = x \mid \theta^{(n)}, x^{(n)}) = p(\theta^{(n)}, x^{(n)}; \theta, x), \quad n \in \mathbb{N},$$

with the convention $(\theta^{(0)}, x^{(0)}) = \theta_0$.

The sequence $(\theta_n, x_n)_{n\in\mathbb{N}^*}$ is called a *partially observed random chain.* The simplest example of a partially observed random chain is a sequence of independent identically distributed random variables, the common probability distribution of which depends on an unknown parameter.

Therefore, for a partially observed random chain, at any time n, the statistician knows the observed trajectory $x^{(n)}$. As regards the unobserved trajectory $\theta^{(n)}$, the conditional probability π_n of $\theta^{(n)}$, given $x^{(n)}$, can be computed, $n \in \mathbb{N}^*$. The pair $\sigma_n = (\pi_n, x^{(n)})$ is called *the observer's state* at time $n \in \mathbb{N}^*$. We define $\sigma_0 = (\pi_0, x^{(0)}) = \pi_0$ as the *a priori* probability distribution of θ_0. How does σ_n change when at time $n+1$ the observed outcome x occurs? Clearly $\sigma_{n+1} = (\pi_{n+1}, x^{(n+1)})$, where

$$x^{(n+1)} = x^{(n)}x = (x_1, \ldots, x_n, x)$$

and, by Bayes' formula,

$$\pi_{n+1}(\theta^{(n)}\theta) = \frac{\pi_n(\theta^{(n)})p(\theta^{(n)}, x^{(n)}; \theta, x)}{\sum_{\tilde{\theta}^{(n)} \in \Theta^{n+1}, \tilde{\theta} \in \Theta} \pi_n(\tilde{\theta}^n)p(\tilde{\theta}^{(n)}, x^{(n)}; \tilde{\theta}, x)},$$

where $\theta^{(n)}\theta = (\theta_0, \ldots, \theta_n, \theta)$. If the denominator of the fraction above equals 0, then π_{n+1} can be arbitrarily given. Thus π_{n+1} appears to be a function S_x of $\pi_n, n \in \mathbb{N}^*$. Clearly, the probability of the transition from π_n to $\pi_{n+1} = S_x\pi_n$ equals

$$P(\sigma_n, x) = \sum_{\theta^{(n)} \in \Theta^{n+1}, \theta \in \Theta} \pi_n(\theta^{(n)})p(\theta^{(n)}, x^{(n)}; \theta, x),$$

where $\sigma_n = (\pi_n, x^{(n)}), n \in \mathbb{N}$.

It follows from the considerations above that the sequence $(\sigma_n)_{n \in \mathbb{N}}$ is equivalent to the MC $(\zeta_n)_{n \in \mathbb{N}}, \zeta_0 = \pi_0$, associated with the RSCC $\{(W_n, \mathscr{W}_n)_{n \in \mathbb{N}}, (X, \mathscr{X}), u, P\}$, where

$$W_0 = M_0, \quad W_n = M_n \times \mathfrak{X}^n, \quad n \in \mathbb{N}^*,$$

$$\mathscr{W}_0 = \mathscr{M}_0, \quad \mathscr{W}_n = \mathscr{M}_n \otimes \mathscr{P}(\mathfrak{X}^n), \quad n \in \mathbb{N}^*,$$

$$X = \mathfrak{X}, \mathscr{X} = \mathscr{P}(\mathfrak{X}),$$

$$u(\cdot, x) = S_x, \quad x \in \mathfrak{X},$$

$$P(w, x) = P(\sigma_n, x) \quad \text{if} \quad w = \sigma_n \in W_n, \quad n \in \mathbb{N}.$$

Here M_n stands for the set of probability distributions on Θ^{n+1} (and therefore is a subset of $\mathbb{R}_+^{\Theta^{n+1}}$), while $\mathscr{M}_n$ is the σ-algebra of Borel sets in $M_n, n \in \mathbb{N}$.

Example 12 *Controlled partially observed random chains* (*Dynkin* (1965) *Širjaev* (1967)). Generalizing the set-up of the preceding example, let us assume that the transition probabilities $p(\theta^{(n)}, x^{(n)}; \theta, x) = p^d(\theta^{(n)}, x^{(n)}; \theta, x)$ depend on a parameter $d \in D$ (a denumerable set), the value of which should be interpreted as a *decision* or *control* enabling the observer to interfere with the course of the process evolution. More precisely, we shall assume that, at any time, the statistician, using the available information, is able to choose either a value $d \in D$ (*non-randomized* control) or a probability distribution on D (*randomized* control). Let us denote by $d^{(n)} = (d_1, \ldots, d_n)$ the values of the control at times $1, \ldots, n$, and assume that

$$\mathsf{P}(\theta_{n+1} = \theta, x_{n+1} = x, d_{n+1} = d \mid \theta^{(n)}, x^{(n)}, d^{(n)}) = p^d(\theta^{(n)}, x^{(n)}; \theta, x)$$

with the convention $(\theta^{(0)}, x^{(0)}, d^{(0)}) = \theta_0$. Note that, at any time n, the statistician knows both $d^{(n)}$ and the observed trajectory $x^{(n)}$. As regards the unobserved trajectory $\theta^{(n)}$, the conditional probability π_n of $\theta^{(n)}$, given

$x^{(n)}$ and $d^{(n)}$, can be computed, $n\in\mathbb{N}^*$. The triple $\sigma_n=(\pi_n, x^{(n)}, d^{(n)})$ is called *the observer's state* at time $n\in\mathbb{N}^*$. We define $\sigma_0=(\pi_0, x^{(0)}, d^{(0)})=\pi_0$ as the *a priori* probability distribution of θ_0. If at time $n+1$ the observed outcome x occurs and the control d is chosen, then, clearly, $\sigma_{n+1}=(\pi_{n+1}, x^{(n+1)}, d^{(n+1)})$, where $x^{(n+1)}=x^{(n)}x=(x_1,\ldots,x_n,x)$, $d^{(n+1)}=d^{(n)}d=(d_1,\ldots,d_n,d)$, and, by Bayes' formula,

$$\pi_{n+1}(\theta^{(n)}\theta)=\frac{\pi_n(\theta^{(n)})p^d(\theta^{(n)},x^{(n)};\theta,x)}{\sum\limits_{\tilde{\theta}^{(n)}\in\Theta^{n+1},\tilde{\theta}\in\Theta}\pi_n(\tilde{\theta}^{(n)})p^d(\tilde{\theta}^{(n)},x^{(n)},\tilde{\theta},x)},$$

where $\theta^{(n)}\theta=(\theta_0,\ldots,\theta_n,\theta)$. If the denominator of the fraction above equals 0, then π_{n+1} can be arbitrarily given. Thus π_{n+1} appears to be a function $S_{x,d}$ of $\pi_n, n\in\mathbb{N}$. Clearly, the probability of the transition from π_n to $\pi_{n+1}=S_{x,d}\pi_n$ equals

$$P(\sigma_n,x,d)=\sum_{\theta^{(n)}\in\Theta^{n+1},\theta\in\Theta}\pi_n(\theta^{(n)})p^d(\theta^{(n)},x^{(n)},\theta,x),$$

where $\sigma_n=(\pi_n,x^{(n)},d^{(n)})$, $n\in\mathbb{N}$.

It follows from the considerations above that the sequence $(\sigma_n)_{n\in\mathbb{N}}$ is equivalent to the MC $(\zeta_n)_{n\in\mathbb{N}}, \zeta_0=\pi_0$, associated with the RSCC $\{(W_n,\mathscr{W}_n)_{n\in\mathbb{N}},(X,\mathscr{X}),u,P\}$, where

$$W_0=M_0,\quad W_n=M_n\times\mathfrak{X}^n\times D^n,\quad n\in\mathbb{N}^*,$$

$$\mathscr{W}_0=\mathscr{M}_0,\quad \mathscr{W}_n=\mathscr{M}_n\otimes\mathscr{P}(\mathfrak{X}^n)\otimes\mathscr{P}(D^n),\quad n\in\mathbb{N}^*,$$

$$X=\mathfrak{X}\times D,\quad \mathscr{X}=\mathscr{P}(\mathfrak{X})\otimes\mathscr{P}(D),$$

$$u(\cdot,x,d)=S_{x,d},\quad (x,d)\in X,$$

$$P(w,x,d)=P(\sigma_n,x,d)\quad\text{if}\quad w=\sigma_n\in W_n,\quad n\in\mathbb{N}.$$

Here M_n and $\mathscr{M}_n$, $n\in\mathbb{N}$, have the same meaning as in the preceding example.

Example 13 *Decision models and dynamic programming.* As remarked by Hinderer (1970, p. 49) the last two examples can be viewed as special cases of decision models. Moreover, the concept of a decision model is intrinsic in various problems ranging from the classical problem of maximizing a real valued function of a finite number of variables under constraints, to problems of optimal search and optimal stopping. The appropriate references are to be found in Hinderer (1970). The concept of a decision model is in fact the mathematical framework for the theory of dynamic programming under risk. Following Hinderer (1970, Chapter 1) the former consists of a collection $\{S,A,D,(p_n)_{n\in\mathbb{N}},(r_n)_{n\in\mathbb{N}}\}$ of objects, where

(i) S is a non-empty denumerable† set called the *state space*;

† This assumption is not essential. It is made here just for the sake of simplicity.

(ii) A is a non-empty denumerable set called the *action space*;

(iii) $D=(D_n)_{n\in\mathbb{N}^*}$ is a sequence of maps D_n from certain sets $H_n\subset\bar{H}_n=S$ or $S\times A\times\cdots\times S\times A\times S$ $(2n-1$ factors), according as $n=1$ or $n\geqslant 2$, into the class of all non-empty subsets of A such that

$$H_1=S,$$
$$H_{n+1}=\{(h,a,s):h\in H_n, a\in D_n(h), s\in S\},\quad n\in\mathbb{N}^*.$$

Note that $\bar{H}_n$ is called the *set of histories* at time n, H_n the *set of admissible histories* at time n, and $D_n(h)$ the *set of admissible actions* at time n under admissible history h. Denoting $K_n=\{(h,a):h\in H_n, a\in D_n(h)\}$ we can write $H_{n+1}=K_n\times S, n\in\mathbb{N}^*$;

(iv) $\mathbf{p}_0$ is a probability distribution on S called the initial distribution, and p_n is a t.p.f. from K_n to S (i.e. for any $(h,a)\in K_n$, $p_n(h,a,\cdot)$ is a probability distribution on S) called the *transition law* from time n to time $n+1, n\in\mathbb{N}^*$;

(v) r_n is an extended real-valued function on K_n called the *reward* during the time interval $(n,n+1)$.

A collection $\{S,A,D,(p_n)_{n\in\mathbb{N}},(r_n)_{n\in\mathbb{N}}\}$ with the above properties is called a (*stochastic dynamic*) *decision model*. To define a dynamic optimization problem, it is necessary to choose a policy and a criterion of optimality (under suitable assumptions on the r_n). A (*deterministic admissible*) *policy* is a sequence $f=(f_n)_{n\in\mathbb{N}^*}$ of maps $f_n:S^n\to A$ such that

$$f_n(y)\in D_n(h_n^f(y)),\quad n\in\mathbb{N}^*,\quad y\in S^n,$$

where

$$h_n^f(y)=\begin{cases}(s_1), & \text{if } n=1\\ (s_1,f_1(s_1),s_2,f_2(s_1,s_2),\ldots,f_{n-1}(s_1,\ldots,s_{n-1}),s_n), & \text{if } n\geqslant 2\end{cases}$$

denotes the history at time n obtained when the sequence $y=(s_1,\ldots,s_n)$ of states occurred and actions were chosen according to f.

The application of a policy f generates a stochastic process, the *decision process* determined by f, which can be described as follows. Starting at time $n=1$ at some point s_1, selected from S according to the initial probability distribution $\mathbf{p}_0$, one takes action $f_1(s_1)$ whereupon one gets the reward $r_1(s_1,f_1(s_1))$, and the system moves to some $s_2\in S$ selected according to $p_1(s_1,f_1(s_1),\cdot)$. Then action $f_2(s_1,s_2)$ is taken, giving the reward $r_2(s_1,f_1(s_1),s_2,f_2(s_2))$ and the system moves to some $s_3\in S$ selected according to $p_2(s_1,f_1(s_1);s_2,f_2\ (s_1,s_2),\cdot)$, and so on.

In fact, for a given policy f, we are dealing with the RSCC $\{(W_n^f,\mathscr{W}_n^f)_{n\in\mathbb{N}},(X,\mathscr{X}),(u_n^f)_{n\in\mathbb{N}},({}^nP^f)_{n\in\mathbb{N}}\}$, where

$$W_0^f=\{w_0\}\text{ – a one-element set,}$$
$$W_n^f=\{(s_1,f_1(s_1),\ldots,s_n,f_n(s_1,\ldots,s_n)):(s_1,\ldots,s_n)\in S^n\},$$

$$\mathscr{W}_n^f = \mathscr{P}(W_n^f), \quad n \in \mathbb{N}^*,$$
$$X = S, \quad \mathscr{X} = \mathscr{P}(S),$$
$$u_0^f(w_0, x) = (x, f_1(x)), \quad x \in X,$$
$$u_n^f(w, x) = (w, x, f_{n+1}(\mathrm{pr}_{13\ldots(2n-1)} w, x)), \quad n \in \mathbb{N}^*, \quad w \in W_n^f, \quad x \in X,$$
$${}^0P^f = \mathbf{p}_0, \quad {}^nP^f(w, \cdot) = p_n(w, \cdot), \quad n \in \mathbb{N}^*, \quad w \in W_n^f.$$

The sequence of states $(s_n)_{n \in \mathbb{N}^*}$ appears then to be equivalent to the infinite order chain associated with the above RSCC.

If a randomized admissible policy is applied, instead of a deterministic one, we are faced with another situation.

A *randomized admissible policy* is a sequence $\pi = (\pi_n)_{n \in \mathbb{N}}$ of t.p.f.s π_n from H_{n+1} to A, $n \in \mathbb{N}$ (hence π_0 is a t.p.f. from S to A), such that

$$\operatorname{supp} \pi_0(s_1, \cdot) \subseteq D_1(s_1), \quad s_1 \in S,$$
$$\operatorname{supp} \pi_n(s_1, a_1, \ldots, s_n, a_n, s_{n+1}, \cdot) \subseteq D_{n+1}(s_1, a_1, \ldots, s_n, a_n, s_{n+1})$$

for all $n \in \mathbb{N}^*$, $(s_1, \ldots, s_n, s_{n+1}) \in S^{n+1}$, $a_1 \in \operatorname{supp} \pi_0(s_1, \cdot), \ldots, a_n \in \operatorname{supp} \pi_{n-1}(s_1, a_1, \ldots, s_{n-1}, a_{n-1}, s_n, \cdot)$.

In the present case we are led to consider the non-homogeneous GRSCC $\{(W_n^\pi, \mathscr{W}_n^\pi)_{n \in \mathbb{N}}, (X, \mathscr{X}), ({}^n\Pi^\pi)_{n \in \mathbb{N}}, ({}^nP^\pi)_{n \in \mathbb{N}}\}$, where

$$W_0^\pi = \{w_0\} \text{ – a one-element set,}$$
$$W_n^\pi = \{(s_1, a_1, \ldots, s_n, a_n) : (s_1, \ldots, s_n) \in S^n, a_1 \in \operatorname{supp} \pi_0(s_1, \cdot), \ldots$$
$$\ldots, a_n \in \operatorname{supp} \pi_{n-1}(s_1, a_1, \ldots, s_{n-1}, a_{n-1}, s_n, \cdot)\},$$
$$\mathscr{W}_n^\pi = \mathscr{P}(W_n^\pi), \quad n \in \mathbb{N}^*,$$
$$X = S, \quad \mathscr{X} = \mathscr{P}(S),$$
$${}^0\Pi^\pi(w_0, x, \cdot) = \pi_0(x, \cdot), \quad x \in X,$$
$${}^n\Pi^\pi(w, x, \cdot) = \pi_n(s_1, a_1, \ldots, s_n, a_n, s_{n+1}, \cdot), \quad n \in \mathbb{N}^*,$$
$$(\text{for } w = (s_1, a_1, \ldots, s_n, a_n) \in W_n^\pi, \quad x = s_{n+1} \in X),$$
$${}^0P^\pi = \mathbf{p}_0, \quad {}^nP^\pi(w, \cdot) = p_n(w, \cdot), \quad n \in \mathbb{N}^*, \quad w \in W_n^\pi.$$

In this case the decision process can be described as follows. Starting at time $n = 1$ at some point s_1, selected from S according to the initial distribution $\mathbf{p}_0$, one takes action a_1 according to the t.p.f. $\pi_0(s_1, \cdot)$ whereupon one gets the reward $r_1(s_1, a_1)$, and the system moves to some $s_2 \in S$ selected according to $p_1(s_1, a_1, \cdot)$. Then, according to the t.p.f. $\pi_1(s_1, a_1, s_2, \cdot)$, action a_2 is taken, giving the reward $r_2(s_1, a_1, s_2, a_2)$, and the system moves to some $s_3 \in S$ selected according to $p_2(s_1, a_1, s_2, a_2, \cdot)$, and so on.

A *Markov* randomized policy is a policy $\pi = (\pi_n)_{n \in \mathbb{N}}$ for which π_0 is a t.p.f. from S to A, while π_n, $n \in \mathbb{N}^*$, is a t.p.f. from $A \times S$ to A. In other words, at a given moment, a decision is taken with a probability depending only on the state of the system at that moment and on the last decision taken (and not depending on the entire history of the system up to that moment). Considering only Markov randomized policies, we obtain the so-called Markov decision model due to Blackwell (1965). This model is also a non-homogeneous GRSCC $\{(W, \mathscr{W}), (X, \mathscr{X}), ({}^n\Pi)_{n\in\mathbb{N}}, ({}^nP)_{n\in\mathbb{N}}\}$, where

$$W = A, \quad \mathscr{W} = \mathscr{P}(A),$$
$$X = S, \quad \mathscr{X} = \mathscr{P}(S),$$
$${}^n\Pi = \pi_n, \quad {}^nP = p_n, \quad n\in\mathbb{N}.$$

Finally, we note that the study of decision models under uncertainty – for which the sequence $(p_n)_{n\in\mathbb{N}}$ is not completely known to the observer (including the extreme case of complete uncertainty, when nothing is known about the p_n) – can be reduced to the study of suitably chosen decision models. See Rieder (1975) and the references therein.

Example 14 *A dynamic stochastic production and inventory model* (*Hinderer* (1973)). Let us first define the elements of this model.

Z is a denumerable set called the set of the external states of the system; we assume that its elements are not observable;

$q = (q_{x,y})_{x,y\in Z}$ is a stochastic matrix;

A is a finite set called the set of the internal states of the system: an internal state $a\in A$ is a vector (c, l, v), whose components represent the production capacity, the stock of goods, and the stocking capacity, respectively.

At the beginning of the time interval $(t, t+1)$, $t\in\mathbb{N}^*$, a manager is faced with an internal state (c_t, l_t, v_t) of the system and with an external state ζ_t. He is supposed to choose a policy leading to a new internal state $(c_{t+1}, l_{t+1}, v_{t+1})$ such that the production and stocking costs be minimized. In Hinderer's model the vector $(c_{t+1}, l_{t+1}, v_{t+1})$ may be chosen from a subset $D(a_t, \zeta_t)$ of $A \times Z$, which is defined by imposing some natural (from the economic point of view) restrictions on the model. The policy chosen by the manager is assumed to be randomized, i.e. a_{t+1} is chosen according to a t.p.f. Ψ from $A \times Z$ to A. Mathematically speaking, the restrictions mentioned above lead to the fact that the support of the probability $\Psi(a_t, \zeta_t, \cdot)$ coincides with $D(a_t, \zeta_t)$. On the other hand, at the beginning of the time interval $(t+1, t+2)$, $t\in\mathbb{N}^*$, the probability that the external state ζ_{t+1} is y, given that $\zeta_t = x$, is $q_{x,y}$.

The model just described may be looked at as a homogeneous GRSCC $\{(W, \mathscr{W}), (X, \mathscr{X}), \Pi, P\}$, where

$$W = A \times Z, \quad X = A,$$

$$\mathscr{W} = \mathscr{P}(W), \quad \mathscr{X} = \mathscr{P}(X),$$

$$P((a,x),c) = \Psi(a,x,c), \quad (a,x) \in W, \quad c \in X,$$

$$\Pi((a,x),c,(b,y)) = q_{x,y}\delta(c,b), \quad (a,x),(b,y) \in W, \quad c \in X.$$

Here $\delta(c,b) = 1$ if $c = b$, and $\delta(c,b) = 0$ otherwise.

1.3 Classification problems

1.3.1 A communication relation

The classification of states for a simple homogeneous denumerable MC is a problem which has been solved long since. This is not the case for other discrete parameter stochastic processes. Let us notice for instance that even for simple or multiple non-homogeneous denumerable MCs this problem does not have a complete solution.

In this section, we shall deal with the problem of classifying the elements of the set X of a homogeneous RSCC $\{(W, \mathscr{W}), (X, \mathscr{X}), u, P\}$, in the case where X is denumerable.

Definition 1.3.1. *Element $j \in X$ is said to be* accessible *from $i \in X$ (denoted $i \to j$) iff, for any $w \in W$, there exists $m = m(w) \in \mathbb{N}^*$ such that $P^m(wi, j) > 0$. Elements i and $j \in X$ are said to* communicate *(denoted $i \leftrightarrow j$) iff $i \to j$ and $j \to i$.*

Proposition 1.3.2. *We have $i \to j$ iff, for any $w \in W$, there exists $m = m(w) \in \mathbb{N}^*$ and $i^{(m)} = (i_1, \ldots, i_m) \in X^m$ (also depending on w), with $i_m = j$, for which*

$$P_m(wi, i^{(m)}) > 0.$$

Proof. Since

$$P^m(wi, j) = \sum_{\{i^{(m)} : i^{(m)} \in X^m, i_m = j\}} P_m(wi, i^{(m)}),$$

the assertion above is an immediate consequence of Definition 1.3.1. □

Note that, given $i \in X$, it is possible that there is no $j \in X$ accessible from i (see Problem 10). This fact illustrates an essential difference in comparison with the case of simple homogeneous MCs.

In general, the relation '$\to$' is neither reflexive nor symmetrical, but it is transitive, as stated in Proposition 1.3.3.

Proposition 1.3.3. *If $i \to j$ and $j \to k$, then $i \to k$.*

Proof. We have

$$P^{m+n}(wi,k)=\sum_{i^{(m)}\in X^m} P_m(wi,i^{(m)})P^n(wii^{(m)},k)\geqslant P_m(wi,i^{(m)})P^n(wii^{(m)},k)$$

for all $i^{(m)}\in X^m$. Now, take $i_m=j$ and use Proposition 1.3.2 to conclude the proof. □

Hence the relation '$\leftrightarrow$' is symmetrical and transitive; it is also reflexive on the set $\tilde{X}$ of those $i\in X$ for which $i\leftrightarrow i$. Let us say that two elements $i,j\in X$ belong to the same class iff they communicate. In this way, the relation '$\leftrightarrow$' splits the set $\tilde{X}$ into disjoint subsets, called *classes.*

Let us agree that any element $i\in X-\tilde{X}$ forms a class by itself. A class containing the element $i\in X$ will be denoted C(i). We say that a property which makes sense for all elements in a class, is a *class property* iff its validity for one element in the class implies its validity for all the elements in the class.

Definition 1.3.4. *Element $i\in X$ is said to be* essential *iff, for any $j\in X$ with $i\to j$, also $j\to i$; otherwise it is said to be* inessential.

This definition, together with Proposition 1.3.3, implies the validity of Proposition 1.3.5.

Proposition 1.3.5. *An inessential element is not accessible from any essential element. Therefore, the property of an element of being essential (inessential) is a class property.*

Proposition 1.3.5 enables us to speak of *essential* and *inessential classes.*

To be able to go on we should introduce the following condition.

Condition C. *The natural integer m in Definition 1.3.1 does not depend on w.*

It is obvious that Condition C introduces a new restriction, making it even more difficult to find elements which are accessible from a given element. A simple condition which implies condition C is Condition K_1.

Condition K_1. *There exists a number $\alpha>0$ such that $P(w'i,j)\geqslant\alpha P(w''i,j)$ for all $w',w''\in W$, and $i,j\in X$.*

A condition of this type was introduced by Kolmogorov (1931) to study simple non-homogeneous MCs. Note that α is necessarily $\leqslant 1$ (this is obtained by summing the inequality in the definition over $j\in X$). Obviously,

Condition K_1 holds for simple homogeneous MCs (but not for all multiple MCs!)

Proposition 1.3.6. *Condition* K_1 *implies Condition* C.

Proof. Let $w' \in W$, $i, j \in X$, and assume that $P^m(w'i, j) > 0$ for some $m \in \mathbb{N}^*$. We shall show that $P^m(wi, j) > 0$ for all $w \in W$. Indeed, by Proposition 1.3.2, there exists $i^{(m)} \in X$, with $i_m = j$, such that $P_m(w'i, i^{(m)}) > 0$. Now Condition K_1 yields

$$P^m(wi, j) \geqslant P_m(wi, i^{(m)}) = \begin{cases} P(wi, j), & \text{if } m = 1 \\ P(wi, i_1) \prod_{r=1}^{m-1} P(wii^{(r)}, i_{r+1}), & \text{if } m > 1 \end{cases}$$
$$\geqslant \alpha^m P_m(w'i, j) > 0. \qquad \square$$

We shall write $i \xrightarrow{m} j$ if $P^m(wi, j) > 0$ for all $w \in W$.

Proposition 1.3.7. *If* $i \xrightarrow{m} j$ *and* $j \xrightarrow{n} k$, *then* $i \xrightarrow{m+n} k$.

The proof is identical to that of Proposition 1.3.3.

Throughout the remainder of this subsection we shall assume that Condition C holds.

If $i \to i$, the greatest common divisor of the natural integers n for which $i \xrightarrow{n} i$ will be called the *period* of i and will be denoted d_i.

Theorem 1.3.8. *The property of having period d is a class property.*

Using Proposition 1.3.7 the proof is obtained along the lines of the proof in the Markovian case (see Chung (1967, p. 13)).

Theorem 1.3.8 enables us to speak of the *period* of a class.

Theorem 1.3.9. *For any* $j \in C(i)$ *there exists a unique residue class* r_j *modulo* d_i *such that if* $i \xrightarrow{n} j$ *then* $n \equiv r_j \pmod{d_i}$. *There exists* $n(j)$ *such that if* $n \geqslant n(j)$ *then* $i \xrightarrow{nd_i + r_j} j$.

As for Theorem 1.3.8, the proof follows the ideas in the Markovian case (see Chung (1967, p. 14)).

The set of the elements $j \in C(i)$ which correspond by Theorem 1.3.9 to the same residue class r modulo d_i will be called a subclass of C(i), more precisely the *subclass* $C_r(i)$, $0 \leqslant r \leqslant d_i - 1$. It is convenient to extend this definition to all non-negative integers r, by putting $C_r(i) = C_s(i)$ whenever $r \equiv s (\mathrm{mod}\, d_i)$. It is obvious that the d_i subclasses $C_r(i)$, $0 \leqslant r \leqslant d_i - 1$, are disjoint and that their union is $C(i)$. Moreover, if $j \in C_r(i)$, then $C_s(j) = C_{r+s}(i)$ for all non-negative integers r and s. So, in fact, the subclasses $C_r(i)$, $0 \leqslant r \leqslant d_i - 1$, do not depend on the choice of i in the class $C(i)$ except for a cyclic permutation. Therefore they will simply be denoted C_r, $0 \leqslant r \leqslant d_i - 1$.

Theorem 1.3.10. *Let C be an essential class of period d. If* $i \in C_r$, *then*

$$\sum_{j \in C_{r+n}} P^n(wi, j) = 1$$

for all $n \in \mathbb{N}^*$ *and* $w \in W$; *here* $C_r = C_s$ *if* $r \equiv s (\mathrm{mod}\, d)$.

The proof is similar to that in the Markovian case (see Chung (1967, p. 15)).

A consequence of Theorem 1.3.10 is that if $i \in C_r$ and $i \xrightarrow{1} j$ then $j \in C_{r+1}$. Hence, if $\xi_m \in C_0$ then $\xi_{m+1} \in C_1, \ldots, \xi_{m+d-1} \in C_{d-1}$, $\xi_{m+d} \in C_0$, and so on. This shows that $C_0, \ldots, C_{d-1}$ are all non-empty and justifies calling them the *cyclically moving subclasses* of the essential class C.

Definition 1.3.11. *A non-empty set* $A \subset X$ *is said to be* (stochastically) closed *iff no element* $j \in X - A$ *is accessible from the elements* $i \in A$. *A closed set is said to be* minimal *iff it does not contain a proper closed subset.*

The proofs of the next two theorems are identical to the corresponding ones in the Markovian case (see Chung (1967, p. 16)).

Theorem 1.3.12. *A set is minimal closed iff it is an essential class.*

Theorem 1.3.13. *A finite union of inessential classes is not closed.*

1.3.2 Random systems with complete connections of type (B)

The consideration of this special class of RSCCs is partly motivated by the possibility of proving results analogous to those valid for homogeneous finite MCs.

Definition 1.3.14. *An RSCC for which X is denumerable is said to be* of type (B_m) *if there exists a convergent series* $\Sigma_{n \geqslant m} a_n$ *of subunitary* *non-negative numbers such that*

$$P(w'i^{(n)}, i) = P(w''i^{(n)}, i)(1 + \theta a_n)$$

for all $w', w'' \in W$, $i^{(n)} \in X^n$, $n \geqslant m$, *and* $i \in X$, *with* $|\theta| = |\theta(w', w'', i^{(n)}, i)| \leqslant 1$.

Hence, for an RSCC of type (B_m), the conditional probabilities $\mathbf{P}_{w'}(\xi_{n+1} = i \mid \xi^{(n)} = i^{(n)})$ and $\mathbf{P}_{w''}(\xi_{n+1} = i \mid \xi^{(n)} = i^{(n)})$ are either both equal to 0 or both different from 0 and their ratio tends to 1 as $n \to \infty$. It is obvious that a multiple homogeneous MC of order m is an RSCC of type (B_m) with $a_n = 0$, $n \geqslant m$.

Let us consider an RSCC of type (B_m) for which Condition K_1 holds.† Then, by the results in Subsection 1.3.1, the set X splits into essential and inessential classes. Actually, this decomposition coincides with that induced by a homogeneous MC with transition matrix $(P(wi, j))_{i,j \in X}$, where $w \in W$ is arbitrarily fixed. (For any $w \in W$, these matrices have the same incidence matrix (see Iosifescu (1980b, p. 222)).) If X is finite, an analysis similar to that in the Markovian case results in the following conclusions (cf. Doeblin & Fortet (1937)).

(i) If C is an essential class, then

$$\mathbf{P}_{wi}(\xi_n = i \text{ for infinitely many } n \in \mathbb{N}^*) = 1$$

for all $i \in \mathrm{C}$ and $w \in W$. Hence we can say that all elements in an essential class are recurrent.

(ii) If T denotes the set of inessential elements, then there exist two positive constants $a > 0$, and $0 < b < 1$, such that

$$\mathbf{P}_w(\xi_n \in \mathrm{T}) \leqslant ab^n$$

for all $w \in W$, and $n \in \mathbb{N}^*$. As a consequence, if $i \in \mathrm{T}$, then $\lim_{n \to \infty} P^n(w, i) = 0$ for all $w \in W$.

(iii) If C is an essential class of period $d \geqslant 1$, with subclasses $\mathrm{C}_0, \ldots, \mathrm{C}_{d-1}$, then, for any $0 \leqslant r$, $s \leqslant d - 1$, for all $w \in W$, $i \in \mathrm{C}_r$, and $j \in \mathrm{C}_s$, we have $P^n(wi, j) = 0$ if $n \not\equiv s - r (\operatorname{mod} d)$ and the limits

$$\lim_{n \to \infty} P^{nd+s-r}(wi, j) = p_s(j) > 0$$

independent of $w \in W$ and $i \in \mathrm{C}_r$ do exist; moreover $\sum_{j \in \mathrm{C}_s} p_s(j) = 1$.

(iv) If C is an essential class of period $d \geqslant 1$, with subclasses $\mathrm{C}_0, \ldots, \mathrm{C}_{d-1}$, then, for any $0 \leqslant s \leqslant d - 1$, for all $w \in W$ and $j \in \mathrm{C}_s$ we have

$$\lim_{n \to \infty} \frac{P^n(w, j)}{f_{(s-n)(\operatorname{mod} d)}(w, j)} = p_s(j)$$

† Let us notice that for an RSCC of type (B_1) Condition K_1 holds with $\alpha = 1 - \alpha_1$.

which implies

$$\lim_{n\to\infty} \frac{\sum_{m=1}^{n} P^m(w,j)}{n} = \frac{p_s(j)}{d} \sum_{r=0}^{d-1} f_r(w,j),$$

where $f_r(w,j) = \mathbf{P}_w(\xi_n = j$ for at least one value $n \equiv r(\text{mod } d)), 0 \leqslant r \leqslant d-1$.

Problems

1. Show that the map $u^{(n)}$ is $(\mathscr{W} \otimes \mathscr{X}, \mathscr{W})$-measurable for all $n \in \mathbb{N}^*$. (*Hint*: Use induction w.r.t. n.)

2. In the proof of Theorem 1.1.2, we used the fact that $\mathscr{K}_{[1,n]} = \mathscr{L}_{[1,n]}$. Does this still hold for a homogeneous GRSCC?

3. Given a homogeneous RSCC, prove that
$$\mathbf{P}_{w_0}(\theta^{-n}A) = (U^n P(\cdot, A))(w_0), n \in \mathbb{N}^*,$$
for all $A \in \mathscr{X}$, where θ is the shift operator on $\Omega = X^{\mathbb{N}^*}$ defined by $\theta((x_n)_{n\in\mathbb{N}^*}) = (x_{n+1})_{n\in\mathbb{N}^*}$.
(*Hint*: Use (1.1.7) and a monotone class argument to prove that
$$\mathbf{P}_{w_0}(\theta^{-n}A \mid \xi^{(n)}, \zeta^{(n)}) = \mathbf{P}_{w_0}(\theta^{-n}A \mid \zeta_n) = \mathbf{P}_{\zeta_n}(A), \mathbf{P}_{w_0}\text{-a.s.})$$

4. Prove that any MC may be looked at as the associated MC of a (non-homogeneous's) RSCC for which the t.p.f.'s ${}^tP$, $t \in \mathbb{N}$, do not depend on $w \in W$. (In other words, in this special case, the infinite order chain $(\xi_n)_{n\in\mathbb{N}^*}$ reduces to a sequence of independent random variables.) (*Hint*: See Section A3.2.)

5. Prove Corollary 1.1.4.

6. Prove the existence theorem for a non-homogeneous RSCC (Theorem 1.1.6) and for a GRSCC (Theorem 1.1.9).

7. Prove equations (1.1.21) and (1.1.22).

8. Prove the analogue of the Chapman–Kolmogorov equation for a homogeneous GRSCC.

9. (*Extinction probability for a Galton–Watson–Bienaymé process with random environments*) In the framework of Example 4, let $q(a)$ denote the extinction probability of the process $(Z_n)_{n\in\mathbb{N}}$ assuming it starts with $Z_0 = a$ individuals, i.e.
$$q(a) = \mathbf{P}(Z_n = 0 \text{ for some } n \in \mathbb{N}^* \mid Z_0 = a), \quad a \in \mathbb{N}^*.$$

It is easily seen that

$$q(a) = \lim_{n\to\infty} \mathbf{P}(Z_n = 0 | Z_0 = a) = \lim_{n\to\infty} \mathbf{E}_0(\zeta_n^a)$$

for all $a \in \mathbb{N}^*$.

Prove that ζ_n^a converges weakly under $\mathbf{P}_0$ as $n \to \infty$ to a $[0,1]$-valued random variable V, and $q(a) = \mathbf{E}_0(V^a)$, $a \in \mathbb{N}^*$. (Weissner (1971))

Does this statement remain true when $\mathbf{E}_0(\mathbf{P}_0)$ is replaced by $\mathbf{E}_s(\mathbf{P}_s)$ with $0 < s < 1$?

(*Hint*: The existence of the limits $\lim_{n\to\infty} \mathbf{E}_0(\zeta_n^a)$, $a \in \mathbb{N}^*$, enables us to assert the existence of V. See Feller (1966, p. 244).)

10. Consider the RSCC $\{(W, \mathscr{W}), (X, \mathscr{X}), u, P\}$, where $W = \{\mathbf{p} = (p_1, p_2): p_1, p_2 \geqslant 0,\ p_1 + p_2 = 1\}$, $\mathscr{W} = \mathscr{B}_W$, $X = \{1, 2\}$, $\mathscr{X} = \mathscr{P}(X)$, $u(\mathbf{p}, i) = [\operatorname{sgn} p_1(1,0) + \operatorname{sgn} p_2(0,1)]/(\operatorname{sgn} p_1 + \operatorname{sgn} p_2)$, $P(\mathbf{p}, i) = p_i$, $\mathbf{p} \in W$, $i \in X$. Show that the sets $\{j : i \to j\}$, $i \in X$, are empty.

11. Prove that the RSCC associated with the continued fraction expansion (Example 6 in Section 1.2) is of type (B_1) (and therefore satisfies condition K_1).

12. Prove that for an RSCC of type (B_m) we have

$$P_l(w' i^{(n)}, A) = P_l(w'' i^{(n)}, A)\alpha_n(l)$$

for all $w', w'' \in W$, $n \geqslant m$, $i^{(n)} \in X^n$, $l \in \mathbb{N}^*$ and $A \subset X^l$, where

$$\prod_{r=n}^{n+l-1} (1 - a_r) \leqslant \alpha_n(l) \leqslant \prod_{r=n}^{n+l-1} (1 + a_r).$$

Show that $\lim_{n\to\infty} \alpha_n(l) = 1$ uniformly w.r.t. $l \in \mathbb{N}^*$.

2

Ergodicity

2.1 Implications of ergodicity

2.1.1 Preliminaries

As with an MC, for an RSCC, the natural problem arises of estimating how different the finite dimensional distributions of the sequence $(\xi_n)_{n\in\mathbb{N}^*}$ are, corresponding to two probabilities $\mathbf{P}_{w'}$ and $\mathbf{P}_{w''}$, $w', w'' \in W$ (see Theorem 1.1.2). Equivalently, for arbitrarily given $n, r \in \mathbb{N}^*$, we want to know how different the probability distributions of the random variable $[\xi_n, \ldots, \xi_{n+r-1}]$ are, under $\mathbf{P}_{w'}$ and $\mathbf{P}_{w''}$ respectively. Thus, we are interested in finding an estimate of the quantity $|P_r^n(w', A) - P_r^n(w'', A)|$ for all $n, r \in \mathbb{N}^*$, $w', w'' \in W$, and $A \in \mathscr{X}^r$; moreover, it is important to establish conditions under which there exists a limit as $n \to \infty$ for $P_r^n(w, A)$, which should not depend on the starting point w of the sequence $(\zeta_n)_{n\in\mathbb{N}}$. Answers to these problems are given in Proposition 2.1.2 and Theorem 2.1.3; these results also justify the introduction of the concept of uniform ergodicity (Definition 2.1.4). The fundamental theorem (Theorem 2.1.5), and the comments following it, emphasize the key role that this concept plays in the study of the asymptotic behaviour of a homogeneous RSCC.

To prove the results which follow we shall frequently make use of Lemma 2.1.1.

Lemma 2.1.1. *Let $(E, \mathscr{E}, \mu)$ be a measure space, where μ is a finite signed measure such that $\mu(E) = 0$. Then*

$$\left| \int_E f \, d\mu \right| \leqslant \tfrac{1}{2} \| \mu \| \operatorname{osc} f$$

for any complex-valued bounded $\mathscr{E}$-measurable function f defined on E; here $\operatorname{osc} f = \sup_{u,v\in E} |f(u) - f(v)|$.

Proof. Let $E = E^+ \cup E^-$ be the Hahn decomposition of E w.r.t. μ (see Section A1.4). The condition $\mu(E) = 0$ implies that $\mu(E^+) = -\mu(E^-) = \|\mu\|/2$.

We have

$$\int_E f\,\mathrm{d}\mu = \int_{E^+} f(u)\mu(\mathrm{d}u) + \int_{E^-} f(v)\mu(\mathrm{d}v)$$
$$= \int_{E^+}\left(f(u) - \frac{1}{\mu(E^+)}\int_{E^-} - f(v)\mu(\mathrm{d}v)\right)\mu(\mathrm{d}u).$$

The inequality stated in the lemma follows if we notice that $\operatorname{osc} f$ is the diameter of the closed convex hull $\overline{\mathrm{co}}\, f(E)$ of the range of f, and that

$$\left(\int_{E^-} - f(v)\mu(\mathrm{d}v)\right)\Big/\mu(E^+) \in \overline{\mathrm{co}}\, f(E). \qquad \square$$

Put

$$b_n = \sup |P(w'x^{(n)}, A) - P(w''x^{(n)}, A)|, \quad n \in \mathbb{N}^*,$$

the supremum being taken over all $w', w'' \in W$, $x^{(n)} \in X^n$, and $A \in \mathscr{X}$.

Proposition 2.1.2. *For any $n \in \mathbb{N}^*$ we have*

$$|P_r(w'x^{(n)}, A) - P_r(w''x^{(n)}, A)| \leqslant \sum_{j=n}^{n+r-1} b_j$$

for all $r \in \mathbb{N}^*$, $w', w'' \in W$, $x^{(n)} \in X^n$, and $A \in \mathscr{X}^r$.

Proof. We proceed by induction. For $r = 1$ the inequality is true by the definition of b_n. Assume the statement holds for r and let us prove it for $r+1$. Putting $w_1 = w'x^{(n)}$, $w_2 = w''x^{(n)}$, for any $A \in \mathscr{X}^{r+1}$ we can write

$$P_{r+1}(w_1, A) - P_{r+1}(w_2, A) = \int_{X^{r+1}} P_r(w_1, \mathrm{d}y^{(r)})P(w_1 y^{(r)}, \mathrm{d}z)\chi_A(y^{(r)}, z)$$
$$- \int_{X^{r+1}} P_r(w_2, \mathrm{d}y^{(r)})P(w_2 y^{(r)}, \mathrm{d}z)\chi_A(y^{(r)}, z)$$
$$= \int_{X^r} [P_r(w_1, \mathrm{d}y^{(r)}) - P_r(w_2, \mathrm{d}y^{(r)})] \int_X P(w_1 y^{(r)}, \mathrm{d}z)\chi_A(y^{(r)}, z)$$
$$+ \int_{X^r} P_r(w_2, \mathrm{d}y^{(r)}) \int_X [P(w_1 y^{(r)}, \mathrm{d}z) - P(w_2 y^{(r)}, \mathrm{d}z)]\chi_A(y^{(r)}, z).$$

Using Lemma 2.1.1 twice, by the induction hypothesis, we get

$$|P_{r+1}(w_1, A) - P_{r+1}(w_2, A)| \leqslant \sum_{j=n}^{n+r-1} b_j + b_{n+r}$$

for all $w', w'' \in W$, $x^{(n)} \in X^n$, and $A \in \mathscr{X}^{r+1}$, which completes the proof. $\square$

2.1.2 A special ergodic theorem

The estimate obtained in Proposition 2.1.2 allows us to prove an ergodic theorem, i.e. to show that, imposing certain conditions on the sequence $(b_n)_{n\in\mathbb{N}^*}$ and the t.p.f. P it is possible to approximate (in a certain sense) each t.p.f. P_r^n, $r\in\mathbb{N}^*$, by a probability $\mathbf{P}_r^\infty$ on $\mathscr{X}^r$, which does not depend on $w\in W$.

Theorem 2.1.3. *Assume that the series $\Sigma_{n\in\mathbb{N}^*} b_n$ converges and that there exists a probability $\mathbf{p}$ on $\mathscr{X}$ and a constant a, $0<a\leqslant 1$, such that $P(w,A)\geqslant a\mathbf{p}(A)$ for all $w\in W$ and $A\in\mathscr{X}$. Let $s\in\mathbb{N}^*$ denote the smallest natural integer such that $\Sigma_{j\geqslant s} b_j\leqslant 1/4$. Then for any $r\in\mathbb{N}^*$ there exists a probability $\mathbf{P}_r^\infty$ on $\mathscr{X}^r$ such that*

$$|P_r^n(w,A)-\mathbf{P}_r^\infty(A)|\leqslant \inf_{1\leqslant k<n}\left[\left(1-\frac{a^s}{4}\right)^{n/k-1}+\frac{4}{a^s}\sum_{j\geqslant k} b_j\right]$$

for all $n\in\mathbb{N}^*$, $w\in W$, and $A\in\mathscr{X}^r$.

Proof. Let $\mathbf{p}^n$ denote the product probability $\mathbf{p}_{\{1,\ldots,n\}}$ where $\mathbf{p}_i=\mathbf{p}$, $1\leqslant i\leqslant n$, $n\in\mathbb{N}^*$ (see Section A1.7), and, for $w\in W$ and $A\in\mathscr{X}^r$, let

$$p_r(w,A)=\begin{cases}\mathbf{p}^r(A), & \text{if } r\leqslant s,\\ \displaystyle\int_{X^s}\mathbf{p}^s(\mathrm{d}x^{(s)})\int_{X^{r-s}}P_{r-s}(wx^{(s)},\mathrm{d}y^{(r-s)})\chi_A(x^{(s)},y^{(r-s)}), & \text{if } r>s.\end{cases}$$

Applying Lemma 2.1.1 and Proposition 2.1.2 we get

$$|p_r(w',A)-p_r(w'',A)|\leqslant\sum_{j\geqslant s} b_j\leqslant\frac{1}{4} \tag{2.1.1}$$

for all $r\in\mathbb{N}^*$, $w',w''\in W$, and $A\in\mathscr{X}^r$. On the other hand, the definition of p_r and P_r and the hypothesis made concerning $\mathbf{p}$ enable us to write

$$P_r(w,A)\geqslant a^s p_r(w,A) \tag{2.1.2}$$

for all $r\in\mathbb{N}^*$, $w\in W$, and $A\in\mathscr{X}^r$. Now (2.1.1) and (2.1.2) imply that

$$P_r(w',A)\geqslant a^s\left[p_r(w'',A)-\frac{1}{4}\right]$$

for all $r\in\mathbb{N}^*, w',w''\in W$, and $A\in\mathscr{X}^r$. Keeping w'' fixed in the above inequality, it easily follows that, for any $r\in\mathbb{N}^*$, either $P_r(w',A)\geqslant a^s/4$ for all $w'\in W$, or $P_r(w',X^r-A)\geqslant a^s/4$ for all $w'\in W$. Therefore

$$|P_r(w',A)-P_r(w'',A)|=|P_r(w',X^r-A)-P_r(w'',X^r-A)|\leqslant 1-\frac{a^s}{4} \tag{2.1.3}$$

for all $r\in\mathbb{N}^*, w',w''\in W$, and $A\in\mathscr{X}^r$.

Now, using Corollary 1.1.3, for all $r\in\mathbb{N}^*$, $w',w''\in W$, $A\in\mathscr{X}^r$, and any

$k, n \in \mathbb{N}^*$ such that $1 \leqslant k < n$, we can write

$$
\begin{aligned}
P_r^n(w', A) - P_r^n(w'', A) &= \int_{X^k} P_k(w', dx^{(k)}) P_r^{n-k}(w'x^{(k)}, A) \\
&\quad - \int_{X^k} P_k(w'', dx^{(k)}) P_r^{n-k}(w''x^{(k)}, A) \\
&= \int_{X^k} [P_k(w', dx^{(k)}) - P_k(w'', dx^{(k)})] P_r^{n-k}(w'x^{(k)}, A) \\
&\quad + \int_{X^k} P_k(w'', dx^{(k)}) [P_r^{n-k}(w'x^{(k)}, A) - P_r^{n-k}(w''x^{(k)}, A)] \\
&= J_1 + J_2.
\end{aligned}
$$

Set

$$\underline{P}_r^n(A) = \inf_{w \in W} P_r^n(w, A), \quad \bar{P}_r^n(A) = \sup_{w \in W} P_r^n(w, A).$$

It follows by (1.1.12) that the sequence $(\underline{P}_r^n(A))_{n \in \mathbb{N}^*}$ is non-decreasing while the sequence $(\bar{P}_r^n(A))_{n \in \mathbb{N}^*}$ is non-increasing. Thus both sequences are convergent; we shall prove that they have a common limit.

Lemma 2.1.1 and inequality (2.1.3) imply that

$$|J_1| \leqslant \left(1 - \frac{a^s}{4}\right) [\bar{P}_r^{n-k}(A) - \underline{P}_r^{n-k}(A)],$$

while the same lemma and Proposition 2.1.2 yield

$$|J_2| \leqslant \sum_{j \geqslant k} b_j.$$

Therefore we have

$$\bar{P}_r^n(A) - \underline{P}_r^n(A) \leqslant \left(1 - \frac{a^s}{4}\right) [\bar{P}_r^{n-k}(A) - \underline{P}_r^{n-k}(A)] + \sum_{j \geqslant k} b_j,$$

hence, for any $r \in \mathbb{N}^*$, $A \in \mathscr{X}^r$, and $k, n \in \mathbb{N}^*$ such that $1 \leqslant k < n$,

$$\bar{P}_r^n(A) - \underline{P}_r^n(A) \leqslant \left(1 - \frac{a^s}{4}\right)^l [\bar{P}_r^{n-kl}(A) - \underline{P}_r^{n-kl}(A)] + \sum_{j \geqslant k} b_j \left[\sum_{h=0}^{l-1} \left(1 - \frac{a^s}{4}\right)^h\right]$$

for any $l \in \mathbb{N}^*$ such that $lk < n$. This last inequality implies that $\lim_{n \to \infty} [\bar{P}_r^n(A) - \underline{P}_r^n(A)] = 0$ for all $r \in \mathbb{N}^*$ and $A \in \mathscr{X}^r$, i.e. the sequences $(\bar{P}_r^n(A))_{n \in \mathbb{N}^*}$ and $(\underline{P}_r^n(A))_{n \in \mathbb{N}^*}$ do indeed have a common limit, which we shall denote by $\mathbf{P}_r^\infty(A)$.

The domination claimed in the theorem follows immediately from the above inequality. The fact that $\mathbf{P}_r^\infty$ is a probability on $\mathscr{X}^r$ for any $r \in \mathbb{N}^*$ is a consequence of the Vitali–Hahn–Saks Theorem A1.1. □

Definition 2.1.4. *A homogeneous RSCC (a homogeneous GRSCC, respectively) is said to be* uniformly ergodic *iff for any* $r \in \mathbb{N}^*$ *there exists a probability* $\mathbf{P}_r^\infty$ *on* $\mathscr{X}^r$ *such that*

$$\lim_{n\to\infty} P_r^n(w, A) = \mathbf{P}_r^\infty(A)$$

uniformly w.r.t. $w \in W$, $A \in \mathscr{X}^r$, *and* $r \in \mathbb{N}^*$. *Alternatively, we say that* uniform ergodicity *holds.*

In the light of this definition, Theorem 2.1.3 gives a set of conditions for a homogeneous RSCC to be uniformly ergodic, and also an estimate of the rate of convergence to 0 of the non-increasing sequence $(\varepsilon_n)_{n\in\mathbb{N}^*}$ defined as

$$\varepsilon_n = \sup_{r\in\mathbb{N}^*} \sup_{w\in W, A\in\mathscr{X}^r} |P_r^n(w, A) - \mathbf{P}_r^\infty(A)|.$$

Remark. In the case of an MC, where $W = X$, $\mathscr{W} = \mathscr{X}$, $u(w, x) = x$ (see Example 1 in Section 1.2), the existence of a probability $\mathbf{P}_1^\infty = \mathbf{P}^\infty$ on $\mathscr{X}$ such that

$$\lim_{n\to\infty} P^n(x, A) = \mathbf{P}^\infty(A) \tag{2.1.4}$$

uniformly w.r.t. $x \in X$ and $A \in \mathscr{X}$, ensures the existence of some probabilities $\mathbf{P}_r^\infty$ on $\mathscr{X}^r$, $r \in \mathbb{N}^*$, such that

$$\lim_{n\to\infty} P_r^n(x, A) = \mathbf{P}_r^\infty(A)$$

uniformly w.r.t. to $x \in X$, $A \in \mathscr{X}^r$, and $r \in \mathbb{N}^*$. This statement is easily proved, putting

$$\mathbf{P}_r^\infty(A) = \int_{X^r} \mathbf{P}^\infty(\mathrm{d}x_1) P(x_1, \mathrm{d}x_2) \cdots P(x_{r-1}, \mathrm{d}x_r) \chi_A(x^{(r)})$$

for $r > 1$.

A necessary and sufficient condition for (2.1.4) to hold (in fact for the uniform ergodicity of an MC) is the following: there exists a natural integer n_0 such that

$$\sup_{\substack{x', x'' \in X \\ A \in \mathscr{X}}} |P^{n_0}(x', A) - P^{n_0}(x'', A)| < 1.$$

For a uniformly ergodic MC, the corresponding sequence $(\varepsilon_n)_{n\in\mathbb{N}^*}$ converges exponentially to 0. (See Section A3.2 and Problem 8.)

2.1.3 A fundamental theorem

The result below shows that for a uniformly ergodic homogeneous RSCC the X-valued sequence $(\xi_n)_{n\in\mathbb{N}^*}$ on $(\Omega, \mathscr{K})$, constructed in Theorem 1.1.2, enjoys some mixing properties (see Appendix 3); moreover, it is strictly stationary under a certain probability.

Theorem 2.1.5. *For a uniformly ergodic homogeneous RSCC there exists a probability* $\mathbf{P}_\infty$ *on* $(\Omega, \mathscr{K})$ *such that*

(i) *the sequence* $(\xi_n)_{n\in\mathbb{N}^*}$ *is strictly stationary under* $\mathbf{P}_\infty$;

(ii) *we have*

$$\mathbf{P}_w(A) = \mathbf{P}_\infty(A) \tag{2.1.5}$$

for all $w\in W$, *and any event* A *in the tail* σ-*algebra of the sequence* $(\xi_n)_{n\in\mathbb{N}^*}$;

(iii) *the sequence* $(\xi_n)_{n\in\mathbb{N}^*}$ *is* φ-*mixing under both* $\mathbf{P}_w$, *for any* $w\in W$, *and* $\mathbf{P}_\infty$, *and the* φ-*mixing coefficients satisfy the inequalities*

$$\varphi_{\mathbf{P}_w}(n) = \varphi_w(n) \leqslant \varepsilon_n + \varepsilon_{n+1}, \tag{2.1.6}$$

$$\varphi_{\mathbf{P}_\infty}(n) = \varphi_\infty(n) \leqslant \varepsilon_n,\ n\in\mathbb{N}^*. \tag{2.1.7}$$

Proof. (i) For an arbitrary set E in $\mathscr{K}_{[1,r]} = \sigma((\xi_i)_{1\leqslant i\leqslant r}) = \{E: E = B^{(r)} \times X^{\mathbb{N}^*}, B^{(r)}\in\mathscr{X}^r\}$, $r\in\mathbb{N}^*$, we define $\mathbf{P}_\infty$ by the equation

$$\mathbf{P}_\infty(E) = \mathbf{P}_r^\infty(B^{(r)}).$$

Since $P_{k+r}^n(w, B\times X^r) = P_k^n(w,B)$ for all $n,k,r\in\mathbb{N}^*$, $w\in W$, and $B\in\mathscr{X}^k$, we have

$$\mathbf{P}_{k+r}^\infty(B\times X^r) = \mathbf{P}_k^\infty(B) \tag{2.1.8}$$

for all $k,r\in\mathbb{N}^*$ and $B\in\mathscr{X}^k$. This can be proved by letting $n\to\infty$ in the obvious inequality below

$$|\mathbf{P}_{k+r}^\infty(B\times X^r) - \mathbf{P}_k^\infty(B)| \leqslant |\mathbf{P}_{k+r}^\infty(B\times X^r) - P_{k+r}^n(w, B\times X^r)| + |P_k^n(w,B) - \mathbf{P}_k^\infty(B)| \leqslant 2\varepsilon_n.$$

Equation (2.1.8) shows that $\mathbf{P}_\infty$ is well-defined on the algebra $\mathscr{A} = \bigcup_{r\in\mathbb{N}^*}\mathscr{K}_{[1,r]}$. On the other hand, $\mathbf{P}_\infty$ is an additive probability on $\mathscr{A}$. Since $\mathscr{K} = \sigma(\mathscr{A})$, to prove that $\mathbf{P}_\infty$ uniquely extends to $\mathscr{K}$, it is sufficient to show that, for any decreasing sequence of sets $(E_n)_{n\in\mathbb{N}^*}$ in $\mathscr{A}$, such that $\bigcap_{n\in N^*}E_n = \varnothing$, we have $\lim_{n\to\infty}\mathbf{P}_\infty(E_n) = 0$ (see Section A1.4).

Define, as usual, the shift operator θ on Ω by

$$\theta(\omega) = (x_{n+1})_{n\in\mathbb{N}^*}$$

if $\omega = (x_n)_{n\in\mathbb{N}^*}$. Uniform ergodicity implies that

$$|\mathbf{P}_w(\theta^{-k}(E)) - \mathbf{P}_\infty(E)| \leqslant \varepsilon_{k+1} \tag{2.1.9}$$

for all k, $r\in\mathbb{N}^*$, $w\in W$, and $E\in\mathscr{K}_{[1,r]}$.

Since $\bigcap_{n\in\mathbb{N}^*} E_n = \varnothing$, we also have $\bigcap_{n\in\mathbb{N}^*}\theta^{-k}(E_n) = \varnothing$ for all $k\in\mathbb{N}^*$. But since $\mathbf{P}_w$ is a completely additive probability on $\mathscr{K}$, we have $\lim_{n\to\infty}\mathbf{P}_w(\theta^{-k}(E_n)) = 0$ for all $w\in W$ and $k\in\mathbb{N}^*$. It follows then, by (2.1.9), that $\limsup_{n\to\infty}\mathbf{P}_\infty(E_n) \leqslant \varepsilon_{k+1}$ for all $k\in\mathbb{N}^*$, which in turn implies that $\lim_{n\to\infty}\mathbf{P}_\infty(E_n) = 0$; this completes the proof of the possibility of uniquely extending $\mathbf{P}_\infty$ to the whole of $\mathscr{K}$.

For any $h, n, r\in\mathbb{N}^*$ and $B^{(r)}\in\mathscr{X}^r$, the inequality

$$|\mathbf{P}^\infty_{h+r}(X^h \times B^{(r)}) - \mathbf{P}^\infty_r(B^{(r)})| \leqslant \varepsilon_n + \varepsilon_{n+h},$$

is also an immediate consequence of uniform ergodicity; letting $n\to\infty$ we get

$$\mathbf{P}^\infty_{h+r}(X^h \times B^{(r)}) = \mathbf{P}^\infty_r(B^{(r)})$$

for all $h, r\in\mathbb{N}^*$ and $B^{(r)}\in\mathscr{X}^r$. By the definition of $\mathbf{P}_\infty$, the last equation implies that

$$\mathbf{P}_\infty([\xi_{h+1},\ldots,\xi_{h+r}]\in B^{(r)}) = \mathbf{P}^\infty_r(B^{(r)})$$

for all $h, r\in\mathbb{N}^*$ and $B^{(r)}\in\mathscr{X}^r$, i.e. the strict stationarity of the sequence $(\xi_n)_{n\in\mathbb{N}^*}$ under $\mathbf{P}_\infty$.

(ii) Let us first show that for any $n\in\mathbb{N}^*$ the inequality

$$|\mathbf{P}_w(A) - \mathbf{P}_\infty(A)| \leqslant \varepsilon_n \tag{2.1.10}$$

holds for all $A\in\mathscr{K}_{[n,\infty)} = \sigma((\xi_i)_{i\geqslant n})$. For any $r\in\mathbb{N}^*$, by uniform ergodicity, for any set A in $\mathscr{K}_{[n,n+r-1]} = \sigma((\xi_i)_{n\leqslant i\leqslant n+r-1})$, $A = X^{n-1}\times A^{(r)}\times X^{\mathbb{N}^*}$, $A^{(r)}\in\mathscr{X}^r$, we have

$$\begin{aligned}|\mathbf{P}_w(A) - \mathbf{P}_\infty(A)| &= |P^n_r(w, A^{(r)}) - \mathbf{P}^\infty_{n+r}(X^n\times A^{(r)})|\\ &= |P^n_r(w, A^{(r)}) - \mathbf{P}^\infty_r(A^{(r)})| \leqslant \varepsilon_n.\end{aligned}$$

Hence (2.1.10) holds for all the sets in $\mathscr{K}_{[n,n+r-1]}$, $r\in\mathbb{N}^*$, and therefore for all the sets in $\mathscr{A}_n = \bigcup_{r\in\mathbb{N}^*}\mathscr{K}_{[n,n+r-1]}$. It is easy to see that the collection of all the sets A satisfying (2.1.10) is a monotone class which includes $\mathscr{A}_n$ and hence $\sigma(\mathscr{A}_n)$; but $\sigma(\mathscr{A}_n) = \mathscr{K}_{[n,\infty)}$, which concludes the proof of the validity of (2.1.10) for any $A\in\mathscr{K}_{[n,\infty)}$.

Now for a set A in the tail σ-algebra $\bigcap_{n\in\mathbb{N}^*}\mathscr{K}_{[n,\infty)}$ of $(\xi_n)_{n\in\mathbb{N}^*}$, inequality (2.1.10) holds for any $n\in\mathbb{N}^*$. Hence, by letting $n\to\infty$, we obtain (2.1.5).

(iii) Let us first prove that the sequence $(\xi_n)_{n\in\mathbb{N}^*}$ is φ-mixing under $\mathbf{P}_w$ for any $w\in W$.

By Theorem 1.1.2 we have

$$\mathbf{P}_w([\xi_{l+n},\ldots,\xi_{l+n+r-1}]\in A\,|\,\xi^{(l)}) = P^n_r(w\xi^{(l)}, A),\ \mathbf{P}_w\text{-a.s.},$$
$$\mathbf{P}_w([\xi_{l+n},\ldots,\xi_{l+n+r-1}]\in A) = P^{l+n}_r(w, A)$$

for all $w\in W$, $l, n, r\in\mathbb{N}^*$, and $A\in\mathscr{X}^r$. Then uniform ergodicity implies that

$$|\mathbf{P}_w([\xi_{l+n},\ldots,\xi_{l+n+r-1}]\in A\,|\,\xi^{(l)}) - \mathbf{P}_w([\xi_{l+n},\ldots,\xi_{l+n+r-1}]\in A)| \leqslant \varepsilon_n + \varepsilon_{n+l}.$$

For any $l, n, r \in \mathbb{N}^*$, putting $A_1 = \{\xi^{(l)} \in B\}$ for an arbitrarily given $B \in \mathscr{X}^l$, $A_2 = \{[\xi_{l+n}, \ldots, \xi_{l+n+r-1}] \in A\}$, and using the last inequality, we get

$$|\mathbf{P}_w(A_1 \cap A_2) - \mathbf{P}_w(A_1)\mathbf{P}_w(A_2)| \leqslant (\varepsilon_n + \varepsilon_{n+l})\mathbf{P}_w(A_1) \qquad (2.1.11)$$

for all $A_1 \in \mathscr{K}_{[1,l]}$, $A_2 \in \mathscr{K}_{[l+n, l+n+r-1]}$, and $w \in W$. If we put (see Section A3.1)

$$\varphi_w(n) = \sup |\mathbf{P}_w(A_2 | A_1) - \mathbf{P}_w(A_2)|, n \in \mathbb{N}^*,$$

the supremum being taken over all $A_1 \in \mathscr{K}_{[1,l]}$ for which $\mathbf{P}_w(A_1) \neq 0$, $A_2 \in \mathscr{K}_{[l+n, l+n+r-1]}$, and all $l, r \in \mathbb{N}^*$, then (2.1.11) yields

$$\varphi_w(n) \leqslant \varepsilon_n + \varepsilon_{n+1},$$

which shows that $\lim_{n\to\infty} \varphi_w(n) = 0$ for any $w \in W$. Therefore the sequence $(\xi_n)_{n \in \mathbb{N}^*}$ is φ-mixing under $\mathbf{P}_w$ for any $w \in W$.

We shall now prove that $(\xi_n)_{n \in \mathbb{N}^*}$ is also φ-mixing under $\mathbf{P}_\infty$. For any $l, n, r \in \mathbb{N}^*$ and $l_0 \leqslant l$ let us put $A_1 = \{[\xi_{l-l_0+1}, \ldots, \xi_l] \in A^{(l_0)}\}$, $A_2 = \{[\xi_{l+n}, \ldots, \xi_{l+n+r-1}] \in A^{(r)}\}$ for arbitrarily given $A^{(l_0)} \in \mathscr{X}^{l_0}$ and $A^{(r)} \in \mathscr{X}^r$. Writing (2.1.11) for these sets, we get

$$\begin{aligned}
&|\mathbf{P}_w(A_1 \cap A_2) - \mathbf{P}_w(A_1)\mathbf{P}_w(A_2)| \\
&\quad = |P^{l-l_0+1}_{l_0+n+r-1}(w, A^{(l_0)} \times X^{n-1} \times A^{(r)}) - P^{l-l_0+1}_{l_0}(w, A^{(l_0)}) P^{l+n}_r(w, A^{(r)})| \\
&\quad \leqslant (\varepsilon_n + \varepsilon_{n+l}) P^{l-l_0+1}_{l_0}(w, A^{(l_0)}).
\end{aligned}$$

Letting $l \to \infty$ we get

$$|\mathbf{P}^\infty_{l_0+n+r-1}(A^{(l_0)} \times X^{n-1} \times A^{(r)}) - \mathbf{P}^\infty_{l_0}(A^{(l_0)})\mathbf{P}^\infty_r(A^{(r)})| \leqslant \varepsilon_n \mathbf{P}^\infty_{l_0}(A^{(l_0)})$$

for all $A^{(l_0)} \in \mathscr{X}^{l_0}$, $A^{(r)} \in \mathscr{X}^r$, and $l_0, n, r \in \mathbb{N}^*$, or, equivalently,

$$|\mathbf{P}_\infty(A_1 \cap A_2) - \mathbf{P}_\infty(A_1)\mathbf{P}_\infty(A_2)| \leqslant \varepsilon_n \mathbf{P}_\infty(A_1)$$

for all $A_1 \in \mathscr{K}_{[1,l_0]}$, $A_2 \in \mathscr{K}_{[l_0+n, l_0+n+r-1]}$, and $l_0, n, r \in \mathbb{N}^*$. Therefore we have $\varphi_\infty(n) \leqslant \varepsilon_n$ for all $n \in \mathbb{N}^*$, hence $\lim_{n\to\infty} \varphi_\infty(n) = 0$; here

$$\varphi_\infty(n) = \sup |\mathbf{P}_\infty(A_2 | A_1) - \mathbf{P}_\infty(A_2)|, n \in \mathbb{N}^*,$$

the supremum being taken over all $A_1 \in \mathscr{K}_{[1,l_0]}$ for which $\mathbf{P}_\infty(A_1) \neq 0$, $A_2 \in \mathscr{K}_{[l_0+n, l_0+n+r-1]}$, and all $l_0, r \in \mathbb{N}^*$. □

Remark. Theorem 2.1.5 still holds for a uniformly ergodic homogeneous GRSCC. The proof above has to be only slightly modified to suit this case.

As is well known, an extensive literature exists on the asymptotic behaviour of strictly stationary φ-mixing sequences (central limit theorem, law of large numbers, law of the iterated logarithm, etc.). That is why Theorem 2.1.5 should be considered fundamental. Not only does it show that the sequence $(\xi_n)_{n \in \mathbb{N}^*}$ associated with a uniformly ergodic homogeneous RSCC is φ-mixing and strictly stationary under $\mathbf{P}_\infty$, but it also shows that there is a close relationship between the asymptotic behaviour of the sequence under $\mathbf{P}_w$, $w \in W$, and its behaviour under $\mathbf{P}_\infty$. This will enable us to prove repeatedly that a certain property holds under $\mathbf{P}_\infty$, and then makes it relatively easy to show that it also holds under $\mathbf{P}_w$, for any $w \in W$.

The key role played by the concept of uniform ergodicity makes it worthwhile to look for necessary and sufficient conditions for an RSCC to be uniformly ergodic. Such theorems will be presented in the next subsections.

As far as the uniform ergodicity of a GRSCC is concerned, the situation is quite different. Le Calvé & Theodorescu (1971) state that the study of the uniform ergodicity of a homogeneous GRSCC, for which (W, d) is a metric compact space, can be reduced to the study of the uniform ergodicity of a certain homogeneous RSCC associated with it. If such were the case, it would follow that to each ergodic theorem for a homogeneous RSCC there corresponds an ergodic theorem for a homogeneous GRSCC, and the only problem to be solved would be to see if it is possible to transform the conditions imposed on the associated RSCC into conditions on the GRSCC under consideration. By the remarks following Proposition 1.1.10, the above assertions prove not to be true, so we have to look for another way of approaching the study of the asymptotic behaviour of a GRSCC.

2.2 The homogeneous case

2.2.1 Preliminaries

We start with a formal definition.

Definition 2.2.1. *A non-homogeneous RSCC is said to be* uniformly weak-ergodic *iff*

$$\lim_{n\to\infty} |{}^tP_r^n(w', A) - {}^tP_r^n(w'', A)| = 0$$

uniformly w.r.t. $t \in \mathfrak{T}$, w', $w'' \in W$, $A \in \mathscr{X}^r$, *and* $r \in \mathbb{N}^*$.

A non-homogeneous RSCC is said to be uniformly strong-ergodic *iff for any* $r \in \mathbb{N}^*$ *there exists a probability* $\mathbf{P}_r^\infty$ *on* $\mathscr{X}^r$ *such that*

$$\lim_{n\to\infty} {}^tP_r^n(w, A) = \mathbf{P}_r^\infty(A)$$

uniformly w.r.t. $t \in \mathfrak{T}$, $w \in W, A \in \mathscr{X}^r$ *and* $r \in \mathbb{N}^*$. *Alternatively, we say that* uniform weak-(strong-)ergodicity *holds.*

It is immediate that uniform strong-ergodicity implies uniform weak-ergodicity. The proposition below shows that the two concepts are equivalent in the case of a homogeneous RSCC. This is why in the homogeneous case we shall simply speak of uniform ergodicity (see also Definition 2.1.4).

Proposition 2.2.2. *For a homogeneous RSCC uniform strong-ergodicity is implied by uniform weak-ergodicity.*

Proof. For all $r \in \mathbb{N}^*$ and $A \in \mathscr{X}^r$ put (*cf.* the proof of Theorem 2.1.3)

$$\underline{P}_r^n(A) = \inf_{w \in W} P_r^n(w, A), \quad \bar{P}_r^n(A) = \sup_{w \in W} P_r^n(w, A).$$

Take $s = 1$ in (1.1.12) to conclude that the sequence $(\underline{P}_r^n(A))_{n \in \mathbb{N}^*}$ is non-decreasing, while the sequence $(\bar{P}_r^n(A))_{n \in \mathbb{N}^*}$ is non-increasing. Uniform weak-ergodicity implies that

$$\lim_{n \to \infty} [\bar{P}_r^n(A) - \underline{P}_r^n(A)] = 0$$

uniformly w.r.t. $A \in \mathscr{X}^r$ and $r \in \mathbb{N}^*$; therefore the two sequences have a common limit, say, $\mathbf{P}_r^\infty(A)$. Since

$$\lim_{n \to \infty} P_r^n(w, A) = \mathbf{P}_r^\infty(A)$$

uniformly w.r.t. $w \in W$, $A \in \mathscr{X}^r$, and $r \in \mathbb{N}^*$, it follows that $\mathbf{P}_r^\infty$ is a probability on $\mathscr{X}^r$ as a consequence of the Vitali–Hahn–Saks Theorem A1.1. □

2.2.2 Condition $FLS(A_0, v)$

From now on this section is devoted to studying only homogeneous RSCCs, namely to establishing necessary and sufficient conditions for them to be uniformly ergodic.

First, we introduce a few conditions which will be used in the study of the asymptotic behaviour of an RSCC.

Put $Y = \bigcup_{r \in \mathbb{N}^*} X^r$, and for any fixed $A \in \mathscr{X}$ let $Y_n(A)$ be the set of the elements of Y which contain among their components at least n which belong to A.

Condition FLS (A_0, v). *Let $A_0 \in \mathscr{X}$ and $v \in \mathbb{N}^*$ be fixed.*

(i) *There exists $\gamma > 0$ such that $P_1^v(w, A_0) \geq \gamma$ for all $w \in W$.*

(ii) *If we set*

$$a_n = \sup |P(w'x^{(r)}, A) - P(w''x^{(r)}, A)|, n \in \mathbb{N}^*,$$

where the supremum is taken over all $w', w'' \in W$, $x^{(r)} \in Y_n(A_0)$, *and* $A \in \mathscr{X}$, *then*

$$\sum_{n \in \mathbb{N}^*} a_n < \infty.$$

We note that (ii) implies that the sequence $(a_n)_{n \in \mathbb{N}^*}$ is non-increasing.

Proposition 2.2.3. *If Condition* FLS(A_0, v) *holds, then for any* $n \in \mathbb{N}^*$ *we have*

$$|P_s(w'x^{(r)}, A) - P_s(w''x^{(r)}, A)| \leqslant (2v/\gamma) \sum_{j=n}^{n+[(s-1)/v]} a_j$$

for all $s \in \mathbb{N}^*$, $x^{(r)} \in Y_n(A_0)$, $w', w'' \in W$, *and* $A \in \mathscr{X}^s$.

If we replace s by $s+h-1$ and A by $X^{h-1} \times A$, $A \in \mathscr{X}^s$ in the inequality above, we get the result given by Corollary 2.2.4.

Corollary 2.2.4. *If Condition* FLS(A_0, v) *holds, then for any* $n \in \mathbb{N}^*$, *we have*

$$|P_s^h(w'x^{(r)}, A) - P_s^h(w''x^{(r)}, A)| \leqslant (2v/\gamma) \sum_{j=n}^{n+[(s+h-2)/v]} a_j$$

for all $s, h \in \mathbb{N}^*$, $x^{(r)} \in Y_n(A_0)$, $w', w'' \in W$, *and* $A \in \mathscr{X}^s$.

To prove Proposition 2.2.3, as well as other results below, we need Lemma 2.2.5.

Lemma 2.2.5. *Let* $(\alpha_{m,n})_{m,n \in \mathbb{N}^*}$ *and* $(\beta_n)_{n \in \mathbb{N}^*}$ *be sequences of real numbers. If there exist* $v \in \mathbb{N}^*$ *and* $\gamma \in (0, 1]$ *such that*

$$\alpha_{m,n} \leqslant \beta_m + \gamma \alpha_{m+1, n-v} + (1-\gamma)\alpha_{m,n-v}$$

for all $m \in \mathbb{N}^*$ *and* $n > v$, *then*

$$\alpha_{m,n} \leqslant \frac{1}{\gamma} \sum_{r=0}^{p-1} \beta_{m+r} + \sum_{r=0}^{p} \binom{p}{r} \gamma^r (1-\gamma)^{p-r} \alpha_{m+r, n-pv}$$

for all $m \in \mathbb{N}^*$ *and* $n > v$, *where* p *is the greatest natural integer such that* $pv < n$.

Proof. Iterating the inequality assumed, we get

$$\alpha_{m,n} \leqslant \sum_{r=0}^{p-1} A_r \beta_{m+r} + \sum_{r=0}^{p} \binom{p}{r} \gamma^r (1-\gamma)^{p-r} \alpha_{m+r, n-pv},$$

where

$$A_r = \gamma^r \left[1 + \sum_{s \in \mathbb{N}} \binom{r+s+1}{r} (1-\gamma)^{s+1} \right], r \in \mathbb{N}.$$

We have

$$A_0 = 1 + \sum_{s\in\mathbb{N}} (1-\gamma)^{s+1} = \frac{1}{\gamma}.$$

Let us prove that $A_{r+1} = A_r$, $r\in\mathbb{N}$. We have indeed

$$\begin{aligned} A_r + \frac{1-\gamma}{\gamma} A_{r+1} &= \gamma^r\left[1 + \sum_{s\in\mathbb{N}}\left[\binom{r+s+1}{r} + \binom{r+s+1}{r+1}\right](1-\gamma)^{s+1}\right] \\ &= \gamma^r\left[1 + \sum_{s\in\mathbb{N}}\binom{r+s+2}{r+1}(1-\gamma)^{s+1}\right] = \frac{1}{\gamma}A_{r+1}, \end{aligned}$$

whence $A_r = A_{r+1}$, $r\in\mathbb{N}$. □

Proof of Proposition 2.2.3. Let $s, n\in\mathbb{N}^*$, $x^{(r)}\in Y_n(A_0)$, $w', w''\in W$, and $A\in\mathscr{X}^s$ be arbitrarily given and put $w_1 = w'x^{(r)}$, $w_2 = w''x^{(r)}$; then

$$\begin{aligned} &|P_s(w_1, A) - P_s(w_2, A)| \\ &= \left|\int_X P(w_1, dy_1)\int_X P(w_1y_1, dy_2)\cdots\int_X P(w_1(y_1,\ldots,y_{s-1}), dy_s)\chi_A(y^{(s)})\right. \\ &\quad \left. - \int_X P(w_2, dy_1)\int_X P(w_2y_1, dy_2)\cdots\int_X P(w_2(y_1,\ldots,y_{s-1}), dy_s)\chi_A(y^{(s)})\right| \\ &\leqslant \sum_{k=1}^{s} |J_{k-1} - J_k|, \end{aligned} \tag{2.2.1}$$

where

$$J_0 = \int_X P(w_1, dy_1)\int_X P(w_1y_1, dy_2)\cdots\int_X P(w_1(y_1,\ldots,y_{s-1}), dy_s)\chi_A(y^{(s)}),$$

$$J_s = \int_X P(w_2, dy_1)\int_X P(w_2y_1, dy_2)\cdots\int_X P(w_2(y_1,\ldots,y_{s-1}), dy_s)\chi_A(y^{(s)}),$$

$$\begin{aligned} J_k = &\int_X P(w_2, dy_1)\int_X P(w_2y_1, dy_2)\cdots\int_X P(w_2(y_1,\ldots,y_{k-1}), dy_k) \\ &\times\int_X P(w_1(y_1,\ldots,y_k), dy_{k+1})\int_X P(w_1(y_1,\ldots,y_{k+1}), dy_{k+2})\cdots \\ &\times\int_X P(w_1(y_1,\ldots,y_{s-1}), dy_s)\chi_A(y^{(s)}), \quad k = 1,2,\ldots,s-1. \end{aligned}$$

Note that

$$\begin{aligned} |J_{k-1} - J_k| \leqslant &\int_X P(w_2, dy_1)\int_X P(w_2y_1, dy_2)\cdots \\ &\times\int_X P(w_2(y_1,\ldots,y_{k-1}), dy_k)|h_k(y^{(k)})|, k = 2,\ldots,s-1, \end{aligned}$$

where

$$h_k(y^{(k)}) = \int_X [P(w_1(y_1,\ldots,y_k), \mathrm{d}y_{k+1}) - P(w_2(y_1,\ldots,y_k), \mathrm{d}y_{k+1})]$$

$$\times \int_X P(w_1(y_1,\ldots,y_{k+1}), \mathrm{d}y_{k+2}) \cdots$$

$$\times \int_X P(w_1(y_1,\ldots,y_{s-1}), \mathrm{d}y_s)\chi_A(y^{(s)}).$$

Using Lemma 2.1.1, we obtain $|h_k(y^{(k)})| \leqslant a_n$, whence

$$|J_{k-1} - J_k| \leqslant a_n, \quad 2 \leqslant k \leqslant s-1.$$

It is easy to see that the above inequality also holds for $k=1$ and $k=s$. Then

$$P_s(w, A) = \int_{X^\nu} P_\nu(w, dx^{(\nu)}) \int_{X^{s-\nu}} P_{s-\nu}(wx^{(\nu)}, \mathrm{d}y^{(s-\nu)})\chi_A(x^{(\nu)}, y^{(s-\nu)});$$

therefore, as before, putting $w_1 = w'x^{(r)}$, $w_2 = w''x^{(r)}$, and

$$\Lambda_s(w', w'', A) = P_s(w', A) - P_s(w'', A),$$

we can write

$$|\Lambda_s(w_1, w_2, A)|$$

$$\leqslant \left| \int_{X^\nu} \Lambda_\nu(w_1, w_2, \mathrm{d}x^{(\nu)}) \int_{X^{s-\nu}} P_{s-\nu}(w_1x^{(\nu)}, \mathrm{d}y^{(s-\nu)})\chi_A(x^{(\nu)}, y^{(s-\nu)}) \right|$$

$$+ \left| \int_{X^{\nu-1}\times A_0} P_\nu(w_2, \mathrm{d}x^{(\nu)}) \int_{X^{s-\nu}} \Lambda_{s-\nu}(w_1x^{(\nu)}, w_2x^{(\nu)}, \mathrm{d}y^{(s-\nu)})\chi_A(x^{(\nu)}, y^{(s-\nu)}) \right|$$

$$+ \left| \int_{X^{\nu-1}\times(X-A_0)} P(w_2, \mathrm{d}x^{(\nu)}) \int_{X^{s-\nu}} \Lambda_{s-\nu}(w_1x^{(\nu)}, w_2x^{(\nu)}, \mathrm{d}y^{(s-\nu)})\chi_A(x^{(\nu)}, y^{(s-\nu)}) \right|.$$

By (2.2.2) and Lemma 2.1.1, the first term on the right in the inequality above is dominated by νa_n. To get upper bounds for the other two terms put

$$\alpha_{n,s} = \sup |\Lambda_s(w'x^{(r)}, w''x^{(r)}, A)|,$$

the supremum being taken over all $w', w'' \in W$, $x^{(r)} \in Y_n(A_0)$, and $A \in \mathscr{X}^s$. Then by Condition FLS(A_0, ν) and Lemma 2.1.1, for fixed s and w_2, the second and third terms are dominated by $P_1^\nu(w_2, A_0)\alpha_{n+1,s-\nu}$ and $[1 - P_1^\nu(w_2, A_0)]\alpha_{n,s-\nu}$, respectively. Since for fixed s, $(\alpha_{n,s})_{n\in\mathbb{N}^*}$ is a non-increasing sequence, we have

$$P_1^\nu(w, A_0)(\alpha_{n+1,s-\nu} - \alpha_{n,s-\nu}) \leqslant \gamma(\alpha_{n+1,s-\nu} - \alpha_{n,s-\nu})$$

for all $w \in W$, which finally leads to

$$\alpha_{n,s} \leqslant v a_n + \gamma \alpha_{n+1,s-v} + (1-\gamma)\alpha_{n,s-v}.$$

Now, a resort to Lemma 2.2.5 completes the proof. □

To conclude this subsection, we note that, in the special case where the set A_0 occurring in Condition FLS (A_0, v) coincides with the whole of X, requirement (i) is automatically satisfied with $\gamma = 1$ and any $v \in \mathbb{N}^*$ (we should actually consider that $v = 1$), while requirement (ii) amounts to

$$\sum_{n \in \mathbb{N}^*} b_n < \infty$$

with

$$b_n = \sup |P(w'x^{(n)}, A) - P(w''x^{(n)}, A)|, \quad n \in \mathbb{N}^*,$$

where the supremum is taken over all $w', w'' \in W$, $x^{(n)} \in X^n$, and $A \in \mathscr{X}$.

It is easy to see that for this special case of Condition FLS$(X, 1)$ the corresponding version of Proposition 2.2.3 is Proposition 2.1.2. Actually the domination appearing in the latter can be obtained from the final inequality in the proof of the former, using Lemma 2.2.5.

2.2.3 *Necessary and sufficient conditions for uniform ergodicity*

We shall now introduce Conditions M(n_0) and F(n_0), which are analogous to those used by A. A. Markov and M. Fréchet, respectively, in studying the asymptotic behaviour of MCs.

Condition M(n_0). *Let $n_0 \in \mathbb{N}^*$ be fixed; there exists $\delta > 0$ such that*

$$|P_r^{n_0}(w', A) - P_r^{n_0}(w'', A)| \leqslant 1 - \delta$$

for all $r \in \mathbb{N}^*$, $w', w'' \in W$, and $A \in \mathscr{X}^r$.

Condition F(n_0). *Let $n_0 \in \mathbb{N}^*$ be fixed; there exists $\delta > 0$ such that, for every $r \in \mathbb{N}^*$ and every partition $A_1 \cup A_2 = X^r$, $A_1, A_2 \in \mathscr{X}^r$, we have either*

$$P_r^{n_0}(w, A_1) \geqslant \delta \textit{ for all } w \in W,$$

or

$$P_r^{n_0}(w, A_2) \geqslant \delta \quad \textit{for all} \quad w \in W.$$

Proposition 2.2.6. *Conditions* M(n_0) *and* F(n_0) *are equivalent.*

Proof. For every $r \in \mathbb{N}^*$ and every partition $A_1 \cup A_2 = X^r$, $A_1, A_2 \in \mathscr{X}^r$, we have

$$|P_r^{n_0}(w',A_1)-P_r^{n_0}(w'',A_1)|=|P_r^{n_0}(w',A_2)-P_r^{n_0}(w'',A_2)|.$$

Hence it is clear that Condition F(n_0) implies Condition M(n_0).

Now suppose that Condition M(n_0) holds. Put

$$f_r(A)=\inf_{w\in W} P_r^{n_0}(w,A), \quad A\in\mathscr{X}^r.$$

Then

$$\sup_{w\in W} P_r^{n_0}(w,A)=\sup_{w\in W}[1-P_r^{n_0}(w,X^r-A)]=1-f_r(X^r-A).$$

Condition M(n_0) implies that

$$1-f_r(X^r-A)-f_r(A)\leqslant 1-\delta,$$

thus

$$f_r(X^r-A)+f_r(A)\geqslant\delta,$$

whence either $f_r(X^r-A)\geqslant\delta/2$ or $f_r(A)\geqslant\delta/2$. Therefore Condition $f(n_0)$ holds (with $\delta/2$ playing the part of δ). □

The theorem below gives necessary and sufficient conditions for a homogeneous RSCC to be uniformly ergodic.

Theorem 2.2.7. *Condition* M(n_0) *for any* $n_0\in\mathbb{N}^*$ *large enough is necessary for a homogeneous RSCC to be uniformly ergodic. If Condition* FLS(A_0,ν) *holds, then Condition* M(n_0) *for some* $n_0\in\mathbb{N}^*$ *is also sufficient for uniform ergodicity. Moreover, for any* $n>n_0$, *we have*

$$|P_s^n(w,A)-\mathbf{P}_s^\infty(A)|$$
$$\leqslant \inf_{\substack{p,r\\ p_0\nu<p\nu<r\nu<n}}\left[\frac{\left(3\sum_{k=0}^{p-1}\binom{r}{k}(1-\gamma)^{r-k}+\frac{6\nu}{\gamma}\sum_{j\geqslant p-p_0}a_j\right)}{\delta}+(1-\delta)^{(n/\nu r)-1}\right]$$

for all $s\in\mathbb{N}^*$, $w\in W$, *and* $A\in\mathscr{X}^s$; *here* p_0 *is the greatest non-negative integer such that* $p_0\nu<n_0$.

Proof. The necessity of Condition M(n_0) follows from the definition of uniform ergodicity.

To prove sufficiency, for $n,s\in\mathbb{N}^*$, $w',w''\in W$, and $A\in\mathscr{X}^s$ put

$$\Lambda_s^n(w',w'',A)=P_s^n(w',A)-P_s^n(w'',A).$$

It follows from Condition M(n_0) that

$$\tfrac{1}{2}\operatorname{var}\Lambda_s^{n_0}(w',w'',\cdot)\leqslant 1-\delta \tag{2.2.3}$$

for all $s\in\mathbb{N}^*$ and $w',w''\in W$. For all $s\in\mathbb{N}^*$, $A\in\mathscr{X}^s$, $w,w',w''\in W$, and any

$n, r \in \mathbb{N}^*$ such that $n_0 \leqslant vr < n$, we can write

$$\Lambda_s^n(w', w'', A) = \int_{X^{vr-n_0+1}} \Lambda_{vr-n_0+1}^{n_0}(w', w'', dy^{(vr-n_0+1)}) P_s^{n-vr}(wy^{(vr-n_0+1)}, A)$$
$$+ \int_{X^{vr}} P_{vr}(w', dx^{(vr)}) \Lambda_s^{n-vr}(w'x^{(vr)}, w(x_{n_0}, \dots, x_{vr}), A)$$
$$- \int_{X^{vr}} P_{vr}(w'', dx^{(vr)}) \Lambda_s^{n-vr}(w'x^{(vr)}, w(x_{n_0}, \dots, x_{vr}), A)$$
$$+ \int_{X^{vr}} P_{vr}(w'', dx^{(vr)}) \Lambda_s^{n-vr}(w'x^{(vr)}, w''x^{(vr)}, A).$$

This equation can be verified by straightforward computation, taking into account equations (1.1.12) and (1.1.4). According to Lemma 2.1.1 and inequality (2.2.3), for fixed $n, r, s \in \mathbb{N}^*$ and $A \in \mathscr{X}^s$, the modulus of the first integral on the right in the above equation is dominated by $(1-\delta) g_s^{n-vr}(A)$, where

$$g_s^n(A) = \sup_{w', w'' \in W} |\Lambda_s^n(w', w'', A)|.$$

To obtain upper bounds for the other three integrals there, let us split the space X^{vr} into 2^r disjoint sets, each of the form

$$(X^{v-1} \times A_1) \times \cdots \times (X^{v-1} \times A_r),$$

where A_i, $1 \leqslant i \leqslant r$, is either A_0 or $X - A_0$. Consider an integral which is taken over a set in which A_0 appears k times, $k \leqslant p-1$, $p < r$; then the modulus of the integrand is dominated by 1, and therefore by (i) in Condition FLS(A_0, v), the modulus of the integral itself is dominated by $(1-\gamma)^{r-k}$. Since there are $3\binom{r}{k}$ integrals taken over the sets in which A_0 appears k times, $k \leqslant p-1$, it follows that the sum of the moduli of these integrals is dominated by

$$3 \sum_{k=0}^{p-1} \binom{r}{k} (1-\gamma)^{r-k}.$$

By Corollary 2.2.4, the modulus of the sum of the integrals taken over the sets in which A_0 appears at least p times is dominated by

$$\frac{6v}{\gamma} \sum_{j \geqslant p - p_0} a_j.$$

We are thus led to the inequality

$$g_s^n(A) \leqslant (1-\delta) g_s^{n-vr}(A) + 3 \sum_{k=0}^{p-1} \binom{r}{k} (1-\gamma)^{r-k} + \frac{6v}{\gamma} \sum_{j \geqslant p - p_0} a_j,$$

which is valid for all $s \in \mathbb{N}^*$, $A \in \mathscr{X}^s$, and any $n, p, r \in \mathbb{N}^*$, for which $p_0 v < pv < rv < n$. Hence, for any $n > n_0$, we have

$$g_s^n(A) \leqslant \inf_{\substack{p,r \\ p_0 v < pv < rv < n}} \left[\frac{3 \sum_{k=0}^{p-1} \binom{r}{k} (1-\gamma)^{r-k} + \frac{6v}{\gamma} \sum_{j \geqslant p - p_0} a_j}{\delta} + (1-\delta)^{(n/vr)-1} \right],$$

for all $s \in \mathbb{N}^*$ and $A \in \mathscr{X}^s$. Let first $n \to \infty$, then $r \to \infty$, and finally $p \to \infty$, to conclude that the non-decreasing sequence $(\inf_{w \in W} P_s^n(w, A))_{n \in \mathbb{N}^*}$ and the non-increasing sequence $(\sup_{w \in W} P_s^n(w, A))_{n \in \mathbb{N}^*}$ have a common limit, say $\mathbf{P}_s^\infty(A)$. As a consequence of the Vitali–Hahn–Saks Theorem A1.1, $\mathbf{P}_s^\infty$ is a probability on $\mathscr{X}^s$ for any $s \in \mathbb{N}^*$, and the inequality to be proved is immediately obtained from the last inequality. □

2.2.4 *A sufficient condition for ergodicity*

Now we shall introduce another condition, which together with Condition $\mathrm{FLS}(A_0, v)$ ensures the uniform ergodicity of a homogeneous RSCC.

For $k \in \mathbb{N}$ let us set

$$W_k = \begin{cases} W, & \text{if } k = 0, \\ \{wx^{(k)} : w \in W, x^{(k)} \in X^k\}, & \text{if } k \in \mathbb{N}^*. \end{cases}$$

Condition $\mathrm{K}_1(A_0, k)$. *Let $A_0 \in \mathscr{X}$ and $k \in \mathbb{N}$ be fixed; there exist a probability* $\mathbf{p}$ *on $A_0 \cap \mathscr{X}$ and a number $\alpha > 0$ such that*

$$P(w, A \cap A_0) \geqslant \alpha \mathbf{p}(A \cap A_0)$$

for all $w \in W_k$ and $A \in \mathscr{X}$.

Proposition 2.2.8. *Conditions $\mathrm{K}_1(A_0, k)$ and $\mathrm{FLS}(A_0, v)$ are sufficient for a homogeneous RSCC to be uniformly ergodic.*

Proof. We shall prove that, under the assumptions made, Condition $\mathrm{F}(k+1)$ holds with $\delta = \alpha^s/4$, where $s \in \mathbb{N}^*$ is chosen such that

$$\sum_{j \geqslant s} a_j \leqslant \frac{\gamma}{8v}.$$

The result stated will then follow from Proposition 2.2.6 and Theorem 2.2.7.

For $n \in \mathbb{N}^*$ let us denote by $\mathbf{p}^n$ the product probability $\mathbf{p}_{\{1,\ldots,n\}}$, where $\mathbf{p}_i = \mathbf{p}$, $1 \leqslant i \leqslant n$ (see Section A1.7), and, for $w \in W$, $r \in \mathbb{N}^*$, and $A \in \mathscr{X}^r$ define

$$p_r(w, A) = \begin{cases} \mathbf{p}^r(A \cap A_0^r), & \text{if } r \leqslant s, \\ \displaystyle\int_{A_0^s} \mathbf{p}^s(dx^{(s)}) \int_{X^{r-s}} P_{r-s}(wx^{(s)}, dy^{(r-s)}) \chi_A(x^{(s)}, y^{(r-s)}), & \text{if } r > s. \end{cases}$$

(Here $A_0^n = \underbrace{A_0 \times \cdots \times A_0}_{n \text{ times}}$.) Using Lemma 2.1.1 and Proposition 2.2.3 we get

$$|p_r(w', A) - p_r(w'', A)| \leqslant \frac{2v}{\gamma} \sum_{j \geqslant s} a_j \leqslant \frac{1}{4}$$

for all $r \in \mathbb{N}^*$, $w', w'' \in W$, and $A \in \mathscr{X}^r$. On the other hand, a straightforward computation yields

$$P_r(w', A) \geqslant \alpha^s p_r(w', A)$$

for all $r \in \mathbb{N}^*$, $w' \in W_k$, and $A \in \mathscr{X}^r$. From the last two inequalities we obtain

$$P_r(w', A) \geqslant \alpha^s p_r(w'', A) - \frac{\alpha^s}{4}$$

for all $r \in \mathbb{N}^*$, $w' \in W_k$, $w'' \in W$, and $A \in \mathscr{X}^r$. Use the equation

$$P_r^{k+1}(w, A) = \int_{X^k} P_k(w, dx^{(k)}) P_r(wx^{(k)}, A)$$

(where $wx^{(k)} \in W_k$!), and the last inequality in which w'' is kept fixed, to deduce that, for any $r \in \mathbb{N}^*$, we have

$$P_r^{k+1}(w, A) \geqslant \alpha^s [p_r(w'', A) - \tfrac{1}{4}]$$

for all $w \in W$ and $A \in \mathscr{X}^r$. Write this inequality replacing A by $X^r - A$ to conclude that Condition F$(k+1)$ holds with $\delta = \alpha^s/4$, since either $p_r(w'', A) \geqslant \frac{1}{2}$ or $p_r(w'', X^r - A) \geqslant \frac{1}{2}$. □

2.3 The non-homogeneous case

2.3.1 *Weak-ergodicity*

By Proposition 2.2.2, for a homogeneous RSCC, uniform weak-ergodicity and uniform strong-ergodicity are equivalent. This is not the case for a non-homogeneous RSCC. To establish necessary and sufficient conditions for a non-homogeneous RSCC to be uniformly strong-(weak-) ergodic, we need conditions analogous to those in the previous subsection, but, of course, stronger.

Condition FLS″(A_0, v). *Let $A_0 \in \mathscr{X}$ and $v \in \mathbb{N}^*$ be fixed.*

(i″) *There exists $\gamma > 0$ such that ${}^{t}P_1^v(w, A_0) \geqslant \gamma$ for all $w \in W$ and $t \in \mathfrak{T}$.*

(ii″) *If we set*

$$a_n'' = \sup |{}^{t'+r}P((w'x^{(r)})_{t'}, A) - {}^{t''+r}P((w''x^{(r)})_{t''}, A)|, \quad n \in \mathbb{N}^*,$$

where the supremum is taken over all $t', t'' \in \mathfrak{T}$, $w', w'' \in W$, $x^{(r)} \in Y_n(A_0)$, and $A \in \mathscr{X}$, then

$$\sum_{n \in \mathbb{N}^*} a_n'' < \infty.$$

Condition FLS$'(A_0, \nu)$. *Let* $A_0 \in \mathscr{X}$ *and* $\nu \in \mathbb{N}^*$ *be fixed.*

(i') *There exists* $\gamma > 0$ *such that* ${}^tP_1^\nu(w, A_0) \geqslant \gamma$ *for all* $w \in W$, *and* $t \in \mathfrak{T}$.

(ii') *If we set*

$$a'_n = \sup |{}^{t+r}P((w'x^{(r)})_t, A) - {}^{t+r}P((w''x^{(r)})_t, A)|, \quad n \in \mathbb{N}^*,$$

where the supremum is taken over all $t \in \mathfrak{T}$, $w', w'' \in W$, $x^{(r)} \in Y_n(A_0)$, *and* $A \in \mathscr{X}$, *then*

$$\sum_{n \in \mathbb{N}^*} a'_n < \infty.$$

It is obvious that the first condition is stronger than the second, and that both of them reduce to Condition FLS(A_0, ν) in the case of a homogeneous RSCC. Similar remarks apply for Conditions M$''(n_0)$ and M$'(n_0)$, F$''(n_0)$ and F$'(n_0)$, respectively, to be introduced below.

Condition M$''(n_0)$. *Let* $n_0 \in \mathbb{N}^*$ *be fixed; there exists* $\delta > 0$ *such that*

$$|{}^{t'}P_r^{n_0}(w', A) - {}^{t''}P_r^{n_0}(w'', A)| \leqslant 1 - \delta$$

for all $t', t'' \in \mathfrak{T}$, $r \in \mathbb{N}^*$, $w', w'' \in W$, *and* $A \in \mathscr{X}^r$.

Condition M$'(n_0)$. *Let* $n_0 \in \mathbb{N}^*$ *be fixed; there exists* $\delta > 0$ *such that*

$$|{}^tP_r^{n_0}(w', A) - {}^tP_r^{n_0}(w'', A)| \leqslant 1 - \delta$$

for all $t \in \mathfrak{T}$, $r \in \mathbb{N}^*$, $w', w'' \in W$, *and* $A \in \mathscr{X}^r$.

Condition F$''(n_0)$. *Let* $n_0 \in \mathbb{N}^*$ *be fixed; there exists* $\delta > 0$ *such that, for every* $r \in \mathbb{N}^*$, *and every partition* $A_1 \cup A_2 = X^r$, $A_1, A_2 \in \mathscr{X}^r$, *we have either*

$${}^tP_r^{n_0}(w, A_1) \geqslant \delta \quad \textit{for all} \quad t \in \mathfrak{T} \textit{ and } w \in W,$$

or

$${}^tP_r^{n_0}(w, A_2) \geqslant \delta \quad \textit{for all} \quad t \in \mathfrak{T} \textit{ and } w \in W.$$

Condition F$'(n_0)$. *Let* $n_0 \in \mathbb{N}^*$ *be fixed; there exists* $\delta > 0$ *such that, for every* $r \in \mathbb{N}^*$, $t \in \mathfrak{T}$, *and every partition* $A_1 \cup A_2 = X^r$, $A_1, A_2 \in \mathscr{X}^r$, *we have either*

$${}^tP_r^{n_0}(w, A_1) \geqslant \delta \quad \textit{for all} \quad w \in W,$$

or

$${}^tP_r^{n_0}(w, A_2) \geqslant \delta \quad \textit{for all} \quad w \in W.$$

As in the homogeneous case, it is easy to prove Proposition 2.3.1.

Proposition 2.3.1. *Conditions* M$'(n_0)$ *and* F$'(n_0)$, M$''(n_0)$ *and* F$''(n_0)$, *respectively, are equivalent.*

The results which follow, concerning the uniform weak-ergodicity of a non-homogeneous RSCC, are analogous to those in the previous subsection; their proofs are similar and hence are left to the reader.

Proposition 2.3.2. *If Condition* $\mathrm{FLS}'(A_0, v)$ *holds, then, for any* $n \in \mathbb{N}^*$, *we have*

$$|{}^{\mathfrak{t}+r}P_s((w'x^{(r)})_{\mathfrak{t}}, A) - {}^{\mathfrak{t}+r}P_s((w''x^{(r)})_{\mathfrak{t}}, A)| \leqslant (2v/\gamma) \sum_{j=n}^{n+[(s-1)/v]} a'_j$$

for all $\mathfrak{t} \in \mathfrak{T}$, $w', w'' \in W$, $x^{(r)} \in Y_n(A_0)$, $s \in \mathbb{N}^*$ *and* $A \in \mathscr{X}^s$.

Corollary 2.3.3. *If Condition* $\mathrm{FLS}'(A_0, v)$ *holds, then, for any* $n \in \mathbb{N}^*$, *we have*

$$|{}^{\mathfrak{t}+r}P^h_s((w'x^{(r)})_{\mathfrak{t}}, A) - {}^{\mathfrak{t}+r}P^h_s((w''x^{(r)})_{\mathfrak{t}}, A)| \leqslant (2v/\gamma) \sum_{j=n}^{n+[(s+h-2)/v]} a'_j$$

for all $\mathfrak{t} \in \mathfrak{T}$, $w', w'' \in W$, $x^{(r)} \in Y_n(A_0)$, $s, h \in \mathbb{N}^*$, *and* $A \in \mathscr{X}^s$.

These preliminary results make it possible to establish necessary and sufficient conditions for a non-homogeneous RSCC to be uniformly weak-ergodic.

Theorem 2.3.4. *Condition* $\mathrm{M}'(n_0)$ *for any* $n_0 \in \mathbb{N}^*$ *large enough is necessary for a non-homogeneous RSCC to be uniformly weak-ergodic. If Condition* $\mathrm{FLS}'(A_0, v)$ *holds, then Condition* $\mathrm{M}'(n_0)$ *for some* $n_0 \in \mathbb{N}^*$ *is also sufficient for uniform weak-ergodicity. Moreover, for any* $n > n_0$, *we have*

$$|{}^{\mathfrak{t}}P^n_s(w', A) - {}^{\mathfrak{t}}P^n_s(w'', A)|$$

$$\leqslant \inf_{\substack{p,r \\ p_0 v < pv < rv < n}} \left[\frac{3\sum_{k=0}^{p-1} \binom{r}{k} (1-\gamma)^{r-k} + \frac{6v}{\gamma} \sum_{j \geqslant p-p_0} a'_j}{\delta} + (1-\delta)^{(n/vr)-1} \right]$$

for all $s \in \mathbb{N}^*$, $\mathfrak{t} \in \mathfrak{T}$, $w', w'' \in W$, *and* $A \in \mathscr{X}^s$; *here* p_0 *is the greatest non-negative integer such that* $p_0 v < n_0$.

To state a result analogous to that in Proposition 2.2.8, we need a condition similar to Condition $\mathrm{K}_1(A_0, k)$. For this purpose we first define

$$W_{\mathfrak{t},n} = \{(wx^{(n)})_{\mathfrak{t}} : w \in W, x^{(n)} \in X^n\}$$

for all $\mathfrak{t} \in \mathfrak{T}$ and $n \in \mathbb{N}^*$; then, for $k \in \mathbb{N}$, set

$$W_k = \begin{cases} W, & \text{if } k = 0, \\ \bigcup_{\mathfrak{t} \in \mathfrak{T}} W_{\mathfrak{t},k}, & \text{if } k \in \mathbb{N}^*. \end{cases}$$

Condition $K_1'(A_0, k)$. *Let $A_0 \in \mathscr{X}$ and $k \in \mathbb{N}$ be fixed; there exists a family $({}^t\mathbf{p})_{t \in \mathfrak{T}}$ of probabilities on $A_0 \cap \mathscr{X}$ and a number $\alpha > 0$ such that*

$$ {}^tP(w, A \cap A_0) \geqslant \alpha\, {}^t\mathbf{p}(A \cap A_0) $$

for all $w \in W_k$, $A \in \mathscr{X}$, and $t \in \mathfrak{T}$.

Proposition 2.3.5. *Conditions $K_1'(A_0, k)$ and $FLS'(A_0, v)$ are sufficient for a non-homogeneous RSCC to be uniformly weak-ergodic.*

In fact, similarly to the proof of Proposition 2.2.8, we can prove that, under the assumptions made, the RSCC considered also satisfies Condition $F'(k+1)$ with $\delta = \alpha^s/4$, where $s \in \mathbb{N}^*$ is chosen such that

$$ \sum_{j \geqslant s} a_j' \leqslant \frac{\gamma}{8v}. $$

2.3.2 *Strong-ergodicity*

Results concerning the uniform strong-ergodicity of a non-homogeneous RSCC can be proved by adapting the proofs in Section 2.2 to the non-homogeneous case. Nonetheless there is another much more elegant approach, which essentially consists in reducing the study of the non-homogeneous case to its study in the homogeneous case (Iosifescu (1965b)). This idea, which we shall present in this subsection, makes it unnecessary to prove all the results stated below, which are given for the sake of completeness.

With a non-homogeneous RSCC $\{(W, \mathscr{W}), (X, \mathscr{X}), (u_t)_{t \in \mathfrak{T}}, ({}^tP)_{t \in \mathfrak{T}}\}$ we associate a homogeneous RSCC $\{(\bar{W}, \bar{\mathscr{W}}), (X, \mathscr{X}), \bar{u}, \bar{P}\}$, where

$$ \bar{W} = W \times \mathfrak{T}, \quad \bar{\mathscr{W}} = \mathscr{W} \otimes \mathscr{P}(\mathfrak{T}), $$

$$ \bar{u}(\bar{w}, x) = \bar{u}((w, t), x) = ((wx)_t, t+1), $$

$$ \bar{P}(\bar{w}, A) = \bar{P}((w, t), A) = {}^tP(w, A). $$

Notice that the space $(X, \mathscr{X})$ is the same for both systems.

It is easy to verify that $\bar{u}$ is $(\bar{\mathscr{W}} \otimes \mathscr{X}, \bar{\mathscr{W}})$-measurable and $\bar{P}$ is a t.p.f. from $(\bar{W}, \bar{\mathscr{W}})$ to $(X, \mathscr{X})$. It is also obvious that

$$ \bar{w}x^{(r)} = ((wx^{(r)})_t, t+r) $$

for all $\bar{w} = (w, t) \in \bar{W}$, $r \in \mathbb{N}^*$ and $x^{(r)} \in X^r$, which yields

$$ {}^{t+r}P((wx^{(r)})_t, A) = \bar{P}(\bar{w}x^{(r)}, A) \tag{2.3.1} $$

for all $r \in \mathbb{N}^*$, $t \in \mathfrak{T}$, $w \in W$, $x^{(r)} \in X^r$, and $A \in \mathscr{X}$, where $\bar{w} = (w, t)$. Hence if we set

$$ \bar{a}_n = \sup |\bar{P}(\bar{w}'x^{(r)}, A) - \bar{P}(\bar{w}''x^{(r)}, A)|, \quad n \in \mathbb{N}^*, $$

the supremum being taken over all $\bar{w}', \bar{w}'' \in \bar{W}, x^{(r)} \in Y_n(A_0)$, and $A \in \mathscr{X}$, then

$$\bar{a}_n = a''_n.$$

On the other hand, taking into account the definitions of ${}^{t+r}P_s$ and ${}^{t+r}P^h_s$, and using (2.3.1), we deduce that

$$
\begin{aligned}
{}^{t+r}P_s((wx^{(r)})_t, A) &= \bar{P}_s(\bar{w}x^{(r)}, A),\\
{}^{t+r}P^h_s((wx^{(r)})_t, A) &= \bar{P}^h_s(\bar{w}x^{(r)}, A)
\end{aligned}
$$

for all $r, s, h \in \mathbb{N}^*$, $t \in \mathfrak{T}$, $w \in W$, $x^{(r)} \in X^r$, and $A \in \mathscr{X}^s$, where $\bar{w} = (w, t)$.

From the remarks above, it becomes clear that Proposition 2.3.6 below is an immediate consequence of Proposition 2.2.3; similar remarks are to be made for the other results below.

Proposition 2.3.6. *If Condition* FLS$''(A_0, \nu)$ *holds, then for any* $n \in \mathbb{N}^*$, *we have*

$$|{}^{t'+r}P_s((w'x^{(r)})_{t'}, A) - {}^{t''+r}P_s((w''x^{(r)})_{t''}, A)| \leqslant (2\nu/\gamma) \sum_{j=n}^{n+[(s-1)/\nu]} a''_j$$

for all $t', t'', \in \mathfrak{T}$, $w', w'' \in W$, $x^{(r)} \in Y_n(A_0)$, $s \in \mathbb{N}^*$ *and* $A \in \mathscr{X}^s$.

Corollary 2.3.7. *If Condition* FLS$''$ (A_0, ν) *holds, then for any* $n \in \mathbb{N}^*$, *we have*

$$|{}^{t'+r}P^h_s((w'x^{(r)})_{t'}, A) - {}^{t''+r}P^h_s((w''x^{(r)})_{t''}, A)| \leqslant (2\nu/\gamma) \sum_{j=n}^{n+[(s+h-2)/\nu]} a''_j$$

for all $t', t'', \in \mathfrak{T}$, w', $w'' \in W$, $x^{(r)} \in Y_n(A_0)$, $s, h \in \mathbb{N}^*$, *and* $A \in \mathscr{X}^s$.

Theorem 2.3.8. *Condition* M$''(n_0)$ *for any* $n_0 \in \mathbb{N}^*$ *large enough is necessary for a non-homogeneous RSCC to be uniformly strong-ergodic. If Condition* FLS$''(A_0, \nu)$ *holds, then Condition* M$''(n_0)$ *for some* $n_0 \in \mathbb{N}^*$ *is also sufficient for uniform strong-ergodicity. Moreover, for any* $n > n_0$, *we have*

$$
\begin{aligned}
&|{}^tP^n_s(w, A) - \mathbf{P}^\infty_s(A)|\\
&\leqslant \inf_{\substack{p,r\\ p_0\nu < p\nu < r\nu < n}} \left[\frac{3\sum_{k=0}^{p-1}\binom{r}{k}(1-\gamma)^{r-k} + \dfrac{6\nu}{\gamma}\sum_{j \geqslant p-p_0} a''_j}{\delta} + (1-\delta)^{(n/\nu r)-1} \right]
\end{aligned}
$$

for all $t \in \mathfrak{T}$, $s \in \mathbb{N}^*$, $w \in W$, *and* $A \in \mathscr{X}^s$, *where* p_0 *is the greatest non-negative integer such that* $p_0\nu < n_0$.

To give a sufficient condition for a non-homogeneous RSCC to be uniformly strong-ergodic we need Condition K$''_1(A_0, k)$.

Condition $K_1''(A_0, k)$. *Let $A_0 \in \mathscr{X}$ and $k \in \mathbb{N}$ be fixed; there exists a probability* $\mathbf{p}$ *on $A_0 \cap \mathscr{X}$ and a number $\alpha > 0$ such that*

$${}^tP(w, A \cap A_0) \geqslant \alpha \mathbf{p}(A \cap A_0)$$

for all $w \in W_k$, $A \in \mathscr{X}$, and $t \in \mathfrak{T}$.

Proposition 2.3.9. *Conditions* $K_1''(A_0, k)$ *and* $\mathrm{FLS}''(A_0, v)$ *are sufficient for a non-homogeneous RSCC to be uniformly strong-ergodic.*

In fact, similarly to the proof of Proposition 2.2.8, we can prove that, under the assumptions made, the RSCC considered also satisfies Condition $F''(k+1)$ with $\delta = \alpha^s/4$, where s is chosen such that

$$\sum_{j \geqslant s} a_j'' \leqslant \frac{\gamma}{8v}.$$

Remark. Let us set

$$\varepsilon_n = \sup |{}^tP_r^n(w, A) - \mathbf{P}_r^\infty(A)|, \quad n \in \mathbb{N}^*,$$

where the supremum is taken over all $t \in \mathfrak{T}$, $w \in W$, $A \in \mathscr{X}^r$, and $r \in \mathbb{N}^*$. Then it is not difficult to see that Theorem 2.1.5 still holds true for a uniformly strong-ergodic non-homogeneous RSCC. The changes which should be made in the proof in order to show the validity of (ii) and (iii) of the theorem are left to the reader.

2.4 An application to the associated Markov chain

Let $B_{L,A}(W, \mathscr{W})$ denote the collection of functions $f \in B(W, \mathscr{W})$ for which, for any $n \in \mathbb{N}^*$,

$$|f(w'x^{(r)}) - f(w''x^{(r)})| \leqslant l_n$$

for all w', $w'' \in W$, and $x^{(r)} \in Y_n(A)$; here $L = (l_n)_{n \in \mathbb{N}^*}$ is an arbitrary sequence of positive numbers.

The results below describe the asymptotic behaviour of the sequence $(U^n f)_{n \in \mathbb{N}^*}$ with $f \in B_{L,A}(W, \mathscr{W})$ in the case where the homogeneous RSCC considered satisfies various conditions previously introduced.

Proposition 2.4.1. *If Condition* $\mathrm{FLS}(A_0, v)$ *holds then, for any $n \in \mathbb{N}^*$, we have*

$$|U^h f(w'x^{(r)}) - U^h f(w''x^{(r)})| \leqslant \frac{2v}{\gamma} (\operatorname{osc} f) \sum_{j=n}^{n+[(h-1)/v]} a_j + l_n$$

for all $h \in \mathbb{N}^*$, $w', w'' \in W, x^{(r)} \in Y_n(A_0)$, and $f \in B_{L,A_0}(W, \mathscr{W})$.

Proof. We have

$$U^h f(w) = \int_{X^h} P_h(w, \mathrm{d}x^{(h)}) f(wx^{(h)}).$$

Using an inequality analogous to (2.2.1) and applying Lemma 2.1.1, we obtain for any $h \leqslant v$

$$|U^h f(w'x^{(r)}) - U^h f(w''x^{(r)})| \leqslant (v-1)a_n(\operatorname{osc} f) + l_n \qquad (2.4.1)$$

for all w', $w'' \in W$, $x^{(r)} \in Y_n(A_0)$, and $f \in B_{L,A_0}(W, \mathscr{W})$; it follows that the proposition is true for $h \leqslant v$.

If $h > v$ we can write

$$U^h f(w) = \int_{X^v} P_v(w, \mathrm{d}x^{(v)}) U^{h-v} f(wx^{(v)}).$$

Let us put $w_1 = w'x^{(r)}$, $w_2 = w''x^{(r)}$, and

$$\tilde{U}^h f(w', w'') = U^h f(w') - U^h f(w'');$$

then we get

$$\begin{aligned}|\tilde{U}^h f(w_1, w_2)| \leqslant & \left| \int_{X^v} [P_v(w_1, \mathrm{d}x^{(v)}) - P_v(w_2, \mathrm{d}x^{(v)})] U^{h-v} f(w_1 x^{(v)}) \right| \\ & + \left| \int_{X^{v-1} \times A_0} P_v(w_2, \mathrm{d}x^{(v)}) \tilde{U}^{h-v} f(w_1 x^{(v)}, w_2 x^{(v)}) \right| \\ & + \left| \int_{X^{v-1} \times (X - A_0)} P_v(w_2, \mathrm{d}x^{(v)}) \tilde{U}^{h-v} f(w_1 x^{(v)}, w_2 x^{(v)}) \right|.\end{aligned}$$

From this inequality, by using (2.2.2) and Lemma 2.1.1 as in the proof of Proposition 2.2.3, we obtain $\alpha_{n,h} \leqslant v a_n(\operatorname{osc} f) + \gamma \alpha_{n+1,h-v} + (1-\gamma)\alpha_{n,h-v}$, where $\alpha_{n,h} = \sup |\tilde{U}^h f(w'x^{(r)}, w''x^{(r)})|$, the supremum being taken over all w', $w'' \in W$ and $x^{(r)} \in Y_n(A_0)$. Taking into account (2.4.1) and using Lemma 2.1.1 once again, we get the stated result. □

Theorem 2.4.2. *If Conditions* M(n_0) *and* FLS(A_0, v) *hold, then for every real-valued* $f \in B_{L,A_0}(W, \mathscr{W})$ *with* $\lim_{n\to\infty} l_n = 0$ *there exists a constant* $U^\infty f$ *such that, for any* $n > n_0$, *we have*

$$\begin{aligned}&|U^n f - U^\infty f| \\ &\leqslant \inf_{\substack{p,r \\ p_0 v < pv < rv < n}} \left[\frac{\left(3 \sum_{k=0}^{p-1} \binom{r}{k} (1-\gamma)^{r-k} + \frac{6v}{\gamma} \sum_{j \geqslant p - p_0} a_j \right) \operatorname{osc} f + 3 l_{p-p_0}}{\delta} \right. \\ &\qquad \left. + (1-\delta)^{(n/vr)-1} \operatorname{osc} f \right]\end{aligned}$$

where p_0 is the greatest non-negative integer such that $p_0 v < n_0$.

Proof. Put

$$\Lambda_s^n(w', w'', A) = P_s^n(w', A) - P_s^n(w'', A)$$

for all $n, s \in \mathbb{N}^*$, $w', w'' \in W$, and $A \in \mathscr{X}^s$. It follows from Condition M(n_0) that

$$\tfrac{1}{2}\operatorname{var} \Lambda_s^{n_0}(w', w'', \cdot) \leqslant 1 - \delta$$

for all $s \in \mathbb{N}^*$ and w', $w'' \in W$. Using the same notation as in Proposition 2.4.1, for all $w, w', w'' \in W$ and any $n, r \in \mathbb{N}^*$ such that $n_0 \leqslant vr < n$, we can write

$$\begin{aligned}\tilde{U}^n f(w', w'') = & \int_{X^{vr-n_0+1}} \Lambda_{vr-n_0+1}^{n_0}(w', w'', dy^{(vr-n_0+1)}) U^{n-vr} f(w' y^{(vr-n_0+1)}) \\ & + \int_{X^{vr}} P_{vr}(w', dx^{(vr)}) \tilde{U}^{n-vr} f(w' x^{(vr)}, w'(x_{n_0}, \ldots, x_{vr})) \\ & - \int_{X^{vr}} P_{vr}(w'', dx^{(vr)}) \tilde{U}^{n-vr} f(w' x^{(vr)}, w''(x_{n_0}, \ldots, x_{vr})) \\ & + \int_{X^{vr}} P_{vr}(w'', dx^{(vr)}) \tilde{U}^{n-vr} f(w' x^{(vr)}, w'' x^{(vr)}).\end{aligned}$$

From this equation, proceeding as in the proof of Theorem 2.2.7, we deduce that for any $p \in \mathbb{N}^*$ such that $p_0 < p < r$ we have

$$\begin{aligned}\operatorname{osc} U^n f \leqslant & (1-\delta) \operatorname{osc} U^{n-vr} f + 3(\operatorname{osc} f) \sum_{k=0}^{p-1} \binom{r}{k} (1-\gamma)^{r-k} \\ & + \frac{6v}{\gamma} (\operatorname{osc} f) \sum_{j \geqslant p-p_0} a_j + 3 l_{p-p_0}.\end{aligned}$$

Since the sequences $(\sup_{w \in W} U^n f(w))_{n \in \mathbb{N}^*}$ and $(\inf_{w \in W} U^n f(w))_{n \in \mathbb{N}^*}$ are non-increasing and non-decreasing, respectively, we can easily prove the existence of $U^\infty f$ as well as the claimed domination. □

Remark. The domination in Theorem 2.4.2 is a generalization of that in Theorem 2.2.7 since, by Proposition 2.2.3, for all $s \in \mathbb{N}^*$ and $A \in \mathscr{X}^s$, the function $P_s(\cdot, A)$ belongs to $B_{L, A_0}(W, \mathscr{W})$, the sequence $(l_n)_{n \in \mathbb{N}^*}$ being given by

$$l_n = (2v/\gamma) \sum_{j=n}^{n+[(s-1)/v]} a_j, \quad n \in \mathbb{N}^*. \tag{2.4.2}$$

Theorem 2.4.2 enables us to prove that, under suitable assumptions, there exists an asymptotic distribution for a class of functionals defined on $(\zeta_n)_{n \in \mathbb{N}}$, the MC associated with the homogeneous RSCC considered.

Let f_j, $1 \leqslant j \leqslant m$, be real-valued functions belonging to $B_{L,A_0}(W, \mathscr{W})$ with $\lim_{n\to\infty} l_n = 0$. Put

$$\eta_{n,j} = f_j(\zeta_n), \quad 1 \leqslant j \leqslant m,$$

$$\Psi_n(w_0, \mathbf{u}) = \mathsf{E}_{w_0}\left\{\exp\left[i\sum_{j=1}^{m} u_j\eta_{n,j}\right]\right\}, \quad n \in \mathbb{N}^*,$$

where $\mathbf{u} = (u_1, \ldots, u_m) \in \mathbb{R}^m$.

Now, we are able to prove Theorem 2.4.3.

Theorem 2.4.3. *If Conditions* M(n_0) *and* FLS(A_0, ν) *hold, then there exists a characteristic function* Ψ, *independent of* $w_0 \in W$, *such that, for any* $n > n_0$, *we have*

$$|\Psi_n(w_0, \mathbf{u}) - \Psi(\mathbf{u})|$$

$$\leqslant 2|\mathbf{u}| \inf_{\substack{p,r \\ p_0\nu < p\nu < r\nu < n}} \left[\frac{3\left\{l_{p-p_0} + c_m\left[\sum_{k=0}^{p-1}\binom{r}{k}(1-\gamma)^{r-k} + \frac{2\nu}{\gamma}\sum_{j\geqslant p-p_0} a_j\right]\right\}}{\delta} + (1-\delta)^{(n/\nu r)-1} c_m\right]$$

for all $w_0 \in W$ *and* $\mathbf{u} \in \mathbb{R}^m$; *here* p_0 *is the greatest non-negative integer such that* $p_0\nu < n_0$, *and*

$$|\mathbf{u}| = \sum_{j=1}^{m} |u_j|, \quad c_m = \max_{1\leqslant j\leqslant m} \operatorname{osc} f_j.$$

Proof. It is easy to verify that

$$\Psi_{n+1}(\cdot, \mathbf{u}) = U(\Psi_n(\cdot, \mathbf{u})), \quad n \in \mathbb{N},$$

with

$$\Psi_0(w_0, \mathbf{u}) = \exp\left[i\sum_{j=1}^{m} u_j f_j(w_0)\right].$$

Since, for any $n \in \mathbb{N}^*$, the functions

$$\left|\sin\sum_{j=1}^{m} u_j f_j(w'x^{(r)}) - \sin\sum_{j=1}^{m} u_j f_j(w''x^{(r)})\right|,$$

$$\left|\cos\sum_{j=1}^{m} u_j f_j(w'x^{(r)}) - \cos\sum_{j=1}^{m} u_j f_j(w''x^{(r)})\right|$$

are dominated by

$$\sum_{j=1}^{m} |u_j|\,|f_j(w'x^{(r)}) - f_j(w''x^{(r)})| \leqslant l_n|u|$$

for all w', $w'' \in W$, $x^{(r)} \in Y_n(A_0)$, and $\mathbf{u} \in \mathbb{R}^m$, to conclude the proof it only remains to apply Theorem 2.4.2. □

The theorem above shows that, under the assumptions made, the random vector $(\eta_{n,1},\ldots,\eta_{n,m})$ possesses an asymptotic distribution $(n\to\infty)$ independent of w_0.

An important special case is obtained by taking

$$f_j(\cdot)=P_s(\cdot, A_j),\quad 1\leqslant j\leqslant m,$$

for fixed $s\in\mathbb{N}^*$ and $A_j\in\mathscr{X}^s$, $1\leqslant j\leqslant m$, when (2.4.2) and Condition FLS(A_0,v) imply that $\lim_{n\to\infty} l_n=0$.

Of course, it is possible to prove analogous theorems for a non-homogeneous RSCC. For the sake of completeness we formulate below the analogue of Theorem 2.4.2 in this case.

Let $B'_{L,A}(W,\mathscr{W})$ denote the collection of functions $f\in B(W,\mathscr{W})$ for which for any $n\in\mathbb{N}^*$

$$|f((w'x^{(r)})_{t'})-f((w''x^{(r)})_{t''})|\leqslant l_n$$

for any $w', w''\in W$, $t', t''\in\mathfrak{T}$, and $x^{(r)}\in Y_n(A)$; here $L=(l_n)_{n\in\mathbb{N}^*}$ is an arbitrary sequence of positive numbers.

Theorem 2.4.4. *If Conditions* M$''(n_0)$ *and* FLS$''(A_0,v)$ *hold, then, for any* $f\in B'_{L,A_0}(W,\mathscr{W})$ *with* $\lim_{n\to\infty} l_n=0$, *there exists a constant* $U^\infty f$ *such that for any* $n>n_0$ *we have*

$$|{}^tU^nf-U^\infty f|$$
$$\leqslant \inf_{\substack{p,r\\ p_0v<pv<rv<n}}\left[\frac{\left\{3\sum_{k=0}^{p-1}\binom{r}{k}(1-\gamma)^{r-k}+\frac{6v}{\gamma}\sum_{j\geqslant p-p_0}a''_j\right\}\operatorname{osc} f+3l_{p-p_0}}{\delta}\right.$$
$$\left.+(1-\delta)^{(n/vr)-1}\operatorname{osc} f\right]$$

for all $t\in\mathfrak{T}$, *where* p_0 *is the greatest non-negative integer such that* $p_0v<n_0$.

Problems

1. Prove that the hypotheses in Theorem 2.1.3 hold for the RSCC associated with the continued fraction expansion (Example 6 in Section 1.2).
(*Hint*: Take $\mathbf{p}$ to be the probability concentrated at the point 2.)

2. Let us consider the following condition.
Condition $\mathrm{K}_1(A_0,k,\mu,r)$. *Let* $A_0\in\mathscr{X}$, $k\in\mathbb{N}$, *and* $\mu, r\in\mathbb{N}^*$ *be fixed;*

there exists a probability $\mathbf{p}_r$ *on* $A_0^r \cap \mathscr{X}^r$ *and a number* $\alpha_r > 0$ *such that*

$$P_r^\mu(w, A \cap A_0^r) \geqslant \alpha_r \mathbf{p}_r(A \cap A_0^r)$$

for all $w \in W_k$ *and* $A \in \mathscr{X}^r$. *Here*

$$W_k = \begin{cases} W, & \text{if } k = 0 \\ \{wx^{(k)} : w \in W, x^{(k)} \in X^k\}, & \text{if } k \in \mathbb{N}^*, \end{cases}$$

and $A_0^r = \underbrace{A_0 \times \cdots \times A_0}_{r \text{ times}}$.

Prove that, if Condition $\mathrm{K}_1(A_0, 0)$ holds, then Condition $\mathrm{K}_1(A_0, 0, 1, r)$ holds, too.

3. Prove that if Conditions $\mathrm{K}_1(A_0, k, \mu, r)$ and $\mathrm{FLS}(A_0, \nu)$ hold with $\mu > 1$ and r such that

$$\sum_{j \geqslant r} a_j \leqslant \frac{\gamma \alpha_r}{8\nu(\alpha_r + 1)},$$

then Condition $\mathrm{F}(\mu + k)$ also holds with $\delta = \alpha_r/4$, and hence uniform ergodicity holds.
(*Hint*: The proof is similar to that of Proposition 2.2.8, by considering, for $s > r$, the t.p.f. p^s from $(W, \mathscr{W})$ to $(X^s, \mathscr{X}^s)$ defined as

$$p^s(w, A) = \int_{A_0^r} \mathbf{p}_r(dx^{(r)}) \int_{X^{s-r}} P_{s-r}(wx^{(r)}, dy^{(s-r)}) \chi_A(x^{(r)}, y^{(s-r)}).$$

4. (*Open problem*) Except for the case $P(w, A) = P(A)$, for all $w \in W$, $A \in \mathscr{X}$, are there uniformly ergodic homogeneous RSCCs such that $\mathbf{P}_r^\infty$ is the product probability of r probabilities identical to $\mathbf{P}_1^\infty$ for all $r \in \mathbb{N}^*$?

5. For $A \in \mathscr{X}$ let $Y_n(A \times A)$ be the set of elements $(x^{(r)}, y^{(r)})$ in $\bigcup_{r \in \mathbb{N}^*}(X^r \times X^r)$ for which there exists $I_n = \{i_1, \ldots, i_n\} \subset \mathbb{N}^*, i_1 < i_2 < \cdots < i_n \leqslant r$ such that $x_i = y_i$ for $i \notin I_n$, and $(x_i, y_i) \in A \times A$ for $i \in I_n$, $1 \leqslant i \leqslant r$. Let us consider Condition $\mathrm{BL}(A_0)$ for a homogeneous RSCC.

Condition $\mathrm{BL}(A_0)$. *Let* $A_0 \in \mathscr{X}$ *be fixed.*

(i) *There exists* $\gamma > 0$ *such that* $P(w, A_0) \geqslant \gamma$ *for all* $w \in W$.

(ii) *If we set*

$$b_n = \sup |P(w'x^{(r)}, A) - P(w''y^{(r)}, A)|$$

the supremum being taken over all $w', w'' \in W$, $(x^{(r)}, y^{(r)}) \in Y_n(A_0 \times A_0)$, *and* $A \in \mathscr{X}$, *then* $\sum_{n \in \mathbb{N}^*} b_n < \infty$.

Show that if Condition BL(A_0) holds, then Condition FLS(A_0, 1) also holds.

6. Prove that, if Condition BL(A_0) holds, then for all $r \in \mathbb{N}^*$ there exists a probability P_r^∞ on $\mathscr{X}^r$ such that

$$\lim_{n\to\infty} \sup_{w\in W, A\in\mathscr{X}^r} |P_r^n(w, A) - \mathsf{P}_r^\infty(A)| = 0.$$

Notice that the convergence here might not be uniform with respect to $r \in \mathbb{N}^*$. (Bert (1968), Le Calvé (1967))

7. Consider a double MC with state space $X = \{1, 2\}$ and transition probabilities

$$P((1,1),1) = a, \quad P((1,2),1) = 1,$$
$$P((2,1),1) = b, \quad P((2,2),1 = 0$$

where $0 < a, b < 1$. Show that $P^2((i,j),2) > 0$ for all $i,j \in X$, but nevertheless the chain is not uniformly ergodic. (Iosifescu (1963b)) (*Hint*: Construct the simple MC with state space $\{(1,1), (1,2), (2,1), (2,2)\}$ and transition matrix

$$\begin{array}{c} \\ (1,1) \\ (1,2) \\ (2,1) \\ (2,2) \end{array} \begin{array}{c} \begin{array}{cccc} (1,1) & (1,2) & (2,1) & (2,2) \end{array} \\ \begin{pmatrix} a & 1-a & 0 & 0 \\ 0 & 0 & 1 & 0 \\ b & 1-b & 0 & 0 \\ 0 & 0 & 0 & 1 \end{pmatrix} \end{array}$$

in order to prove that

$$\lim_{n\to\infty} P^n((i,j),2) = \begin{cases} \dfrac{1-a}{2+b-2a}, & \text{if } (i,j) = (1,1), (1,2) \text{ or } (2,1) \\ 1, & \text{if } (i,j) = (2,2). \end{cases}$$

8. Prove that a homogeneous MC of order k, $k \in \mathbb{N}^*$, is uniformly ergodic iff Condition $\mathrm{M}_k(n_0)$ below holds.

Condition $\mathrm{M}_k(n_0)$. *Let $n_0 \in \mathbb{N}^*$ be fixed; there exists $\delta > 0$ such that*

$$|P_k^{n_0}(x^{(k)}, A) - P_k^{n_0}(y^{(k)}, A)| \leqslant 1 - \delta$$

for all $x^{(k)}, y^{(k)} \in X^k$ and $A \in \mathscr{X}^k$.

Using this result explain the result in the preceding problem. Show also that if the MC is uniformly ergodic, then the rate of convergence to the limiting probabilities is exponential.
(*Hint*: Condition $\mathrm{M}_k(n_0)$ implies Condition $\mathrm{M}(n_0)$.)

9. (*Open problem*) How should Condition FLS(A_0, v) be strengthened in order to ensure that a homogeneous RSCC verifying the strengthened condition be uniformly ergodic iff there exist $n_0, r_0 \in \mathbb{N}^*$ and $\delta > 0$ such that

$$|P_{r_0}^{n_0}(w', A) - P_{r_0}^{n_0}(w'', A)| \leqslant 1 - \delta$$

for all $w', w'' \in W$ and $A \in \mathscr{X}^{r_0}$? (Of course, the case of multiple MC is ruled out.)

10. Formulate and prove the ergodic theorems in Section 2.3 in the case of multiple non-homogeneous MCs. (Iosifescu (1966a))

11. Prove Propositions 2.3.2 and 2.3.5, and Theorem 2.3.4.

12. Consider the RSCC $\{(W, \mathscr{W}), (X, \mathscr{X}), u, P\}$, where

$$W = [0, 1], \quad \mathscr{W} = \mathscr{B}_W, \quad X = \{1, 2\}, \quad \mathscr{X} = \mathscr{P}(X),$$

$$u(w, 1) = (1 - \theta_1)w + \theta_1, \quad u(w, 2) = (1 - \theta_2)w, \quad 0 < \theta_1, \theta_2 < 1,$$

$$P(w, 1) = h(w), \quad P(w, 2) = 1 - h(w), \quad w \in W,$$

where h is a Lipschitz function on $[0, 1]$ such that $0 < h(w) < 1$, $w \in [0, 1]$.

Prove that the distribution function $F_n(w, \theta_1, \theta_2, h) = \mathsf{P}(\zeta_n \leqslant w)$, $w \in [0, 1]$, $n \in \mathbb{N}$, converges weakly to a distribution function $F(w, \theta_1, \theta_2, h)$ which is continuous in θ_1 and θ_2, and does not depend on F_0. (Karlin (1953))
(*Hint*: Apply Theorem 2.4.3.)

3
The associated Markov chain

3.1 Associated operators

3.1.1 Preliminaries

We have already seen that in many cases (e.g. Examples 2, 4, 9 and 14 in Section 1.2) we are especially interested in the MC associated with either an RSCC or a GRSCC. Moreover, as shown by the two propositions below, the study of the asymptotic behaviour of the associated MC provides information about the asymptotic behaviour of the chain of infinite order in the case of both a homogeneous RSCC and a homogeneous GRSCC. This fact will be used in Section 3.4 to obtain sufficient conditions for the ergodicity of a homogeneous RSCC or of a homogeneous GRSCC, respectively, by letting the associated MC satisfy some topological properties.

To state the propositions below we need some preliminary definitions. Let $(R^{(n)})_{n \in \mathbb{N}^*}$ be a sequence of t.p.f.s from the measurable space $(Y, \mathscr{Y})$ to the measurable space $(Z, \mathscr{Z})$. Then it is possible to define the following types of convergence:

(1) there exists a t.p.f. R^{∞} from $(Y, \mathscr{Y})$ to $(Z, \mathscr{Z})$ such that

$$\lim_{n \to \infty} \frac{1}{n} \sum_{k=1}^{n} R^{(k)}(y, A) = R^{\infty}(y, A)$$

for all $y \in Y$ and $A \in \mathscr{Z}$.

(2) there exists a t.p.f. R^{∞} from $(Y, \mathscr{Y})$ to $(Z, \mathscr{Z})$ such that

$$\lim_{n \to \infty} R^{(n)}(y, A) = R^{\infty}(y, A)$$

for all $y \in Y$ and $A \in \mathscr{Z}$.

(3) there exists a probability $\mathbf{R}^{\infty}$ on $\mathscr{Z}$ such that

$$\lim_{n \to \infty} \frac{1}{n} \sum_{k=1}^{n} R^{(k)}(y, A) = \mathbf{R}^{\infty}(A)$$

for all $y \in Y$ and $A \in \mathscr{Z}$.

(4) there exists a probability $\mathbf{R}^{\infty}$ on $\mathscr{Y}$ such that

$$\lim_{n\to\infty} R^{(n)}(y, A) = \mathbf{R}^{\infty}(A)$$

for all $y \in Y$ and $A \in \mathscr{Y}$.

Proposition 3.1.1. *Let Q^n, $n \in \mathbb{N}^*$, denote the n-step t.p.f. of the MC associated with a homogeneous RSCC. Assume that the sequence $(Q^n)_{n\in\mathbb{N}^*}$ converges in one of the four ways mentioned above. Then for any $r \in \mathbb{N}^*$ the sequence $(P_r^n)_{n\in\mathbb{N}^*}$ converges in the same way.*

Proof. By the remark at the end of Corollary 1.1.3, for any $r \in \mathbb{N}^*$, we can write

$$P_r^n(\cdot, A) = U^{n-1}(P_r(\cdot, A)) = \int_W Q^{n-1}(\cdot, \mathrm{d}w) P_r(w, A)$$

for all $A \in \mathscr{X}^r$. To conclude the proof it is sufficient to define the limit t.p.f. as

$$P_r^{\infty}(w, A) = \int_W Q^{\infty}(w, \mathrm{d}w') P_r(w', A)$$

for all $w \in W$ and $A \in \mathscr{X}^r$, in the case of a convergence of type (1) or (2), and the limit probability as

$$\mathbf{P}_r^{\infty}(A) = \int_W \mathbf{Q}^{\infty}(\mathrm{d}w') P_r(w', A)$$

for all $A \in \mathscr{X}^r$, in the case of a convergence of type (3) or (4). □

Using the remark following equation (1.1.33), we can prove Proposition 3.1.2 in a similar way.

Proposition 3.1.2. *Let Q^n, $n \in \mathbb{N}^*$, denote the n-step t.p.f. of the MC associated with a homogeneous GRSCC. Assume that the sequence $(Q^n)_{n\in\mathbb{N}^*}$ converges in one of the four ways mentioned above. Then for any $r \in \mathbb{N}^*$ the sequence $(P_r^n)_{n\in\mathbb{N}^*}$ converges in the same way.*

Now it becomes obvious that it is important to study the sequences $(Q^n)_{n\in\mathbb{N}^*}$ and $(\sum_{k=1}^{n} Q^k/n)_{n\in\mathbb{N}^*}$, where Q is the t.p.f. of an MC with an arbitrary state space $(W, \mathscr{W})$. For the special case of finite MCs, obtaining results concerning these two sequences is closely related to establishing properties like those quoted below. (See Iosifescu (1980b) for the definition of the concepts and for the proofs of the following results.)

(1) The state space is a union of $l+1$ pairwise disjoint sets: T (the set of non-recurrent states) which the chain leaves after a finite number of

steps with probability 1, and l recurrent classes $E_1, \dots, E_l$. Once entered, a recurrent class cannot be left. Hence l is equal to the dimension of the eigenspace corresponding to the eigenvalue 1 of the transition matrix.

(2) Each recurrent class is a (finite) union of subclasses through which the chain moves cyclically.

(3) For each recurrent class there exists a stationary probability for the chain, the support of which is the class itself. Moreover, the set of these stationary probabilities is a basis for the set of all probabilities which are stationary for the chain.

Generally, the MC associated with an RSCC or a GRSCC, respectively, is not finite. However, we shall prove that there are important classes of such systems, for which the associated MC is of finite type, i.e. it enjoys all the properties quoted above as well as others which are characteristic of finite MCs.

This chapter is devoted to isolating such classes of RSCCs and to establishing the properties of the MCs associated with them. In the last section we shall show how these properties can be used to derive conditions for an RSCC, and a GRSCC, respectively, to be ergodic.

For the notation used in the rest of the chapter, we refer the reader to Section A2.4.

If Q is a t.p.f. on the measurable space $(W, \mathscr{W})$, we shall write U for the Markov operator associated with Q, i.e. for the transition operator associated with the MC with state space $(W, \mathscr{W})$ and t.p.f. Q. The operator U is defined on $B(W, \mathscr{W})$ by the equation

$$Uf(\cdot) = \int_W Q(\cdot, dw')f(w') \tag{3.1.1}$$

(see Section A3.2) and its iterates are given by

$$U^n f(\cdot) = \int_W Q^n(\cdot, dw')f(w'), \quad n \in \mathbb{N}^*, \tag{3.1.2}$$

where Q^n is the n-step t.p.f. The operator V defined as

$$V\mu(\cdot) = \int_W \mu(dw)Q(w, \cdot) \tag{3.1.3}$$

is a linear operator of norm 1 on $ba(W, \mathscr{W})$.[†] Its iterates are given by

$$V^n\mu(\cdot) = \int_W \mu(dw)Q^n(w, \cdot), \quad n \in \mathbb{N}^* \tag{3.1.4}$$

for all $\mu \in ba(W, \mathscr{W})$.

[†] In Section A3.2, the operator V has been restricted to the space $ca(W, \mathscr{W})$. For the integration w.r.t. a finitely additive signed measure (see Dunford & Schwartz (1958, Theorem IV 5.1).

The proposition below, due to Šidák (1962), shows the relationship between the operators U and V.

Proposition 3.1.3. *The operator V on* $\mathrm{ba}(W, \mathscr{W})$ *is the adjoint of the operator U on $B(W, \mathscr{W})$.*

Proof. We have to prove that

$$(\mu, Uf) = (V\mu, f) \tag{3.1.5}$$

for all $f \in B(W, \mathscr{W})$, and $\mu \in \mathrm{ba}(W, \mathscr{W})$, where

$$(\mu, f) = \int_W f(w)\mu(dw)$$

is an isometric isomorphism between $\mathrm{ba}(W, \mathscr{W})$ and the dual of $B(W, \mathscr{W})$ (see Section A2.3). By (3.1.1), equation (3.1.5) becomes

$$\int_W \left[\int_W f(w')Q(w, dw') \right] \mu(dw) = \int_W f(w') \int_W Q(w, dw')\mu(dw). \tag{3.1.6}$$

Obviously, this equation holds for indicators χ_B, $B \in \mathscr{W}$, and hence its validity for an arbitrary function $f \in B(W, \mathscr{W})$ is obtained by a standard argument. □

Remark. As noticed in Section A3.2, V is a linear operator of norm 1 on $\mathrm{ca}(W, \mathscr{W})$, and (3.1.6) (i.e. Proposition 3.1.3) obviously holds in this context.

We shall define

$$\bar{Q}_n(w, A) = \frac{1}{n} \sum_{k=1}^{n} Q^k(w, A) \tag{3.1.7}$$

for all $n \in \mathbb{N}^*$, $w \in W$, and $A \in \mathscr{W}$. It is clear that $\bar{Q}_n$ is a t.p.f. on $(W, \mathscr{W})$; we shall denote the Markov operator associated with $\bar{Q}_n$ by $\bar{U}_n$, and its adjoint by $\bar{V}_n$, $n \in \mathbb{N}^*$.

Definition 3.1.4. *Let $(H, \|\cdot\|_H)$ be a normed space of real-(complex-)valued functions belonging to $B(W, \mathscr{W})$.*

(i) *The operator U is said to be* orderly *w.r.t. H iff there exists a bounded linear operator U^∞ on H such that*

$$\lim_{n \to \infty} \| \bar{U}_n - U^\infty \|_H = 0.$$

(ii) *The operator U is said to be* aperiodic *w.r.t. H iff there exists a bounded linear operator U^∞ on H such that*

$$\lim_{n \to \infty} \| U^n - U^\infty \|_H = 0.$$

(iii) *The operator U is said to be* ergodic *w.r.t. H iff it is orderly and the range $U^\infty(H)$ is one-dimensional.*

(iv) *The operator U is said to be* regular *w.r.t. H iff it is ergodic and aperiodic.*

By extension we shall use the same nomenclature for the MC itself.

3.1.2 Doeblin's condition and (quasi-) compact operators

It is an immediate consequence of (3.1.2) and (3.1.4) that, for the operators U and V associated with an MC, we have $|U^n| = 1$, $\| V^n \| = 1$, $n \in \mathbb{N}^*$. Therefore, letting these operators be (quasi-) compact, we can use the mean ergodic theorem or the uniform ergodic theorem (see Section A2.8), to get information about the asymptotic behaviour of the sequence $(U^n)_{n\in\mathbb{N}^*}$, and hence of the sequence $(Q^n)_{n\in\mathbb{N}^*}$.

Letting the operator V be quasi-compact on $\mathrm{ca}(W, \mathscr{W})$, Yosida & Kakutani (1941) proved several results, the most important of which are mentioned below.

Theorem 3.1.5. *Assume the operator V is quasi-compact on* $\mathrm{ca}(W, \mathscr{W})$. *Then the following results hold.*

(i) *there exists a t.p.f. Q^∞ on $(W, \mathscr{W})$, such that*

$$\lim_{n\to\infty} \sup_{\substack{w\in W \\ A\in\mathscr{W}}} \left| \frac{1}{n} \sum_{k=1}^{n} Q^k(w, A) - Q^\infty(w, A) \right| = 0; \tag{3.1.8}$$

(ii) *there exist a finite collection of sets $\mathrm{E}_1, \ldots, \mathrm{E}_l \in \mathscr{W}$ (called ergodic kernels), some probabilities $\pi_1, \ldots, \pi_l$ on $\mathscr{W}$, and $\mathscr{W}$-measurable real-valued functions $g_1, \ldots, g_l$ defined on W, such that:*

(a) *we have*

$$Q^\infty(w, A) = \sum_{j=1}^{l} g_j(w)\pi_j(A) \tag{3.1.9}$$

for all $w \in W$ and $A \in \mathscr{W}$.

(b) *the probabilities $\pi_1, \ldots, \pi_l$ are stationary for the chain; moreover, the set $\{\pi_1, \ldots, \pi_l\}$ is a basis for the set $\{\mu \in \mathrm{ca}(W, \mathscr{W})\colon V\mu = \mu\}$ and $\pi_j(\mathrm{E}_k) = \delta_{jk}$, $1 \leqslant j, k \leqslant l$.*

(c) *the set $\{g_1, \ldots, g_l\}$ is a basis for the eigenspace corresponding to the eigenvalue 1 of the operator U; moreover, $g_j(w) \geqslant 0$, and $\sum_{j=1}^{l} g_j(w) = 1$ for all $w \in W$.*

(iii) *let $\mathrm{E} = \bigcup_{j=1}^{l} \mathrm{E}_j$; there exists $M > 0$ and $\alpha \in (0, 1)$ such that*

$$\sup_{w\in W} Q^n(w, \mathrm{E}^c) \leqslant M\alpha^n$$

for all $n \in \mathbb{N}^*$.

(iv) *for each* $1 \leqslant j \leqslant l$, *there exist a natural integer* d_j *and disjoint subsets* E_j^p, $1 \leqslant p \leqslant d_j$, *of* E_j *(called subergodic kernels), such that* $Q(w, \mathrm{E}_j^{p+1}) = 1$ *for any* $w \in \mathrm{E}_j^p$ *(here we write* $\mathrm{E}_j^p = \mathrm{E}_j^q$ *whenever* $p \equiv q \pmod{d_j}$*)*.

(v) *denoting by* N_0 *the least common multiple of the natural integers* $d_1, \ldots, d_l$, *we have*

$$\lim_{n\to\infty} \sup_{\substack{w\in W\\ A\in\mathscr{W}}} |Q^{nN_0}(w,A) - Q^{\infty}(w,A)| = 0. \tag{3.1.10}$$

Remark. For any fixed $w \in W$, the ergodic and subergodic kernels in the theorems above are unique up to a $Q^{\infty}(w,\cdot)$-null set, and, generally, $\bigcup_{p=1}^{d_j} \mathrm{E}_j^p \neq \mathrm{E}_j$.

The results by Yosida & Kakutani are more complete than those by Doeblin (1937), who had essentially obtained the ergodic and subergodic decomposition in Theorem 3.1.5 letting the chain fulfil the following condition.

Condition (D) (Doeblin's condition). *There exist a finite measure* φ *on* $(W, \mathscr{W})$, *with* $\varphi(W) > 0$, *a natural integer* p, *and two constants* $\varepsilon > 0$, $\eta > 0$, *such that* $Q^p(w, A) \leqslant 1 - \eta$ *for all* $w \in W$ *and* $A \in \mathscr{W}$ *with* $\varphi(A) \leqslant \varepsilon$.

Condition (D) always holds for finite MCs. In fact, it is a generalization of several conditions previously used by different authors to study the asymptotic behaviour of an MC. We cite below these conditions, in which q is a $\mathscr{W}^2$-measurable real-valued function defined on W^2.

Condition 1 (Hostinsky (1931)). *The t.p.f. of the chain is of the form*

$$Q(w,A) = \int_A q(w,t)\nu(dt), \quad w \in W, \quad A \in \mathscr{W}, \tag{3.1.11}$$

where ν *is a probability on* $\mathscr{W}$ *and* q *is continuous in both arguments (assuming, of course, that* W *is a topological space).*

Condition 2 (Fréchet (1934a, b)). *The t.p.f. is given by* (3.1.11), *where* q *is bounded.*

Condition 3 (Doob (1938)). *The t.p.f. is given by* (3.1.11), *where* $q(w,t)$ *is integrable in* t, *uniformly w.r.t.* w, *i.e. for any* $\varepsilon > 0$, *there exists* $\eta > 0$ *such that* $\int_A q(w,t)\nu(dt) < \varepsilon$ *for all* $w \in W$ *and all* $A \in \mathscr{W}$ *with* $\nu(A) < \eta$.

Neveu (1964, pp. 162–167) proved that the conclusions of Theorem 3.1.5 are still valid if we assume the operator U to be quasi-compact on $B(W, \mathscr{W})$.

As a consequence of the previously presented facts, it is natural to ask whether there is a relationship between Condition (D) and the quasi-compactness of V and U, respectively. Theorems 3.1.6 and 3.1.7 below answer this question.

Theorem 3.1.6. *Condition* (D) *holds iff the operator V is quasi-compact on* $\mathrm{ca}(W, \mathscr{W})$.

Proof. The fact that Condition (D) implies the quasi-compactness of V on $\mathrm{ca}(W, \mathscr{W})$ has been proved by Jacobs (1960, pp. 45–46) and Herkenrath (1977).

To prove the converse, note that by the quasi-compactness of V, Theorem 3.1.5 holds. Hence we shall use the notation of Theorem 3.1.5 and prove that Condition (D) holds, the parts of φ, ε and η being played by

$$\varphi(\cdot) = \frac{1}{l} \sum_{j=1}^{l} \pi_j(\cdot), \quad \varepsilon = \frac{2l-1}{2l^2}, \quad \eta = \frac{1}{4l}.$$

Indeed, if $\varphi(A) \leqslant \varepsilon$, we have

$$\max_{1 \leqslant j \leqslant l} \pi_j(A) \leqslant \sum_{j=1}^{l} \pi_j(A) \leqslant \frac{2l-1}{2l} = 1 - 2\eta.$$

Now, by (3.1.10), since $\sum_{j=1}^{l} g_j(w) = 1$ for all $w \in W$, there exists $n_0 \in N^*$ such that

$$\sup_{w \in W} Q^{n_0 N_0}(w, A) \leqslant 1 - \eta$$

thus completing the proof. □

Theorem 3.1.7. *Condition* (D) *holds iff the operator U is quasi-compact on* $B(W, \mathscr{W})$.

Proof. Suppose first that Condition (D) holds; then Theorem 3.1.5 holds too. Using the t.p.f. Q^∞, we can define a Markov operator U^∞ on $B(W, \mathscr{W})$ by

$$U^\infty f(w) = \int_W f(w') Q^\infty(w, dw').$$

Then, by (3.1.9), we get

$$U^\infty f(w) = \sum_{j=1}^{l} g_j(w) \int_W f(w') \pi_j(dw'),$$

which shows that U^∞ has a finite-dimensional range, hence it is compact (see Section A2.6). Using Lemma 2.1.1, we obtain

$$|U^{nN_0} - U^{\infty}| = \sup_{\substack{f \in B(W, \mathscr{W}) \\ |f| \leq 1}} \sup_{w \in W} \left| \int_W f(w')[Q^{nN_0}(w, dw') - Q^{\infty}(w, dw')] \right|$$

$$\leq \sup_{w \in W} \| Q^{nN_0}(w, \cdot) - Q^{\infty}(w, \cdot) \|$$

$$= 2 \sup_{\substack{w \in W \\ A \in \mathscr{W}}} |Q^{nN_0}(w, A) - Q^{\infty}(w, A)|. \tag{3.1.12}$$

In turn, this inequality and (3.1.10) enable us to conclude that the operator U is quasi-compact.

To prove the converse, we note that, as proved in Neveu (1964, Ch V-3), the quasi-compactness of the operator U on $B(W, \mathscr{W})$ implies the conclusions of Theorem 3.1.5. Then Condition (D) follows along the lines of the proof of the preceding theorem. □

Now it is clear that it is important to find out what conditions are to be imposed on an RSCC in order to ensure that Condition (D) holds for its associated MC. This will be done in the next propositions.

Proposition 3.1.8. *Let $\{(W, \mathscr{W}), (X, \mathscr{X}), u, P\}$ be an RSCC for which:*

(i) *$(W, \mathscr{W}, \nu)$ and $(X, \mathscr{X}, \mu)$ are finite measure spaces.*

(ii) *there exists an $\alpha > 0$ such that $\mu(u_w^{-1}(A)) \leq \alpha \nu(A)$ for all $w \in W$ and $A \in \mathscr{W}$, where $u_w(\cdot) = u(w, \cdot)$.*

(iii) *there exists a $\beta > 0$ such that $P(w, B) \leq \beta \mu(B)$ for all $w \in W$ and $B \in \mathscr{X}$. Then Condition* (D) *holds for the associated MC.*

Proof. Since $Q(w, A) = P(w, u_w^{-1}(A))$ for all $w \in W$ and $A \in \mathscr{W}$, it follows that $Q(w, A) \leq \beta \mu(u_w^{-1}(A)) \leq \beta \alpha \nu(A)$. Condition (D) holds with $\varphi(\cdot) = \beta \alpha \nu(\cdot)$, $p = 1$, and $\varepsilon = \eta = \frac{1}{2}$. □

There are classes of RSCCs for which some hypotheses in the above proposition are automatically fulfilled.

For instance, if $P(w, \cdot)$ is absolutely continuous w.r.t. μ for all $w \in W$, and if the corresponding densities $dP(w, \cdot)/d\mu(\cdot)$ are uniformly bounded, then (iii) holds.

Another special case – which is important in the theory of stochastic economic growth – is obtained by taking $W = [a, b] \subset \mathbb{R}$, $X = [c, d] \subset \mathbb{R}$, $\nu = \mu = \lambda$(Lebesgue measure), and

$$u_w(x) = \alpha(w)x + \beta(w), w \in W, x \in X,$$

where $\alpha, \beta: W \to W$ are measurable functions. Obviously, we should impose some restrictions on these functions, namely the graph of the straight line given by the equation above should be in $[c,d] \times [a,b]$ for every $w \in W$, and its slope should not degenerate; this last restriction means that there exists an $\alpha > 0$ such that $|\alpha(w)| \geqslant \alpha$ for all $w \in W$. Then it follows that

$$\lambda(u_w^{-1}(A)) = \frac{1}{|\alpha(w)|}\lambda(A) \leqslant \frac{1}{\alpha}\lambda(A)$$

for all $w \in W$ and $A \in \mathscr{W}$, which shows that (ii) holds.

It is also easy to see that (ii) holds in the case where the function u_w is piecewise linear for any $w \in W$.

The next proposition refers to the class of RSCCs $\{(W, \mathscr{W}), (X, \mathscr{X}), u, P\}$, for which $u(w,x) = u(w), w \in W, x \in X$, and the σ-algebra $\mathscr{W}$ contains all one-element sets $\{w\}$, $w \in W$. In this case, the t.p.f. of the associated MC is given by

$$Q(w, A) = \delta(u(w), A), \quad w \in W, \quad A \in \mathscr{W}. \tag{3.1.13}$$

Although this proposition gives information on a very special class of RSCCs, it is important both for finding instances in which Condition (D) holds for the MC associated with an RSCC, and for constructing some counter-examples (see Problems 3 and 4).

Proposition. 3.1.9. *Assume that the t.p.f. is given by* (3.1.13). *Then Condition* (D) *holds iff there exists a natural integer p for which $u_p(W)$ is a finite set. Here u_p is the pth iterate of u.*

Proof. To prove sufficiency, assume that $u_p(W) = \{w_1, \ldots, w_l\}$ Let us define $\varphi(\cdot) = \sum_{j=1}^{l} \delta(w_j, \cdot)$, and take $\varepsilon = \frac{1}{2}$. Then for any $A \in \mathscr{W}$ with $\varphi(A) \leqslant \frac{1}{2}$, we have $Q^p(w, A) = \delta(u_p(w), A) = 0$ for all $w \in W$, i.e. Condition (D) holds.

To prove necessity, suppose that, for any $n \in \mathbb{N}^*$, $u_n(W)$ is not a finite set. Let $\varepsilon > 0$ and φ (a finite measure) be those occurring in Condition (D). Then there is a $w_0 \in W$ such that $\varphi\{u_p(w_0)\} \leqslant \varepsilon$, since, if not, we would have $\varphi(W) = \infty$. But $Q^p(w_0, \{u_p(w_0)\}) = \delta(u_p(w), \{u_p(w)\}) = 1$, which leads to a contradiction. □

See Problems 5 and 6 for classes of GRSCCs for which Condition (D) holds for their associated MCs.

3.1.3 Doeblin–Fortet operators

As was noted in the preceding subsection, an MC for which the operator U is quasi-compact on $B(W, \mathscr{W})$ possesses an ergodic and subergodic decomposition. On the other hand, Propositions 3.1.8 and 3.1.9, as well as Problems 5 and 6, isolate classes of RSCCS, and GRSCCs, respectively, for which Condition (D) – which is equivalent to the quasi-compactness of U on $B(W, \mathscr{W})$ – holds for the associated MC. However, the hypotheses in these cases are, obviously, quite restrictive. Therefore it becomes necessary to find more natural (in this context) hypotheses implying the validity of an ergodic and subergodic decomposition for the MC associated with an RSCC and a GRSCC, respectively. The next two subsections are devoted to studying the properties of the Markov operator U on function spaces different from $B(W, \mathscr{W})$.

In the remainder of this subsection, we introduce and study such a function space, namely the space of bounded Lipschitz functions.

We shall assume that (W, d) is a metric space. Put

$$s(f) = \sup_{w' \neq w''} \frac{|f(w') - f(w'')|}{d(w', w'')},$$

$$\| f \|_L = |f| + s(f)$$

for any Lipschitz (real- or complex-valued) function defined on W,and $L(W) = \{f: W \to \mathbb{C}: \| f \|_L < \infty\}$, $L_r(W) = \{f: W \to \mathbb{R}: \| f \|_L < \infty\}$. It is easy to prove that $\| \cdot \|_L$ is a norm, and that $L(W)$ is a normed linear space under both the supremum- and $\| \cdot \|_L$-norm. The theorem below gives stronger results. To formulate it, we need Lemma 3.1.10.

Lemma 3.1.10. *If $(f_n)_{n \in \mathbb{N}^*} \subset L(W)$, and $\lim_{n \to \infty} |f_n - f| = 0$, then $s(f) \leqslant \liminf_{n \to \infty} s(f_n)$.*

Proof. Put $s = \liminf_{n \to \infty} s(f_n)$. The case $s = \infty$ is trivial. Thus, assume $s < \infty$. Then there exists a subsequence $(f_{n_j})_{j \in \mathbb{N}^*}$ such that $s = \lim_{j \to \infty} s(f_{n_j})$. Let $\varepsilon > 0$ be arbitrarily chosen; there exists a natural integer $j(\varepsilon)$ such that $s(f_{n_j}) \leqslant s + \varepsilon$ for all $j \geqslant j(\varepsilon)$. Then, for any $w', w'' \in W, w' \neq w''$, we have

$$\frac{|f(w') - f(w'')|}{d(w', w'')} = \lim_{j \to \infty} \frac{|f_{n_j}(w') - f_{n_j}(w'')|}{d(w', w'')},$$

whence $s(f) \leqslant s + \varepsilon$. Since ε is arbitrary, the proof is complete. □

We omit the simple proof of the following corollary.

Corollary 3.1.11. *Let $(f_n)_{n \in \mathbb{N}^*} \subset L(W)$. If there exists $c > 0$ such that $\| f_n \|_L < c$ for all $n \in \mathbb{N}^*$, and $\lim_{n \to \infty} |f_n - f| = 0$, then $f \in L(W)$, and $\| f \|_L < c$.*

Theorem 3.1.12. *The space $(L(W), \| \cdot \|_L)$ is a Banach algebra with unit 1.*

Proof. The proof is immediate except for the assertions '$(L(W), \|\cdot\|_L)$ is complete', '$L(W)$ is closed under multiplication', and 'the Banach algebra property of the norm holds'.

To prove the first assertion, let us take a sequence $(f_n)_{n\in\mathbb{N}^*} \subset L(W)$ for which $\lim_{n,p\to\infty} \|f_n - f_p\|_L = 0$. Then, we have both $\lim_{n,p\to\infty} |f_n - f_p| = 0$ and $\lim_{n,p\to\infty} s(f_n - f_p) = 0$. Hence there exists $f \in C(W)$ such that $\lim_{n\to\infty} |f_n - f| = 0$. It remains to show that $f \in L(W)$, and $\lim_{n\to\infty} \|f_n - f\|_L = 0$. Let $\varepsilon > 0$ be arbitrary; there exists a natural integer $n(\varepsilon)$ such that $s(f_n - f_r) < \varepsilon$ for all $n, r \geqslant n(\varepsilon)$. If we fix r, we get $\lim_{n\to\infty} (f_n - f_r) = f - f_r = 0$ and, by Lemma 3.1.10,

$$s(f - f_r) \leqslant \liminf_{n\to\infty} s(f_n - f_r) < \varepsilon.$$

Hence $f - f_r$ is in $L(W)$ and so is f. Moreover, since $r \geqslant n(\varepsilon)$ is arbitrary, we also have $\lim_{r\to\infty} s(f - f_r) = 0$, so that the space $(L(W), \|\cdot\|_L)$ is complete.

Next, the obvious inequality

$$s(fg) \leqslant |f| s(g) + |g| s(f)$$

implies that $fg \in L(W)$ and $\|fg\|_L \leqslant \|f\|_L \|g\|_L$ whenever $f, g \in L(W)$.

□

Remarks. 1. If (W, d) is compact, then, by the Stone–Weierstrass theorem, $(L(W), |\cdot|)$ is dense in $C(W)$.

2. By Corollary 3.1.11, Condition (ITM_1) holds for the spaces $X = C(W)$ and $Y = L(W)$ (see Section A2.9).

We shall also need the following result due to M. D. Kirszbraun & E. J. McShane (see Araujo & Giné (1980, pp. 3 and 35)).

Lemma 3.1.13. *If W' is a subset of the metric space (W, d), and $f' \in L'_r = L_r(W')$, then there exists a function $f \in L_r = L_r(W)$ such that $f|_{W'} = f'$ and $\|f\|_{L_r} = \|f'\|_{L'_r}$.*

Now, let us introduce a definition.

Definition 3.1.14. *The transition operator U of an MC with state space (W, d) is said to be a* Doeblin–Fortet operator *iff U takes $L(W)$ into $L(W)$ boundedly w.r.t. $\|\cdot\|_L$, and there exist $k \in \mathbb{N}^*$, $r \in [0, 1)$, and $R < \infty$, such that*

$$s(U^k f) \leqslant r s(f) + R|f| \tag{3.1.14}$$

for all $f \in L(W)$. Alternatively, the MC itself is said to be a Doeblin–Fortet chain.

Remark. The concept of a Doeblin–Fortet chain is entirely different from that of an MC satisfying Condition (D), as illustrated by Problem 7.

The definition below isolates a class of RSCCs, called RSCCs with contraction, for which the associated MCs are Doeblin–Fortet chains.

An RSCC with contraction might be defined as an RSCC $\{(W,\mathscr{W}), (X,\mathscr{X}), u, P\}$, where (W,d) is a metric space and the distance between $w'x$ and $w''x$ is less than that between w' and w'', e.g. there exists $r<1$ such that $d(w'x,w''x)\leqslant rd(w',w'')$ for all $x\in X$ and w', $w''\in W$. A more comprehensive definition was suggested (in a particular context) by Isaac (1962), who assumed that the RSCC satisfies

$$\int_X P(w',\mathrm{d}x)\,d(w'x,w''x)\leqslant rd(w',w'')$$

for all $w',w''\in W$, where $r<1$. The definition below which is even more comprehensive, is due to Norman (1972). Let us put

$$r_j=\sup_{w'\neq w''}\int_{X^j} P_j(w',\mathrm{d}x^{(j)})\frac{d(w'x^{(j)},w''x^{(j)})}{d(w',w'')} \tag{3.1.15}$$

$$R_j=\sup_{A\in\mathscr{X}^j}\sup_{w'\neq w''}\frac{|P_j(w',A)-P_j(w'',A)|}{d(w',w'')}=\sup_{A\in\mathscr{X}^j}\mathrm{s}(P_j(\cdot,A)),\quad j\in\mathbb{N}^*. \tag{3.1.16}$$

Definition 3.1.15. *An RSCC $\{(W,\mathscr{W}),(X,\mathscr{X}),u,P\}$ is said to be an RSCC* with contraction *iff (W,d) is a separable metric space, $r_1<\infty$, $R_1<\infty$, and there exists a natural integer k such that $r_k<1$.*

Remarks. 1. For earlier, less comprehensive, variants of Definition 3.1.15, see Problems 19 and 20.

2. The separability of the space (W,d) ensures the $\mathscr{X}^j$-measurability of the map $x^{(j)}\to d(w'x^{(j)},w''x^{(j)})$ for all w', $w''\in W$, so that the integrals in the definition of r_j make sense for all $j\in\mathbb{N}^*$.

3. A still weaker version of the concept of an RSCC with contraction was introduced by Kaijser (1981b). His setting and requirements are as follows. Let (W,d) be a compact metric space. An RSCC $\{(W,d),(X,\mathscr{X}), u,P\}$ is said to be an RSCC with *weak contraction* iff:

(i) denoting by $(\zeta_n^{(w')})_{n\in\mathbb{N}}$, $\zeta_0^{(w')}=w'$ and $(\zeta_n^{(w'')})_{n\in\mathbb{N}}$, $\zeta_0^{(w'')}=w''$, the associated Markov chains starting from $w'\in W$ and $w''\in W$, respectively, for any $\varepsilon>0$ there exists $n_0\in\mathbb{N}^*$ such that

$$\inf_{w',w''\in W}\mathsf{P}_{w'}\times\mathsf{P}_{w''}(d(\zeta_{n_0}^{(w')},\zeta_{n_0}^{(w'')})<\varepsilon)>0$$

(ii) there exist $k\in\mathbb{N}^*$ and real numbers $\alpha<0$ and $c>0$ such that, if

$d(w', w'') \leqslant c$, $w' \neq w''$, then

$$\min\left(\int_{X^k} P_k(w', \mathrm{d}x^{(k)}) \log \frac{d(w'x^{(k)}, w''x^{(k)})}{d(w', w'')},\right.$$
$$\left.\int_{X^k} P_k(w'', \mathrm{d}x^{(k)}) \log \frac{d(w'x^{(k)}, w''x^{(k)})}{d(w', w'')}\right) \leqslant \alpha.$$

(iii) we have

$$\int_0^\infty t^{-1} \sup_{d(w',w'') \leqslant t} \| Q^k(w', \cdot) - Q^k(w'', \cdot) \| \, \mathrm{d}t < \infty,$$

where $\| \cdot \|$ is the total variation norm (see Section A1.4).

For an application of results obtained by Kaijser for RSCCs with weak contraction see Góra (1985).

Theorem 3.1.16. *The MC associated with an RSCC with contraction is a Doeblin–Fortet chain.*

To prove this theorem we need the following lemma.

Lemma 3.1.17. *For any $f \in L(W)$ and $j \in \mathbb{N}^*$, we have*

$$|U^j f(w') - U^j f(w'')| \leqslant r_j \mathrm{s}(f) \, d(w', w'') + \operatorname{osc} f \, \| P_j(w', \cdot) - P_j(w'', \cdot) \|$$

for all w', $w'' \in W$.

Proof. The inequality follows from the equation

$$U^j f(w') - U^j f(w'') = \int_{X^j} P_j(w', \mathrm{d}x^{(j)}) [f(w'x^{(j)}) - f(w''x^{(j)})]$$
$$+ \int_{X^j} f(w''x^{(j)}) [P_j(w', \mathrm{d}x^{(j)}) - P_j(w'', \mathrm{d}x^{(j)})]$$

by applying Lemma 2.1.1. □

Proof of Theorem 3.1.16. By Lemma 3.1.17, we have

$$\mathrm{s}(U^j f) \leqslant r_j \mathrm{s}(f) + 2R_j |f| \tag{3.1.17}$$

for all $f \in L(W)$ and $j \in \mathbb{N}^*$. Taking $j = 1$, we get

$$\| Uf \|_L \leqslant r_1 \mathrm{s}(f) + (2R_1 + 1) |f| \leqslant \bar{R} \| f \|_L,$$

since $|Uf| \leqslant |f|$; here $\bar{R} = \max\{r_1, 2R_1 + 1\}$. This last inequality shows that U is bounded on $L(W)$. Taking $j = k$ in (3.1.17) yields (3.1.14) with $r = r_k < 1$ and $R = 2R_k < \infty$. □

For some learning models the space W is compact, while the space X is finite (see Problem 21); in such a special case it is possible to establish a simpler criterion for verifying that the RSCC under consideration is an RSCC with contraction (see Problem 20).

The next auxiliary result will often be used throughout this chapter.

Lemma 3.1.18. *Let $(a_n)_{n\in\mathbb{N}^*}$ be a sequence of non-negative numbers for which $a_{n+j}\leqslant a_n a_j$ for all $n,j\in\mathbb{N}^*$. Assume that there exists $k\in\mathbb{N}^*$ such that $a_k<1$. Then $\lim_{n\to\infty}(a_n)^{1/n}<1$ exists. Moreover, there exist $a>0$ and $\alpha\in(0,1)$ such that $a_n\leqslant a\alpha^n$ for all $n\in\mathbb{N}^*$.*

Proof. For any $n\geqslant j\geqslant 1$, by iterating the asserted inequality we get $a_n\leqslant a_j^q a_m$; here $q=[n/j]$ and $m=n-jq<j$, with the convention $a_0=1$. Therefore, we have

$$a^* \limsup_{n\to\infty}(a_n)^{1/n}\leqslant (a_j)^{1/j}.$$

So $a^*\leqslant(a_j)^{1/j}$ for any fixed j. Taking $j=k$ yields $a^*<1$; next, letting $j\to\infty$, we get $a^*\leqslant\liminf_{j\to\infty}(a_j)^{1/j}$, hence the existence of the limit $\lim_{n\to\infty}(a_n)^{1/n}$ is proved. It also follows that there exists $n_0\in\mathbb{N}^*$ such that $(a_n)^{1/n}<1$ for all $n\geqslant n_0$. Hence there exists $\alpha\in(0,1)$ such that $a_n\leqslant\alpha^n$ for all $n\geqslant n_0$. The inequality $a_n\leqslant a\alpha^n$ for all $n\in\mathbb{N}^*$ follows by a suitable choice of $a>0$. □

Remark. In fact, Lemma 3.1.18 is a special case of a subadditive ergodic theorem due to J. F. C. Kingman (see Walters (1982, p. 231)).

We shall now consider Markov (especially Doeblin–Fortet) operators acting on $L(W)$, which are either orderly, aperiodic, or ergodic. Let us first remark that, since $1\in L(W)$, for an ergodic Markov operator U on $L(W)$ we have $U^\infty 1=1$, hence the ergodicity of U is equivalent to $U^\infty f$ being a constant (depending on f) for all $f\in L(W)$.

The lemma below is valid for an arbitrary Markov operator acting on a normed space.

Lemma 3.1.19. *Assume that the Markov operator U is aperiodic w.r.t. a normed space, the norm of which is $\|\cdot\|$. Put $T=U-U^\infty$. Then, we have*

$$U^\infty U^\infty=U^\infty, \tag{3.1.18}$$

$$TU^\infty=U^\infty T=0, \tag{3.1.19}$$

$$U^n=U^\infty+T^n,\quad n\in\mathbb{N}^*. \tag{3.1.20}$$

Moreover, there exist positive constants $\alpha<1$ and c such that

$$\|T^n\|\leqslant c\alpha^n,\quad n\in\mathbb{N}^*. \tag{3.1.21}$$

Proof. Since $U^{n+1}=U^nU=UU^n$, letting $n\to\infty$ yields

$$U^\infty=U^\infty U=UU^\infty. \tag{3.1.22}$$

Thus $U^nU^\infty = U^\infty$ by iterating the second equation, and (3.1.18) follows by letting $n \to \infty$; next, (3.1.19) is a consequence of (3.1.18) and (3.1.22). For $n = 1$, (3.1.20) is simply the definition of T, and, for $n > 1$, (3.1.20) is obtained by induction on n, using (3.1.18) and (3.1.19). Since $\| T^{n+k} \| \leqslant \| T^n \| \| T^k \|$ for all $n, k \in \mathbb{N}^*$, and $\lim_{n\to\infty} \| T^n \| = \lim_{n\to\infty} \| U^n - U^\infty \| = 0$, we obtain (3.1.21) by applying Lemma 3.1.18 to the sequence $(\| T^n \|)_{n\in\mathbb{N}^*}$. □

Now we are able to give a sufficient condition for a Doeblin–Fortet chain to be aperiodic.

Theorem 3.1.20. *Assume that U is a Doeblin–Fortet operator, and that there exists a sequence $(l_n)_{n\in\mathbb{N}^*}$, with $\lim_{n\to\infty} l_n = 0$, such that for any $f \in L(W)$ there exists a function (denoted $U^\infty f$) in $B(W)$, satisfying the inequality*

$$|U^n f - U^\infty f| \leqslant l_n \| f \|_L, \quad n \in \mathbb{N}^*. \tag{3.1.23}$$

Then U is aperiodic w.r.t. $L(W)$.

To prove the theorem we need the following lemma.

Lemma 3.1.21. *For any Doeblin–Fortet operator U there exist $r \in [0, 1)$ and $R^* < \infty$ such that*

$$\| U^{mk} f \|_L \leqslant r^m \| f \|_L + R^* |f| \tag{3.1.24}$$

for all $f \in L(W)$ and $m \in \mathbb{N}$; here k is the natural integer specified in Definition 3.1.14. Moreover, $J = \sup_{n\in\mathbb{N}^*} \| U^n \|_L < \infty$.

Proof. By (3.1.14), we have

$$\| U^k f \|_L = s(U^k f) + |U^k f| \leqslant rs(f) + R|f| + |U^k||f|.$$

Hence, since $|U^k| = 1$,

$$\| U^k f \|_L \leqslant r \| f \|_L + R' |f|, \tag{3.1.25}$$

where $R' = R + 1 - r$. By iterating (3.1.25) we get (3.1.24) with $R^* = R(1 - r)^{-1}$. In turn, by (3.1.24), we get $\sup_{m\in\mathbb{N}^*} \| U^{mk} f \|_L < \infty$ for all $f \in L(W)$, and, by the uniform boundedness principle,† we deduce that $c = \sup_{m\in\mathbb{N}^*} \| U^{mk} \|_L < \infty$. But $J = \sup_{n\in\mathbb{N}^*} \| U^n \|_L \leqslant c(\max_{i=0,\ldots,k-1} \| U^i \|_L)$, so that $J < \infty$. □

† The uniform boundedness principle reads as follows. Let $(U_n)_{n\in\mathbb{N}^*}$ be a sequence of bounded linear operators from X into Y, where X and $(Y, \|\cdot\|)$ are Banach spaces. Then the two statements below are equivalent (for the notation see Section A2.6)

$$\sup_{n\in\mathbb{N}} ||| U_n ||| < \infty;$$

$$\sup_{n\in\mathbb{N}} \| U_n x \|_Y < \infty \quad \text{for all } x \in X.$$

Proof of Theorem 3.1.20. By Lemma 3.1.21, $\| U^n f \|_L \leqslant J \| f \|_L$ for all $n \in \mathbb{N}^*$. By (3.1.23) and Corollary 3.1.11, we get $U^\infty f \in L(W)$ and $\| U^\infty f \|_L \leqslant J \| f \|_L$. It is obvious that U^∞ is a linear operator on $L(W)$. Put $T = U - U^\infty$. The validity of (3.1.19) and (3.1.20) follows by arguments similar to those in the proof of Lemma 3.1.19, using (3.1.23) instead of the aperiodicity assumption. So it remains to prove that $\lim_{n \to \infty} \| T^n \|_L = 0$. Since $U^i T^n = T^{i+n}$, $i, n \in \mathbb{N}^*$, we replace f by $T^n f$, $n \in \mathbb{N}^*$, in (3.1.24), to get

$$\| T^{mk+n} f \|_L \leqslant r^m \| T^n f \|_L + R^* | T^n f | \leqslant 2Jr^m \| f \|_L + R^* l_n \| f \|_L.$$

Hence

$$\| T^{mk+n} \|_L \leqslant 2Jr^m + R^* l_n.$$

This shows that $\| T^{mk+n} \|_L < 1$ for all sufficiently large m and n. To conclude the proof, we apply Lemma 3.1.18 to the sequence $(\| T^n \|_L)_{n \in \mathbb{N}^*}$. □

The special importance of the space $L_r(W)$ is emphasized by Theorem 3.1.22 below, which is due to Dudley (1966).

For any $v, \mu \in \mathrm{ca}(W)$, let us put

$$\Delta(v, \mu) = \sup_{\substack{f \in L_r(W) \\ \| f \|_{L_r} = 1}} \left| \int_W f(w) v(\mathrm{d}w) - \int_W f(w) \mu(\mathrm{d}w) \right|.$$

It is easy to prove that Δ is a metric on $\mathrm{ca}(W)$, and hence also on the space $\mathrm{pr}(W)$ of all probabilities on $\mathscr{W}$.

Theorem 3.1.22. (i) *If (W, d) is a separable metric space, then the topology of the weak convergence on* $\mathrm{pr}(W)$ *and the topology induced on* $\mathrm{pr}(W)$ *by the metric Δ are equivalent.*

(ii) *If (W, d) is a complete separable metric space, then the metric space* $(\mathrm{pr}(W), \Delta)$ *is also complete.*

We omit the proof of this theorem; instead, we comment upon its consequences.

Let $\bar{U}$ and $\bar{U}'$ be two bounded Markov operators on $L(W)$, corresponding to the t.p.f.s $\bar{Q}$ and $\bar{Q}'$. We have

$$\| \bar{U} - \bar{U}' \|_L \geqslant \sup_{\substack{f \in L(W) \\ \| f \|_L = 1}} | \bar{U} f - \bar{U}' f | = \sup_{\substack{f \in L(W) \\ \| f \|_L = 1}} \sup_{w \in W} | \bar{U} f(w) - \bar{U}' f(w) |$$

and, interchanging the suprema, we obtain

$$\| \bar{U} - \bar{U}' \|_L \geqslant \sup_{w \in W} \Delta(\bar{Q}(w, \cdot), \bar{Q}'(w, \cdot)). \tag{3.1.26}$$

If the operator U^∞ corresponds to a t.p.f. Q^∞, taking $\bar{U}' = U^\infty$ and either

$\bar{U} = U_n$ or $\bar{U} = U^n$ in (3.1.26), we are able to estimate the rate of convergence to 0 of $\sup_{w \in W} \Delta(Q_n(w, \cdot), Q^\infty(w, \cdot))$ and $\sup_{w \in W} \Delta(Q^n(w, \cdot), Q^\infty(w, \cdot))$, in terms of the rate of convergence to 0 of $\| U_n - U^\infty \|_L$ and $\| U^n - U^\infty \|_L$, respectively.

We shall use such an argument in proving Theorem 3.1.24, which illustrates some properties of orderly and ergodic Doeblin–Fortet operators. To prove this theorem, we need the following lemma.

Lemma 3.1.23. *Let A and B be non-empty sets of the metric space (W, d), for which*

$$d(A, B) = \inf_{w' \in A, w'' \in B} d(w', w'') > 0,$$

and let

$$\rho(w) = \frac{d(w, A)}{d(w, A) + d(w, B)}, \, w \in W.$$

Then $\rho \in L(W)$, $0 \leqslant \rho(w) \leqslant 1$ for all $w \in W$, $\rho = 0$ on A, and $\rho = 1$ on B.

Proof. Put $d(w) = d(w, A) + d(w, B)$. Straightforward computations yield

$$\begin{aligned}&[\rho(w') - \rho(w'')] d(w') d(w'') \\ &= [d(w', A) - d(w'', A)] d(w'', B) + d(w'', A)[d(w'', B) - d(w', B)],\end{aligned}$$

whence we obtain

$$|\rho(w') - \rho(w'')| \leqslant \frac{d(w', w'')}{d(A, B)},$$

since, as is easily seen $s(d(\cdot, A)) \leqslant 1$, $s(d(\cdot, B)) \leqslant 1$, and $d(w) \geqslant d(A, B)$. So $\rho \in L(W)$, which concludes the proof, since the other assertions are immediate. □

Theorem 3.1.24. (i) *If the metric space (W, d) is separable and complete and U is an orderly Doeblin–Fortet operator, then there exists a t.p.f. Q^∞ on $(W, \mathscr{W})$ such that*

$$U^\infty f(w) = \int_W Q^\infty(w, dw') f(w'), \quad f \in L(W). \tag{3.1.27}$$

(ii) *We have*

$$\sup_{B \in \mathscr{W}} s(Q^\infty(\cdot, B)) < \infty. \tag{3.1.28}$$

(iii) *For any $w \in W$ and $B \in \mathscr{W}$*

$$\int_W Q^\infty(w, dw') Q(w', B) = \int_W Q(w, dw') Q^\infty(w', B) = Q^\infty(w, B). \tag{3.1.29}$$

(iv) *If, in addition, U is ergodic, then $Q^\infty(w,\cdot)$ does not depend on w and is the unique stationary probability of the chain.*

Proof. (i) Since U is orderly, the sequence $(\|U_n\|_L)_{n\in\mathbb{N}^*}$ is Cauchy, i.e. $\lim_{m,n\to\infty}\|U_n - U_m\|_L = 0$. Because of (3.1.26) we get $\lim_{m,n\to\infty}\Delta(Q_n(w,\cdot), Q_m(w,\cdot)) = 0$. By Theorem 3.1.22 (ii), the space $(\mathrm{pr}(W),\Delta)$ is complete, so that there exists $Q^\infty(w,\cdot)\in\mathrm{pr}(W)$ such that $Q_n(w,\cdot)\Rightarrow Q^\infty(w,\cdot)$ for all $w\in W$, and (3.1.27) obviously holds. The fact that $Q^\infty(\cdot,B)$ is $\mathscr{W}$-measurable for all $B\in\mathscr{W}$ will follow from the proof of the stronger statement in (ii).

(ii) It follows from Lemma 3.1.21, by replacing f by $U^i f, 0\leqslant i\leqslant k-1$ in (3.1.24), that

$$s(U^j f)\leqslant cr^{[j/k]}\|f\|_L + R^*|f|,\quad j\in\mathbb{N},$$

where $c = \max_{0\leqslant i\leqslant k-1}\|U^i\|_L$ and $R^* = R(1-r)^{-1}$. Therefore

$$s(U_n f)\leqslant\frac{1}{n}\sum_{j=0}^{n-1}s(U^j f) = O(n^{-1})\|f\|_L + R^*|f|,$$

whence, letting $n\to\infty$, we get

$$s(U^\infty f)\leqslant R^*|f| \tag{3.1.30}$$

for all $f\in L(W)$.

Let $B\in\mathscr{W}, B\neq W, B\neq\varnothing$, be closed. Then the sets $A_n = \{w: d(w,B)\geqslant 1/n\}$, $n\in\mathbb{N}^*$, are not empty for sufficiently large n. Since $d(A_n,B)\geqslant 1/n > 0$, Lemma 3.1.23 implies that $\rho_n\in L(W)$, where $\rho_n(w) = d(w,A_n)/[d(w,A_n) + d(w,B)]$. Obviously, $\lim_{n\to\infty}\rho_n(w) = \chi_B(w)$ for all $w\in W$, so that, by dominated convergence, $\lim_{n\to\infty}U^\infty\rho_n(w) = Q^\infty(w,B)$ for all $w\in W$. Since $|\rho_n| = 1, n\in\mathbb{N}^*$, by (3.1.30), we also have

$$|Q^\infty(w',B) - Q^\infty(w'',B)| = \lim_{n\to\infty}|U^\infty\rho_n(w') - U^\infty\rho_n(w'')|\leqslant R^* d(w',w'') \tag{3.1.31}$$

for all $w', w''\in W, w'\neq w''$. The inequality (3.1.31) holds trivially for $B = W$ and $B = \varnothing$; hence it is valid for all closed sets $B\in\mathscr{W}$ and all w', $w''\in W$. Let $B\in\mathscr{W}$ be arbitrary. Then the regularity of the measure $\tilde{Q}(\cdot) = Q^\infty(w',\cdot) + Q^\infty(w'',\cdot)$ implies that, for any $\varepsilon > 0$, there exists a closed set $A(\varepsilon, w', w'') = A\subset B$ such that $\tilde{Q}(B - A) < \varepsilon$. Then, using (3.1.30) once again, we obtain

$$\begin{aligned}
&|Q^\infty(w',B) - Q^\infty(w'',B)|\\
&\quad\leqslant|[Q^\infty(w',B) - Q^\infty(w'',B)] - [Q^\infty(w',A) - Q^\infty(w'',A)]|\\
&\qquad + |Q^\infty(w',A) - Q^\infty(w'',A)|\\
&\quad\leqslant\tilde{Q}(B - A) + |Q^\infty(w',A) - Q^\infty(w'',A)|\\
&\quad\leqslant\varepsilon + R^* d(w',w'').
\end{aligned}$$

Since ε is arbitrary, the proof of (ii) is complete.

(iii) By direct computation

$$UU_nf = U_nUf = U_nf + \frac{1}{n}(U^nf - f)$$

for all $n \in \mathbb{N}^*$ and $f \in L(W)$. Noting that the three terms of the above equation converge in $L(W)$ to $UU^\infty f, U^\infty Uf$, and $U^\infty f$, respectively, we obtain

$$UU^\infty f = U^\infty Uf = U^\infty f$$

for all $f \in L(W)$. Taking $f(\cdot) = Q^\infty(\cdot, B)$ with $B \in \mathscr{W}$ arbitrary, yields (3.1.29).

(iv) If U is ergodic, then $U^\infty f(w') = U^\infty f(w'')$ for all $f \in L(W)$ and $w', w'' \in W$. Taking $f(\cdot) = Q^\infty(\cdot, B)$ with $B \in \mathscr{W}$ arbitrary, we conclude that $Q^\infty(w, \cdot) = \mathbf{Q}^\infty(\cdot)$ does not depend on $w \in W$. Now, (3.1.29) shows that $\mathbf{Q}^\infty$ is a stationary probability of the chain. Suppose that there exists another stationary probability μ, i.e. $V\mu = \mu$. Then, putting $V_n = (1/n)\sum_{k=1}^n V^k$, for any $f \in L(W)$, we have

$$(\mu, f) = (V_n\mu, f) = (\mu, U_nf) \to (\mu, U^\infty f) = U^\infty f = (\mathbf{Q}^\infty, f)$$

as $n \to \infty$, hence $\mu = \mathbf{Q}^\infty$. □

Remark. As illustrated by Problems 14, 15 and 16, the various assertions of this theorem can be obtained either under less restrictive assumptions or for Markov operators acting on function spaces different from $L(W)$.

3.1.4 Markov operators on C(W)

In this subsection we shall study Markov operators acting on $C(W)$, where $(W, \mathcal{O})$ is a topological space (in particular, the topology $\mathcal{O}$ may be induced by a metric d). The problem is to find relationships between the quasi-compactness of the operators U and V, on the one hand, and between the quasi-compactness of U on $C(W)$ and Doeblin's condition, on the other. This is done in Theorems 3.1.25 and 3.1.28 below.

Theorem 3.1.25. *Let $(W, \mathcal{O})$ be a topological space. If U is a quasi-compact Markov operator on $B(W)$ and $U(C(W)) \subset C(W)$, then U is also quasi-compact on $C(W)$.*

Proof. By Theorems 3.1.6 and 3.1.7, the quasi-compactness of U on $B(W)$ is equivalent to the quasi-compactness of V on $\mathrm{ca}(W)$. Hence Theorem 3.1.5 applies. Let U^∞ be the Markov operator on $B(W)$ defined as

$$U^\infty f(w) = \int_W f(w')Q^\infty(w, dw').$$

First, we prove that $U^\infty(C(W)) \subset C(W)$. Let $\varepsilon > 0$ and $f \in C(W)$ be arbitrary; without any loss of generality, we may assume that $|f| \leqslant 1$. Then, by (3.1.10) and (3.1.12), there exists $n_0 = n_0(\varepsilon) \in \mathbb{N}^*$, such that

$$\sup_{w \in W} |U^{nN_0} f(w) - U^\infty f(w)| < \tfrac{1}{3}\varepsilon$$

for all $n \geqslant n_0$. But $U^{nN_0} f \in C(W)$, $n \in \mathbb{N}^*$, so that, for any $w_0 \in W$, there exists a neighbourhood $V_\varepsilon(w_0)$ of w_0, such that, for any $w \in V_\varepsilon(w_0)$, we have

$$|U^{nN_0} f(w) - U^{nN_0} f(w_0)| < \tfrac{1}{3}\varepsilon.$$

Hence

$$\begin{aligned} |U^\infty f(w) - U^\infty f(w_0)| &\leqslant |U^\infty f(w) - U^{nN_0} f(w)| + |U^{nN_0} f(w) - U^{nN_0} f(w_0)| \\ &\quad + |U^{nN_0} f(w_0) - U^\infty f(w_0)| < \varepsilon \end{aligned}$$

for all $n \geqslant n_0(\varepsilon)$ and $w \in V_\varepsilon(w_0)$, which shows that $f \in C(W)$ implies $U^\infty f \in C(W)$. Since U^∞ is a compact operator on $B(W)$ (see the proof of Theorem 3.1.7) and $\lim_{n\to\infty} |U^{nN_0} - U^\infty| = 0$, we deduce that U is quasi-compact on $C(W)$. □

Under additional assumptions on the topology of the space W, further related results can also be proved. To this purpose, we use the fact that, for a compact Hausdorff space $(W, \mathcal{O})$, the space $C^*(W)$ of the linear functionals on $C(W)$ and the space $\mathrm{rca}(W)$ are isometrically isomorphic (see Theorem A2.1). To prove Theorem 3.1.28, we need the following two lemmas, the simple proofs of which are left to the reader. (The second one is a simple consequence of Schauder's theorem – see Section A2.6.)

Lemma 3.1.26. *Let $(Y, \|\cdot\|_Y)$, $(Z, \|\cdot\|_Z)$ be isometrically isomorphic normed linear spaces, with isomorphism $J: Y \to Z$, and A a bounded linear operator on Y. Then A induces a bounded linear operator $\hat{A}$ on Z defined as $\hat{A} = JAJ^{-1}$. If A is (quasi-)compact, then $\hat{A}$ is also (quasi-)compact.*

Lemma 3.1.27. *If A is a quasi-compact operator on a normed linear space Y, then its adjoint A^* is also quasi-compact on Y^*.*

Theorem 3.1.28. *Assume that $(W, \mathcal{O})$ is a Hausdorff compact space. Then*
(i) *the operator U is compact on $C(W)$ iff V is compact on* $\mathrm{rca}(W)$;
(ii) *if U is quasi-compact on $C(W)$, then V is quasi-compact on* $\mathrm{rca}(W)$.

Proof. Let $J: \mathrm{rca}(W) \to C^*(W)$ be the isometric isomorphism defined by

$$(J\mu, f) = \int_W f(w)\mu(\mathrm{d}w)$$

see Section A2.7.

By Schauder's theorem and the preceding two lemmas, the proof amounts to showing that $V\mu = J^{-1}U^*J\mu$ for all $\mu \in \text{ca}(W)$, where U^* is the (genuine!) adjoint of U (see Section A2.6). But this equation is successively equivalent to the equations below

$$JV\mu = U^*J\mu, \quad \mu \in \text{ca}(W),$$

$$(JV\mu, f) = (U^*J\mu, f) = (J\mu, Uf), \quad \mu \in \text{ca}(W), \quad f \in C(W),$$

$$\int_W f(w)V\mu(dw) = \int_W Uf(w)\mu(dw) = \int_W \int_W f(w')Q(w, dw')\mu(dw'),$$

the latter holding by the definition of V. □

Remarks. 1. If (W, d) is a compact metric space, Theorem 3.1.28 shows that the statement 'U is compact on $C(W)$' is equivalent to 'V is compact on $\text{ca}(W)$,' and hence any of them implies Doeblin's condition. At the same time, if U is quasi-compact on $C(W)$, then Condition (D) holds. So, in this case, under either (i) or (ii) in Theorem 3.1.28, Theorem 3.1.5 applies.

2. The converse of (ii) is, generally, not true, as shown by Problem 4.

The next step in our analysis is to ensure the (quasi-)compactness of the operator U on $C(W)$ by continuity assumptions on its underlying t.p.f. Q.

Definition 3.1.29. *Let (Y, d) and (Z, d') be two metric spaces, and R a t.p.f. from $(Y, \mathscr{B}_Y)$ to $(Z, \mathscr{B}_Z)$.*

(i) *The t.p.f. R is said to be* weakly continuous *iff the map $y \to R(y, \cdot)$ from (Y, d) to the space* ca(Z) *endowed with the topology of weak convergence is continuous, i.e. iff, for any sequence $(y_n)_{n \in \mathbb{N}^*} \subset Y$ with $\lim_{n \to \infty} y_n = y_0$, we have*

$$\lim_{n \to \infty} \int_Z f(z)R(y_n, dz) = \int_Z f(z)R(y_0, dz)$$

for all bounded functions f in $C(Z)$.

(ii) *The t.p.f. R is said to be* continuous *iff $R(\cdot, A)$ is continuous for all $A \in \mathscr{B}_Z$.*

(iii) *The t.p.f. R is said to be* strongly continuous *iff the map $y \to R(y, \cdot)$ from (Y, d) to the space* ca(Z) *endowed with the topology of strong convergence is continuous, i.e. iff, for any sequence $(y_n)_{n \in \mathbb{N}^*} \subset Y$ with $\lim_{n \to \infty} y_n = y_0$, we have*

$$\lim_{n \to \infty} \| R(y_n, \cdot) - R(y_0, \cdot) \| = 0.$$

Remarks. 1. Obviously, the strong continuity of a t.p.f. implies both its continuity and weak continuity, and the continuity of a t.p.f. implies its weak continuity.

2. The strong continuity of a t.p.f. R is equivalent to the statement 'the function $R(\cdot, A)$ is uniformly continuous w.r.t. $A \in \mathscr{B}_Z$.'

Weak continuity is too weak a concept to ensure the desired properties for the operator U; in turn, continuity and strong continuity imply the quasi-compactness and the compactness of the operator U on $C(W)$, respectively, if, in addition, we let the state space be a compact metric space. To prove this we need the following lemma, the simple proof of which is left to the reader.

Lemma 3.1.30. (i) *For an MC whose state space is a metric space (W, d) and whose t.p.f. Q is weakly continuous, the transition operator U on $C(W)$ is of norm* 1.

(ii) *If the t.p.f. Q is continuous, then $U(B(W)) \subset C(W)$.*

Theorem 3.1.31. *For an MC whose state space is a compact metric space (W, d) and whose t.p.f. Q is continuous, the transition operator U is quasi-compact on $C(W)$.*

Proof. By the preceding lemma, $U(C(W)) \subset C(W)$. Let us define the set function $\rho: \mathscr{W} \to C(W)$ by $\rho(A) = Q(\cdot, A)$, $A \in \mathscr{W}$. We shall prove that ρ is a vector measure, i.e.

$$\left(f^*, \rho\left(\bigcup_{j \in \mathbb{N}^*} A_j\right)\right) = \sum_{j \in \mathbb{N}^*} (f^*, \rho(A_j)) \tag{3.1.32}$$

for all $f^* \in C^*(W)$ and all sequences $(A_j)_{j \in \mathbb{N}^*} \subset \mathscr{W}$ with $A_i \cap A_j = \varnothing$, $i \neq j$. Putting $g_n(\cdot) = \sum_{j=1}^n Q(\cdot, A_j)$, $n \in \mathbb{N}^*$, and $g(\cdot) = \sum_{j \in \mathbb{N}^*} Q(\cdot, A_j)$, we note that $|g_n| = \sup_{w \in W} Q(\cdot, \bigcup_{j=1}^n A_j) \leqslant 1$ for all $n \in \mathbb{N}^*$, and that $\lim_{n \to \infty} g_n(w) = g(w)$ for all $w \in W$. Then, by the Riesz representation Theorem A2.1, the sequence $(g_n)_{n \in \mathbb{N}^*}$ converges weakly to g, i.e. $\lim_{n \to \infty} (f^*, g_n) = (f^*, g)$ for all $f^* \in C^*(W)$. Therefore

$$\begin{aligned}\left(f^*, \rho\left(\bigcup_{j \in \mathbb{N}^*} A_j\right)\right) &= \left(f^*, \sum_{j \in \mathbb{N}^*} \rho(A_j)\right) = (f^*, g) = (f^*, \lim_{n \to \infty} g_n) \\ &= \lim_{n \to \infty} (f^*, g_n) = \lim_{n \to \infty} \sum_{j=1}^n (f^*, \rho(A_j)) \\ &= \sum_{j \in \mathbb{N}^*} (f^*, \rho(A_j)),\end{aligned}$$

which is (3.1.32). In addition, the measure (f^*, ρ) is regular and hence the operator U is weak-compact on $C(W)$ (see Dunford & Schwartz (1958, p. 493)). Then U^2 is compact on $C(W)$ (see Section A2.6); in turn, this implies that U is quasi-compact on $C(W)$. □

Theorem 3.1.32. *For an MC whose state space is a compact metric space (W, d) and whose t.p.f. Q is strongly continuous, the transition operator U is compact on $C(W)$.*

Proof. Using Lemma 2.1.1, we get

$$\sup_{\substack{f\in C(W)\\ |f|\leqslant 1}} |Uf(w') - Uf(w'')| = \sup_{\substack{f\in C(W)\\ |f|\leqslant 1}} \left| \int_W f(w)[Q(w', \mathrm{d}w) - Q(w'', \mathrm{d}w)] \right|$$
$$\leqslant \| Q(w', \cdot) - Q(w'', \cdot) \|$$

for arbitrary $w', w'' \in W$. This shows that the set $\{Uf : f \in C(W), |f| \leqslant 1\}$ is bounded and equicontinuous. The proof is concluded by applying the Arzelà–Ascoli Theorem A1.16. □

Now, the natural problem which arises is to find assumptions on u and P (on Π and P), ensuring that the MC associated with an RSCC (GRSCC) has either a continuous or a strong continuous t.p.f. Q.

For RSCCs, this problem is solved only in a special case and under quite restrictive assumptions (see Problem 17). As a consequence, we will now confine ourselves to the case of a GRSCC, where the continuity concepts reveal themselves to be much more natural.

Let $\{(W, \mathscr{W}), (X, \mathscr{X}), \Pi, P\}$ be a GRSCC. For the sake of clarity, for $w \in W$ fixed, we shall write Π_w for the t.p.f. from $(X, \mathscr{X})$ to $(W, \mathscr{W})$ defined as $\Pi_w(x, A) = \Pi(w, x, A)$, $x \in X$, $A \in \mathscr{W}$. At the same time, for $x \in X$ fixed, we shall write Π_x for the t.p.f. from $(W, \mathscr{W})$ to itself defined as $\Pi_x(w, A) = \Pi(w, x, A)$, $w \in W$, $A \in \mathscr{W}$.

Proposition 3.1.33. *Let $\{(W, d), (X, \mathscr{X}), \Pi, P\}$ be a GRSCC for which*

(i) *the metric space (W, d) is separable;*

(ii) *the t.p.f. P is strongly continuous;*

(iii) *for any $x \in X$, the t.p.f. Π_x is strongly continuous.*

Then the t.p.f. Q of the associated MC is strongly continuous.

Proof. Because of (i), there exists a denumerable algebra $\mathscr{K}$ which generates $\mathscr{W}$. Moreover, if ρ and ν are two probabilities on $\mathscr{W}$, we have

$$\sup_{A\in\mathscr{W}} |\rho(A) - \nu(A)| = \sup_{A\in\mathscr{K}} |\rho(A) - \nu(A)|. \tag{3.1.33}$$

Indeed, putting $c = \sup_{A \in \mathscr{X}} |\rho(A) - \nu(A)|$, the collection $\mathscr{M} = \{A \in \mathscr{W}: |\rho(A) - \nu(A)| \leqslant c\}$ includes $\mathscr{X}$ and is a monotone class, as shown below. Let $(A_n)_{n \in \mathbb{N}^*} \subset \mathscr{M}$ be a non-decreasing sequence of sets and put $A = \lim_{n \to \infty} A_n$. Then, for any $\varepsilon > 0$, there exist $n_0 = n_0(\varepsilon) \in \mathbb{N}^*$ and $k_0 = k_0(\varepsilon) \in \mathbb{N}^*$ such that

$$|\rho(A) - \rho(A_n)| < \varepsilon, \quad n \geqslant n_0,$$
$$|\nu(A) - \nu(A_k)| < \varepsilon, \quad k \geqslant k_0.$$

Thus, for any $n \geqslant \max(n_0, k_0)$, we have

$$|\rho(A) - \nu(A)| \leqslant |\rho(A) - \rho(A_n)| + |\rho(A_n) - \nu(A_n)| + |\nu(A_n) - \nu(A)| \leqslant c + 2\varepsilon,$$

which shows that $A \in \mathscr{M}$, since $\varepsilon > 0$ is arbitrary. Since $\mathscr{X} \subset \mathscr{M} \subset \mathscr{W}$ and $\sigma(\mathscr{X}) = \mathscr{W}$, we conclude that (3.1.33) holds.

We have to prove that

$$\lim_{n \to \infty} \| Q(w_n, \cdot) - Q(w_0, \cdot) \| = 2 \lim_{n \to \infty} \sup_{A \in \mathscr{W}} |Q(w_n, A) - Q(w_0, A)| = 0$$

for all $w_0 \in W$ and all sequences $(w_n)_{n \in \mathbb{N}^*} \subset W$ with $\lim_{n \to \infty} w_n = w_0$ (see Section A1.4); but, by (3.1.33), it is sufficient to prove that

$$\lim_{n \to \infty} \sup_{A \in \mathscr{X}} |Q(w_n, A) - Q(w_0, A)| = 0.$$

As the supremum is taken over a denumerable set, we can interchange the supremum and the integral to get

$$\begin{aligned}\sup_{A \in \mathscr{X}} |Q(w_n, A) - Q(w_0, A)| &\leqslant \sup_{A \in \mathscr{X}} \int_X |\Pi_x(w_n, A) - \Pi_x(w_0, A)| P(w_0, \mathrm{d}x) \\ &\quad + \sup_{A \in \mathscr{X}} \left| \int_X \Pi_x(w_n, A)[P(w_n, \mathrm{d}x) - P(w_0, \mathrm{d}x)] \right| \\ &\leqslant \int_X \sup_{A \in \mathscr{X}} |\Pi_x(w_n, A) - \Pi_x(w_0, A)| P(w_0, \mathrm{d}x) \\ &\quad + \sup_{A \in \mathscr{X}} \| P(w_n, \cdot) - P(w_0, \cdot) \|,\end{aligned}$$

where we have also used Lemma 2.1.1.

By (iii), we have

$$\lim_{n \to \infty} \sup_{A \in \mathscr{X}} |\Pi_x(w_n, A) - \Pi_x(w_0, A)| = 0$$

for any $x \in X$ fixed, hence the last integral above converges to 0 as $n \to \infty$ by the Lebesgue dominated convergence Theorem A1.8.

Since, by (ii), we also have $\lim_{n \to \infty} \| P(w_n, \cdot) - P(w_0, \cdot) \| = 0$, the proof is complete. □

Remark. Taking into account Theorem 3.1.32, Proposition 3.1.33 tells us that the Markov operator U is compact on $C(W)$, in the case where the metric space (W,d) is compact.

In the light of Theorem 3.1.31 the following two propositions give conditions under which the Markov operator U is quasi-compact on $C(W)$, in the case where the metric space (W,d) is compact.

Proposition 3.1.34. *Let $\{(W,d),(X,\mathscr{X}),\Pi,P\}$ be a GRSCC for which*

(i) *the metric space (W,d) is compact;*

(ii) *the t.p.f. P is continuous;*

(iii) *the t.p.f. Π_x is continuous for all $x \in X$.*

Then the t.p.f. of the associated MC is continuous.

Proof. Let $w_0 \in W$ be arbitrary, and consider a sequence $(w_n)_{n\in\mathbb{N}^*} \subset W$, with $\lim_{n\to\infty} w_n = w_0$. For $A \in \mathscr{W}$ fixed, if we define

$$\mu_n(\cdot) = P(w_n,\cdot), \qquad g_n = 1, \quad n\in\mathbb{N}^*,$$
$$\mu(\cdot) = P(w_0,\cdot), \qquad g = 1,$$
$$f_n(\cdot) = \Pi_{\cdot}(w_n, A), \quad n\in\mathbb{N}^*, \quad f(\cdot) = \Pi_{\cdot}(w_0, A),$$

then, by the hypotheses made, $(f_n)_{n\in\mathbb{N}^*}$ and $(g_n)_{n\in\mathbb{N}^*}$ are two sequences of $\mathscr{W}$-measurable functions which converge to f and g, respectively; moreover, $|f_n| \leqslant |g_n|$ for all $n\in\mathbb{N}^*$, $\lim_{n\to\infty}\mu_n(B) = \mu(B)$ for all $B\in\mathscr{W}$, and

$$\lim_{n\to\infty}\int_X g_n\,\mathrm{d}\mu_n = \int_X g\,\mathrm{d}\mu < \infty.$$

Thus all the hypotheses in the generalized Lebesgue theorem[†] hold and so does its conclusion, i.e.

$$\lim_{n\to\infty}\int_X f_n\,\mathrm{d}\mu_n = \int_X f\,\mathrm{d}\mu.$$

This means

$$\lim_{n\to\infty}\int_X \Pi_x(w_n, A)P(w_n,\mathrm{d}x) = \int_X \Pi_x(w_0, A)P(w_0,\mathrm{d}x),$$

or, equivalently, $\lim_{n\to\infty} Q(w_n, A) = Q(w_0, A)$. □

Proposition 3.1.35. *Let $\{(W,d), (X,d'), \Pi, P\}$ be a GRSCC for which*

(i) *the metric space (W,d) is compact;*

[†] The generalized Lebesgue convergence theorem reads as follows. Let $(X,\mathscr{B})$ be a measurable space and $(\mu_n)_{n\in\mathbb{N}^*}$ a sequence of measures on $\mathscr{B}$ which converge setwise to a measure μ. Let $(f_n)_{n\in\mathbb{N}^*}$ and $(g_n)_{n\in\mathbb{N}^*}$ be two sequences of measurable real-valued functions which converge pointwise to f and g. Suppose $|f_n| \leqslant |g_n|$, $n\in\mathbb{N}^*$, and that $\lim_{n\to\infty}\int g_n\,\mathrm{d}\mu_n = \int g\,\mathrm{d}\mu < \infty$. Then $\lim_{n\to\infty}\int f_n\,\mathrm{d}\mu_n = \int f\,\mathrm{d}\mu$ (see Royden (1968, p. 232)).

(ii) *the t.p.f. P is weakly continuous;*
(iii) *the t.p.f. Π_w is continuous for all $w \in W$;*
(iv) *for any $A \in \mathscr{W}$ the collection $\{\Pi_x(\cdot, A), x \in X\}$ of t.p.f.s is equicontinuous.*
Then the t.p.f. of the associated MC is continuous.

Proof. Let $w_0 \in W$ be arbitrary and consider a sequence $(w_n)_{n \in \mathbb{N}^*} \subset W$ with $\lim_{n \to \infty} w_n = w_0$. We have

$$|Q(w_n, A) - Q(w_0, A)| \leqslant \left| \int_X \Pi_{w_0}(x, A)[P(w_n, \mathrm{d}x) - P(w_0, \mathrm{d}x)] \right| + \int_X |\Pi_x(w_n, A) - \Pi_x(w_0, A)| P(w_n, \mathrm{d}x)$$

for all $A \in \mathscr{W}$. The first term on the right converges to 0 as $n \to \infty$, since $P(w_n, \cdot) \Rightarrow P(w_0, \cdot)$ and $\Pi_{w_0}(\cdot, A)$ is a bounded continuous function. The second term on the right is less than

$$\sup_{x \in X} |\Pi_x(w_n, A) - \Pi_x(w_0, A)|,$$

hence it converges to 0 as $n \to \infty$ by (iv). □

3.2 Compact Markov chains

This section is devoted to describing a class of MCs associated with an RSCC which enjoy properties analogous to those of finite MCs. These MCs, called compact MCs, were introduced and studied by M. F. Norman.

Definition 3.2.1. *An MC is said to be* compact *iff its state space is a compact metric space (W, d) and its transition operator is a Doeblin–Fortet operator.*

Problem 5 shows that, in general, a Doeblin–Fortet chain does not fulfil Doeblin's condition and hence the operator U is not quasi-compact on $B(W)$, nor is the operator V quasi-compact on $\mathrm{ca}(W)$. As a consequence, the mean ergodic Theorem A2.2 does not apply in this context. As illustrated by the theorem below, the key role in studying compact MCs is played by the Ionescu Tulcea–Marinescu Theorem A2.4.

Theorem 3.2.2. *The Ionescu Tulcea–Marinescu Theorem* A2.4 *applies to the transition operator of a compact MC, when it is viewed as acting on the spaces $L(W)$ and $C(W)$.*

Proof. By Corollary 3.1.11, Condition (ITM_1) holds for $(X, |\cdot|) = (C(W), |\cdot|)$

and $(Y, \|\cdot\|) = (L(W), \|\cdot\|_L)$, whether or not (X, d) is compact. Since U is a transition operator, Condition (ITM$_2$) holds with $H = 1$. Next, Condition (ITM$_3$) holds by the definition of a Doeblin–Fortet operator. Finally, to verify Condition (ITM$_4$), let $L' \subset L(W)$ be a bounded subset of $L(W)$. Then the set $U^k L'$ is also bounded in $L(W)$ for any $k \in \mathbb{N}^*$; therefore this set is bounded in $C(W)$ and equicontinuous. Since (W, d) is compact, the Arzelà–Ascoli Theorem A1.16 implies that its closure in $C(W)$ is compact.

□

Theorem 3.1.16 provides an important class of RSCCs for which the associated MC is compact as specified in Proposition 3.2.3.

Proposition 3.2.3. *If for an RSCC with contraction the space (W, d) is compact, then the associated MC is compact.*

Another class of compact MCs is that of finite MCs, regardless of the metric on their state space (see Problem 18).

For any compact MC, the sequence $(U_n f)_{n \in \mathbb{N}^*}$ is equicontinuous for any $f \in C(W)$ (see Problem 11). The theory of Markov operators enjoying this property was worked out by Jamison (1964, 1965), Jamison & Sine (1969), and Rosenblatt (1964a, b, 1967). However, Norman's stronger assumptions enabled him to get the stronger results we give below.

First, an immediate consequence of Theorem 3.2.2 is that Theorem 3.1.24 applies to any compact MC, as is shown more precisely in Theorem 3.2.4.

Theorem 3.2.4. *A compact MC is orderly w.r.t. $L(W)$ and there exists a t.p.f. Q^∞ on $(W, \mathscr{W})$ such that*

$$U^\infty f(\cdot) = \int_W Q^\infty(\cdot, dw) f(w), \quad f \in L(W).$$

Moreover, $Q^\infty(\cdot, A) \in E(1)$ for any $A \in \mathscr{W}$, and for any $w \in W$, $Q^\infty(w, \cdot)$ is a stationary probability for the chain.

Proof. Since $U1 = 1$, it follows that 1 is an eigenvalue of U, and any constant function is a corresponding eigenfunction of U. Then (see Section A2.9), we have

$$U_1^n = \frac{1}{n} \sum_{j=0}^{n-1} U^j,$$

which, by Theorem A2.4(iv), can be written as

$$U_1^n = \frac{1}{n} \sum_{j=1}^{n} U^j + \frac{1}{n} I - \frac{1}{n} U^n = U_1 + \frac{1}{n} \sum_{\substack{\sigma \in E \\ \sigma \neq 1}} \frac{\sigma - \sigma^{n+1}}{1 - \sigma} U_\sigma + \frac{1}{n} \sum_{j=0}^{n} T^j - \frac{1}{n} U^n,$$

where E is the (finite) set of the eigenvalues of modulus 1 of the operator U. In turn, this implies that

$$\| U_1^n - U_1 \|_L = O\left(\frac{1}{n}\right)$$

as $n \to \infty$ which shows that the chain is orderly. The other statements in the theorem have been already proved in Theorem 3.1.24. □

The key concept in studying a compact MC (and also a continuous MC, which will be studied in the next section) is that of an ergodic kernel.

Definition 3.2.5. *A set* $\mathrm{E} \in \mathscr{W}$ *is said to be* stochastically closed *(w.r.t. Q) iff* $Q(w, \mathrm{E}) = 1$ *for all* $w \in \mathrm{E}$. *A set* $\mathrm{E} \in \mathscr{W}$ *is said to be an* ergodic kernel *iff it is topologically and stochastically closed and there is no proper subset of it with these properties.*

Using this concept, it is possible to get results about the so-called 'ergodic decomposition' of the state space of a compact MC, as in the theorem below.

Theorem 3.2.6. (i) *There exist a finite number of ergodic kernels* $\mathrm{E}_1, \ldots, \mathrm{E}_l$, *where* $l = \dim E(1)$.

(ii) *The t.p.f.* Q^∞ *is of the form*

$$Q^\infty(w, A) = \sum_{j=1}^{l} g_j(w)\pi_j(A), \quad w \in W, \quad A \in \mathscr{W},$$

where

(a) *the set* $\{g_1, \ldots, g_l\}$ *is a basis for* $E(1)$;

(b) *the probability* π_j *is a stationary probability for the chain,* $\operatorname{supp} \pi_j = \mathrm{E}_j$, $1 \leqslant j \leqslant l$, *and the set* $\{\pi_1, \ldots, \pi_l\}$ *is a basis for* $\{\mu \in \mathrm{ca}(W); V\mu = \mu\}$.

(iii) *We have*

$$g_j(w) = \mathbf{P}_w(\Omega_j), \quad w \in W, \quad 1 \leqslant j \leqslant l, \quad \textit{where} \quad \Omega_j = \left\{\omega \colon \lim_{n \to \infty} d(\zeta_n(\omega), \mathrm{E}_j) = 0\right\}.$$

Moreover, for any $w \in W$, *we have* $\mathbf{P}_w\left(\bigcup_{j=1}^{l} \Omega_j\right) = 1$.

(iv) *If* $f \in L(W)$ *and* $f(w) = 0$ *for all* $w \in \bigcup_{j=1}^{l} \mathrm{E}_j$, *then the sequence* $(\| U^n f \|_L)_{n \in \mathbb{N}^*}$ *converges exponentially to* 0 *as* $n \to \infty$.

The proof of this theorem, as well as the proofs of all the results below, are not given here since they can be found in Norman (1972, pp. 50–63).

Remarks. 1. Statement (ii) of Theorem 3.2.6 gives the structure of the t.p.f. Q^∞ to which the sequence $(Q_n)_{n\in\mathbb{N}^*}$ converges weakly; this makes it possible to derive necessary and sufficient conditions for a compact MC to be ergodic.

2. Statement (iii) shows that a compact MC converges 'in distance' with probability 1 to one of its ergodic kernels, regardless of its starting point.

3. The proof is based upon a series of lemmas among which two are valid for any MC, whose state space is a compact metric space. These lemmas (3.3.5 and 3.3.7) are to be found in the next section, since they will also be used in the theory of continuous MCs.

To speak of the 'subergodic decomposition' of an MC we first have to introduce the concept of a cycle.

Definition 3.2.7. *A sequence* $\{\mathrm{D}_1,\ldots,\mathrm{D}_p\}$ *of pairwise disjoint sets in* $\mathscr{W}$ *is said to be a* cycle *iff* $Q(w,\mathrm{D}_{j+1})=1$ *for any* $w\in\mathrm{D}_j$, $j\in\mathbb{N}^*$, *with the convention* $\mathrm{D}_s=\mathrm{D}_t$ *iff* $s\equiv t(\mathrm{mod}\,p)$.

A cycle $\{\mathrm{D}_1,\ldots,\mathrm{D}_p\}$ *is said to be* maximal *iff for any other cycle* $\{\mathrm{D}'_1,\ldots,\mathrm{D}'_{p'}\}$, *p is a multiple of p'; and p is said to be the* period *of the maximal cycle.*

Theorem 3.2.8. *Consider a compact MC having the ergodic kernels* E_j, $1\leqslant j\leqslant l$. *Then*

(i) *for each* $1\leqslant j\leqslant l$ *there exists a maximal cycle* $\{\mathrm{E}_j^1,\ldots,\mathrm{E}_j^{p_j}\}$ *of topologically closed sets, the union of which is the ergodic kernel* E_j. *The sets in the maximal cycle are unique up to a cyclic permutation, i.e. if* $\{\mathrm{F}_j^1,\ldots,\mathrm{F}_j^{p_j}\}$ *is another maximal cycle, then there exists* $r\in\mathbb{N}^*$ *such that* $\mathrm{F}_j^s=\mathrm{E}_j^{s+r}$ *for all s,* $1\leqslant s\leqslant p_j$;

(ii) *the set E of the eigenvalues of modulus* 1 *of the operator U can be written as* $E=\bigcup_{j=1}^{l}C_{p_j}$, *where* C_p *is the set of all pth roots of unity.*

Remarks. 1. Statement (i) in Theorem 3.2.8 shows that, if a compact MC enters an ergodic kernel, E_j say, then it moves cyclically within it with some period p_j, which is called the period of the ergodic kernel itself.

2. Consider a compact MC with a state space (W,d). Let W' be a stochastically and topologically closed subset of W, let d' be the restriction of d to W', and define $Q'(w,A)=Q(w,A)$ for all $w\in W'$ and $A\in\mathscr{W}'=W'\cap\mathscr{W}$. It is possible to prove that the MC having the state space (W',d') and the t.p.f. Q' is still a compact MC; this MC is called the restriction of the initial chain to the set W'. Consider the restriction of the Markov operator U to the ergodic kernel E_j, and denote by E_j the set of its eigenvalues of modulus 1, $1\leqslant j\leqslant l$. It is then possible to prove that $E_j=C_{p_j}$, $1\leqslant j\leqslant l$.

3. The results in Theorem 3.2.8 make it possible to derive necessary and sufficient conditions for a compact MC to be aperiodic w.r.t. $L(W)$.

The next theorem, which is a direct consequence of Theorems 3.2.6 and 3.2.8, gives necessary and sufficient conditions for a compact MC to be ergodic, aperiodic, and regular, respectively, w.r.t. $L(W)$.

Theorem 3.2.9. (i) *A compact MC is ergodic w.r.t. $L(W)$ iff it has only one ergodic kernel, or, equivalently, iff it possesses a unique stationary probability.*

(ii) *A compact MC is aperiodic w.r.t. $L(W)$ iff $p_j = 1$, $1 \leqslant j \leqslant l$.*

(iii) *A compact MC is regular w.r.t. $L(W)$ iff it has only one ergodic kernel of period 1. Equivalently, it is regular iff it possesses a unique stationary probability π, and the distribution of ζ_n converges weakly to π, regardless of the initial distribution of the chain.*

Remark. For a regular compact MC, the distribution of ζ_n may not converge to π in the norm of total variation (see Problem 26). That is why, unlike finite and continuous MCs, for a regular compact MC the tail σ-algebra may not be trivial (again, see Problem 26).

A special class of compact MCs, which is of importance in some of the applications of MCs, is that of absorbing compact MCs, defined below.

Definition 3.2.10. *A compact MC is said to be* absorbing *iff all its ergodic kernels are one-element sets, i.e.* $E_j = \{a_j\}$, $a_j \in W$, $1 \leqslant j \leqslant l$. *The* a_j, $1 \leqslant j \leqslant l$, *are called* absorbing points. *An RSCC is said to be* absorbing *iff its associated MC is compact and absorbing.*

The aperiodicity w.r.t. $L(W)$ of an absorbing compact MC is obvious and, by Theorem 3.2.6, ζ_n converges in distance with probability 1 to a random absorbing point $a_j \in W$, $j = j(\omega)$, $1 \leqslant j \leqslant l$. If $l > 1$, then the probability of the convergence to each of the a_j, $1 \leqslant j \leqslant l$, depends on the initial distribution of the chain, and therefore the chain is not regular w.r.t. $L(W)$.

The next theorem gives the characteristic properties of an absorbing compact MC.

Theorem 3.2.11. (i) *For an absorbing compact MC there exists a t.p.f. Q^∞ on $(W, \mathscr{W})$ such that*

$$Q^\infty(w, A) = \sum_{j=1}^{l} g_j(w)\delta(a_j, A), \quad w \in W, \quad A \in \mathscr{W},$$

where the a_j are the absorbing points of the chain and

$$g_j(w) = \mathsf{P}_w\left(\left\{\omega: \lim_{n\to\infty} \zeta_n(\omega) = a_j\right\}\right), \quad 1 \leqslant j \leqslant l.$$

(ii) *If U^∞ is the Markov operator corresponding to Q^∞, then there exist positive constants $\alpha < 1$ and D such that*

$$\| U^n f - U^\infty f \|_L \leqslant D\alpha^n \| f \|_L \tag{3.2.1}$$

for all $f \in L(W)$ and $n \in \mathbb{N}^*$.

(iii) *Let b_j, $1 \leqslant j \leqslant l$, be given real numbers. Then $g(w) = \sum_{j=1}^{l} b_j g_j(w)$ is the unique function in $L(W)$ satisfying the functional equation*

$$\int_X P(w, \mathrm{d}x) g(wx) = g(w) \tag{3.2.2}$$

and having the boundary values $g(a_j) = b_j$, $1 \leqslant j \leqslant l$.

Proof. (i) The statement simply follows from the results in Theorem 3.2.6.

(ii) Since U is aperiodic w.r.t. $L(W)$, by Lemma 3.1.19, there exist positive constants $\alpha < 1$ and D such that

$$\| U^n - U^\infty \|_L \leqslant D\alpha^n, \quad n \in \mathbb{N}^*.$$

This immediately implies the validity of (3.2.1).

(iii) Since $g_j \in L(W)$, $1 \leqslant j \leqslant l$, we also have $g \in L(W)$. Moreover, as a consequence of (i), we have

$$g_j(w) = Q^\infty(w, \{a_j\}), \quad 1 \leqslant j \leqslant l, \quad w \in W,$$

$$g_j(a_k) = \mathsf{P}_{a_k}\left(\left\{\omega: \lim_{n\to\infty} \zeta_n(\omega) = a_j\right\}\right) = \delta_{jk}, \quad 1 \leqslant j, k \leqslant l.$$

It follows that $g(a_k) = b_k$, $1 \leqslant k \leqslant l$. Using equation (3.1.29) we can also write

$$\begin{aligned}\int_X P(w, \mathrm{d}x) g(wx) &= \int_W Q(w, \mathrm{d}w') g(w') \\ &= \sum_{j=1}^{l} b_j \int_W Q(w, \mathrm{d}w') Q^\infty(w', \{a_j\}) = \sum_{j=1}^{l} b_j Q^\infty(w, \{a_j\}).\end{aligned}$$

Hence g satisfies (3.2.2). This functional equation may be written $Ug = g$. Suppose $g' \in L(W)$ is another function satisfying this equation and having the same boundary values. Then we have $U(g - g') = g - g'$ and, by Lemma 3.3.7, the function $g - g'$ attains its maximum on the set $\{a_1, \ldots, a_l\}$. It follows that necessarily $g' \equiv g$. □

The next two theorems are criteria for a compact MC to be absorbing or regular w.r.t. $L(W)$; these criteria are expressed in terms of the supports $\Sigma_n(w)$ of the t.p.f.s $Q^n(w, \cdot)$, $n \in \mathbb{N}^*$.

Theorem 3.2.12. *A compact MC is absorbing iff there are some points $a_1, \ldots, a_l$ in W such that, for any $w \in W$, there exists $j = j(w)$, $1 \leqslant j \leqslant l$, for which we have*

$$\lim_{n\to\infty} d(\Sigma_n(w), a_j) = 0.$$

Theorem 3.2.13. *A compact MC is regular w.r.t. $L(W)$ iff there exists a point w_0 in W such that*

$$\lim_{n\to\infty} d(\Sigma_n(w), w_0) = 0$$

for all $w \in W$.

The application of these criteria is facilitated by the inter-relationship among the sets $\Sigma_n(w)$, $n \in \mathbb{N}^*$, which is given in Lemma 3.2.14.

Lemma 3.2.14. *For all $m, n \in \mathbb{N}$, and $w \in W$, we have*

$$\Sigma_{m+n}(w) = \overline{\bigcup_{w' \in \Sigma_m(w)} \Sigma_n(w')}.$$

Remark. The analogues of Theorem 3.2.13 and Lemma 3.2.14 for continuous MCs are given in Section 3.3. Their proofs parallel the proofs of Norman's results cited here.

3.3 Continuous Markov chains

3.3.1 Preliminaries

Let us start with a formal definition.

Definition 3.3.1. *A Markov chain is said to be* continuous *iff its state space is a compact metric space (W, d), and its t.p.f. Q is continuous.*

A finite Markov chain is a continuous MC with respect to any metric on its state space (see Problem 18). Also, there is no relationship between the notions of a continuous MC and a compact MC (see Problems 22 and 23).

On the other hand, Theorems 3.1.33–3.1.35 isolate classes of GRSCCs for which the associated MC is a continuous MC.

The aim of this section is to establish results concerning the ergodic and subergodic decomposition for a continuous MC. By Theorem 3.1.31, the transition operator U of a continuous MC is quasi-compact on $C(W)$, and, by Remark 1 following Theorem 3.1.28, the operator V is quasi-compact on ca(W). Therefore Theorem 3.1.5 applies in the case of a continuous MC, ensuring its ergodic and subergodic decompositions. However, Theorems 3.3.2–3.3.4 below are more precise and, since their proofs are constructive, some other important information about the chain is obtained. Such information is provided by Lemmas 3.3.9, 3.3.12, 3.3.16, and Proposition 3.3.19.

In this section we will frequently use the following notation:

$$K_A^n(w,B)=\mathbf{P}_w(\zeta_n\in B,\zeta_r\notin A, 1\leqslant r\leqslant n-1),\quad n\geqslant 2,$$

$$K_A^1(w,B)=Q(w,B),\ K_A^n(w,A)=K^n(w,A),\ L(w,A)=\sum_{n\in\mathbb{N}^*}K^n(w,A),$$

$w\in W,\quad A,B\in\mathscr{W}$.

It is clear that $L(w,A)$ is the probability that the chain eventually enters the set A, given that it starts from the point w.

By Lemma 3.1.30, $U(B(W))\subset C(W)$, so $U(C(W))\subset C(W)$, hence the n-step t.p.f. Q^n, $n\in\mathbb{N}^*$, is also continuous. Obviously, 1 is an eigenvalue of the operator $U{:}C(W)\to C(W)$. Let $E(1)$ be the eigenspace corresponding to the eigenvalue 1.

The aim of Subsections 3.3.2–3.3.4 is to prove the theorems stated below.

Theorem 3.3.2. (i) *The operator U is orderly w.r.t. $C(W)$.*

(ii) *We have*

$$U^\infty f(\cdot)=\int_W Q^\infty(\cdot,\mathrm{d}w)f(w),\quad f\in C(W),$$

where

(a) $Q^\infty(\cdot,A)\in E(1)$ *for any* $A\in\mathscr{W}$;

(b) *for any* $w\in W$, $Q^\infty(w,\cdot)$ *is a stationary probability for the chain.*

Theorem 3.3.3. (i) *There exist a finite number of ergodic kernels* $\mathrm{E}_1,\ldots,\mathrm{E}_l$, *and* $l=\dim E(1)$.

(ii) *The t.p.f. Q^∞ is given by the formula*

$$Q^\infty(w,A)=\sum_{j=1}^{l}g_j(w)\pi_j(A)$$

for any $w\in W$ and $A\in\mathscr{W}$, where

(a) *the set of functions $\{g_1,\ldots,g_l\}$ is a basis for $E(1)$;*

(b) *the set of probabilities $\{\pi_1,\ldots,\pi_l\}$ is a basis for $\{\mu\in\mathrm{ca}(W){:}V\mu=\mu\}$.*

Theorem 3.3.4. *We have $g_j(w)=L(w,\mathrm{E}_j)$, $w\in W, 1\leqslant j\leqslant l$ and $\sum_{j=1}^{l}g_j(w)=1$ for any $w\in W$.*

These theorems are analogous to Theorems 3.2.4 and 3.2.6. However, Theorem 3.3.4 is stronger than statement (iii) in 3.2.6, where it is only asserted that the chain is attracted in distance to its ergodic kernels with probability 1.

Let us also remark that a continuous MC is uniformly stable (see Problem 13). The hypotheses of uniform stability and of uniform stability in mean (see Problem 14) were used by Jamison (1964, 1965), Jamison & Sine (1969), and Rosenblatt (1964a, b, 1967) to study the ergodic decomposition of an MC. It is worthy of note that, under these hypotheses, a finite ergodic decomposition cannot be obtained.

In Subsection 3.3.5 the validity of a subergodic decomposition is proved; necessary and sufficient conditions for a continuous MC to be ergodic or regular w.r.t. $C(W)$ are also given. The last subsection is devoted to the study of the structure of the tail σ-algebra of the chain.

3.3.2 *Ergodic kernels and transience*

As mentioned before, the key concept in studying a continuous MC is that of an ergodic kernel, the definition of which was given in the preceding section (see Definition 3.2.5).

The lemma below is valid for both compact and continuous MCs.

Lemma 3.3.5. *If the state space of an MC is a compact metric space (W, d), then there exists at least one ergodic kernel, and any two distinct ergodic kernels are disjoint.*

Proof. Let us denote by $\tilde{\mathrm{E}}$ the collection of subsets of W which are closed both topologically and stochastically. Since $W \in \tilde{\mathrm{E}}$, we have $\tilde{\mathrm{E}} \neq \varnothing$; moreover, if $\mathscr{A} \subset \tilde{\mathrm{E}}$ is a totally ordered system w.r.t. $\subset$, then $\mathscr{A}$ has the finite intersection property and hence, $\mathrm{E}_0 = \bigcap_{\mathrm{E} \in \mathscr{A}} \mathrm{E} \neq \varnothing$. Since (W, d) is compact, by Lindelöf's Theorem A1.13, there exists a denumerable subset $\mathscr{B}$ of $\mathscr{A}$ such that $\mathrm{E}_0 = \bigcap_{\mathrm{E} \in \mathscr{B}} \mathrm{E}$, from which we deduce that E_0 is closed both topologically and stochastically. The existence of an ergodic kernel follows now from Zorn's lemma.

If $\mathrm{E}_1, \mathrm{E}_2$ are two distinct ergodic kernels, we should have $\mathrm{E}_0 = \mathrm{E}_1 \cap \mathrm{E}_2 = \varnothing$, because, if not, E_0 is a topologically and stochastically closed proper subset of both E_1 and E_2, leading to a contradiction. □

Lemma 3.3.6. *A continuous MC possesses only a finite number of distinct ergodic kernels.*

Proof. Assume the statement is not true. Let $(\mathrm{E}_n)_{n\in\mathbb{N}^*}$ be a sequence of distinct ergodic kernels, and choose an arbitrary w_n in E_n, $n\in\mathbb{N}^*$. It follows that there exist $w_0\in W$ and a subsequence $(w_{n_k})_{k\in\mathbb{N}^*}$ such that $\lim_{k\to\infty} d(w_{n_k}, w_0)=0$. Put $\mathrm{E}_m=\bigcup_{k=1}^{m}\mathrm{E}_{n_k}$ and $\mathrm{E}=\bigcup_{m\in\mathbb{N}^*}\mathrm{E}_m$. By the preceding lemma, $\mathrm{E}_m\cap\mathrm{E}_{n_p}=\varnothing$ for all $p\geqslant m+1$, whence $Q(w_{n_p},\mathrm{E}_m)=0$ for all $p\geqslant m+1$, so we get

$$Q(w_0,\mathrm{E})=\lim_{m\to\infty}\lim_{p\to\infty} Q(w_{n_p},\mathrm{E}_m)=0.$$

On the other hand,

$$Q(w_0,\mathrm{E})=\lim_{p\to\infty} Q(w_{n_p},\mathrm{E})\geqslant \lim_{p\to\infty} Q(w_{n_p},\mathrm{E}_{n_p})=1,$$

which leads to a contradiction. □

For a related result see Problem 24.

From now on we shall denote by $\mathrm{E}_1,\ldots,\mathrm{E}_l$ the ergodic kernels, and set also $\mathrm{E}=\bigcup_{j=1}^{l}\mathrm{E}_j$ and $\mathrm{T}=W-\mathrm{E}$.

Lemma 3.3.7. *Consider an MC whose state space is a compact metric space (W,d).*

(i) *If $f:W\to\mathbb{R}$ is an upper semi-continuous function for which $f(w)\leqslant Uf(w)$ for all $w\in W$, then f is constant on each ergodic kernel and it attains its maximum on* E.

(ii) *If $f\in E(1)$, then f is constant on each ergodic kernel. If $|\sigma|=1$ and $f\in E(\sigma)$, then $|f|$ is constant on each ergodic kernel and $f\equiv 0$, if $f(w)=0$ for all $w\in\mathrm{E}$.*

Proof. (i) Put $A_j=\{w\in\mathrm{E}_j: f(w)=\max_{w'\in\mathrm{E}_j} f(w')\}$; since the restriction of f to E_j is upper semi-continuous and E_j is a compact set, we deduce that A_j is a non-empty topologically closed subset of E_j (see Proposition A1.17). On the other hand, for any $w\in A_j$, we have

$$\max_{w'\in W} f(w')=f(w)\leqslant\int_W Q(w,dw'')f(w'')=\int_{\mathrm{E}_j} Q(w,dw'')f(w'')$$
$$=f(w)Q(w,A_j)+\int_{\mathrm{E}_j-A_j} Q(w,dw'')f(w''),$$

which leads to a contradiction if $Q(w,\mathrm{E}_j-A_j)=0$. Therefore $Q(w,A_j)=1$ for all $w\in A_j$ and, since E_j is an ergodic kernel, we have necessarily $A_j=\mathrm{E}_j$.

To prove that f attains its maximum on E, we show in a similar manner that the set $A=\{w\in W: f(w)=\max_{w'\in W} f(w')\}$ is topologically and stochastically closed and hence $A\cap\mathrm{E}\neq\varnothing$.

(ii) If $f\in E(1)$, then $\mathrm{Re} f\in E(1)$ and $\mathrm{Im} f\in E(1)$, so both functions are constant on each ergodic kernel. If $|\sigma| = 1$ and $f \in E(\sigma)$, then $g = |f|$ is continuous and $g \leqslant Ug$, so that we only have to apply (i). □

The next result shows that the chain leaves the 'transient' set T with probability 1; moreover, the probability for the chain to be in the set T after n steps converges exponentially to 0 as $n \to \infty$.

Lemma 3.3.8. *There exist positive constants $\rho < 1$ and a such that*

$$\sup_{w \in W} Q^n(w, \mathrm{T}) \leqslant a\rho^n, \quad n \in \mathbb{N}^*. \tag{3.3.1}$$

Proof. By the definition of T, for any $n, k \in \mathbb{N}^*$,

$$Q^{n+k}(w, \mathrm{T}) = \int_W Q^n(w, \mathrm{d}w')Q^k(w', \mathrm{T}) = \int_{\mathrm{T}} Q^n(w, \mathrm{d}w')Q^k(w', \mathrm{T}), \tag{3.3.2'}$$

$$Q^{n+k}(w, \mathrm{T}) = \int_W Q^k(w, \mathrm{d}w')Q^n(w', \mathrm{T}) = \int_{\mathrm{T}} Q^k(w, \mathrm{d}w')Q^n(w', \mathrm{T}). \tag{3.3.2''}$$

Take $k = 1$ in (3.3.2′) to conclude that $(Q^n(\cdot, \mathrm{T}))_{n \in \mathbb{N}^*}$ is a non-increasing sequence of continuous functions. Hence there exists $R(w, \mathrm{T}) = \lim_{n \to \infty} Q^n(w, \mathrm{T})$ for any $w \in W$; moreover, $R(\cdot, \mathrm{T})$ is upper semi-continuous (see Proposition A1.18). Take $k = 1$ in (3.3.2″), let $n \to \infty$, and use the Lebesgue dominated convergence Theorem A1.8 to get

$$R(w, \mathrm{T}) = \int_W Q(w, \mathrm{d}w')R(w', \mathrm{T}).$$

As $Q^n(w, \mathrm{T}) = 0$ for all $w \in \mathrm{E}$ and $n \in \mathbb{N}^*$, it follows from the preceding lemma that $R(w, \mathrm{T}) \equiv 0$. Moreover, the same Proposition A1.18 implies that

$$\lim_{n \to \infty} \sup_{w \in W} Q^n(w, \mathrm{T}) = 0.$$

Using (3.3.2′) we also have

$$\sup_{w \in W} Q^{n+k}(w, \mathrm{T}) \leqslant \sup_{w \in W} Q^n(w, \mathrm{T}) \sup_{w \in W} Q^k(w, \mathrm{T}).$$

The proof can now be concluded using the last two equations and Lemma 3.1.18. □

The next result gives information about the first entrance time of the chain into one of its ergodic kernels.

Lemma 3.3.9. *We have*

$$\lim_{n \to \infty} \sup_{w \in W} \frac{1}{n} \sum_{r=1}^{n} rK^r(w, \mathrm{E}_j) = 0, \quad 1 \leqslant j \leqslant l.$$

Proof. Set

$$k_j^n(w) = \frac{1}{n} \sum_{r=1}^{n} rK^r(w, \mathrm{E}_j).$$

We have $k_j^n(w) = 1/n$ for all $w \in \mathrm{E}_j$, and $k_j^n(w) = 0$ for all $w \in \mathrm{E}_k$, $k \neq j$. For $w \in \mathrm{T}$, by Lemma 3.3.8, we get

$$K^r(w, \mathrm{E}_j) \leqslant \mathbf{P}_w(\zeta_s \in \mathrm{T}, 1 \leqslant s \leqslant r-1) \leqslant Q^{r-1}(w, \mathrm{T}) \leqslant a\rho^{r-1}, \quad r \geqslant 2,$$

and the desired conclusion follows from the inequality

$$k_j^n(w) \leqslant \frac{a}{n}\left(\rho + \frac{1}{\rho} \sum_{r=1}^{\infty} r\rho^r\right). \qquad \square$$

Remark. The proof actually shows that

$$\lim_{n \to \infty} \sup_{w \in W} \frac{1}{n} \sum_{r=1}^{n} r^s K^r(w, \mathrm{E}_j) = 0, \quad 1 \leqslant j \leqslant l,$$

for any $s \in \mathbb{N}^*$, but in the rest of this section we shall only need the result as stated in Lemma 3.3.9.

Lemma 3.3.10. *For any $n \in \mathbb{N}^*$, $w \in W$, and $A \in \mathscr{W}$ we have*

$$\begin{aligned} Q^{n+1}(w, A) &= \sum_{m=1}^{n} \int_A K_A^m(w, \mathrm{d}w') Q^{n+1-m}(w', A) + K^{n+1}(w, A), \\ Q^{n+1}(w, A) &= \sum_{m=1}^{n} \int_A Q^m(w, \mathrm{d}w') K^{n+1-m}(w', A) + K^{n+1}(w, A). \end{aligned} \tag{3.3.3}$$

Proof. Putting $\tau = \inf\{n \in \mathbb{N}^* : \zeta_n(\omega) \in A\}$, we have

$$\begin{aligned} Q^{n+1}(w, A) &= \sum_{m=1}^{n+1} \mathbf{P}_w(\tau = m, \zeta_{n+1} \in A) \\ &= K^{n+1}(w, A) + \sum_{m=1}^{n} \int_{\{\tau = m\}} \mathbf{P}_w(\zeta_{n+1} \in A \mid \mathscr{K}_{[1,m]})(\omega) \mathbf{P}_w(\mathrm{d}\omega). \end{aligned}$$

Denote by G_w^m the joint distribution of the r.v.s $\zeta_1, \ldots, \zeta_m$, given that the chain starts from w. Using the Markov property, we get

$$\begin{aligned} &\int_{\{\tau = m\}} \mathbf{P}_w(\zeta_{n+1} \in A \mid \mathscr{K}_{[1,m]})(\omega) \mathbf{P}_w(\mathrm{d}\omega) \\ &= \int_{\underbrace{A^c \times \ldots \times A^c}_{(m-1)\ \text{times}} \times A} Q^{n+1-m}(w_m, A) G_w^m(\mathrm{d}w_1, \ldots, \mathrm{d}w_{m-1}, \mathrm{d}w_m) \\ &= \int_A Q^{n+1-m}(w', A) K^m(w, \mathrm{d}w'), \end{aligned}$$

which is the first equation in (3.3.3); the second one can be derived in the same manner. □

Proof of Theorem 3.3.4. For any $n\in\mathbb{N}^*$ and $w\in W$ set

$$g_j^n(w) = \sum_{m=1}^{n} K^m(w, \mathrm{E}_j), \quad 1\leqslant j\leqslant l. \tag{3.3.4}$$

The second equation in (3.3.3) can be written as

$$Q^{n+1}(w,A) = \sum_{m=1}^{n} U^m[\chi_A(\cdot)K^{n+1-m}(\cdot,A)](w) + K^{n+1}(w,A),$$

from which we derive the continuity of $K^m(\cdot, A)$ for all $m\in\mathbb{N}^*$, and therefore the continuity of g_j^n, $1\leqslant j\leqslant l$. As $g_j^{n+1}\geqslant g_j^n$, there exists

$$g_j(w) = \lim_{n\to\infty} g_j^n(w). \tag{3.3.5}$$

The first statement of the theorem is obtained from (3.3.4) and (3.3.5). On the other hand,

$$\sum_{j=1}^{l} g_j^n(w) = 1 - \mathbf{P}_w(\zeta_r\in\mathrm{T}, 1\leqslant r\leqslant n) \geqslant 1 - Q^n(w,\mathrm{T})$$

for any $w\in W$ and $n\in\mathbb{N}^*$, which, by using Lemma 3.3.8, yields the second assertion of the theorem. □

3.3.3 *The case of one ergodic kernel*

We assume throughout this subsection that the MC considered has just one ergodic kernel which coincides with W itself.

Lemma 3.3.11. *If there exists a finite measure ν on $\mathscr{W}$ such that*

$$\nu(\Sigma) \leqslant \int_{\Sigma} \nu(dw)Q(w,\Sigma), \tag{3.3.6}$$

where $\Sigma = \operatorname{supp}\nu$, then $\Sigma = W$.

Proof. Putting $\Sigma' = \{w\in W: Q(w,\Sigma) = 1\}$, we can write

$$\begin{aligned}\nu(\Sigma) &\leqslant \int_{\Sigma} \nu(dw)Q(w,\Sigma) = \nu(\Sigma\cap\Sigma') + \int_{\Sigma\setminus\Sigma'} \nu(dw)Q(w,\Sigma)\\ &< \nu(\Sigma\cap\Sigma') + \nu(\Sigma\setminus\Sigma') = \nu(\Sigma),\end{aligned}$$

if $\nu(\Sigma\setminus\Sigma') > 0$, which is contradictory. So we should have $\nu(\Sigma\setminus\Sigma') = 0$, from which we get $\nu(\Sigma\cap\Sigma') = \nu(\Sigma)$. Since $\Sigma = \operatorname{supp}\nu$ and Σ' is a topologically closed set, we deduce that $\Sigma\cap\Sigma' = \Sigma$. Now, Σ is both topologically and stochastically closed, so it should coincide with W. □

Lemma 3.3.12. (i) *There exists a stationary probability* π *and* $\operatorname{supp}\pi = W$.

(ii) *For any* $w \in W$ *and* $n \in \mathbb{N}^*$ *the probability* $Q^n(w, \cdot)$ *is absolutely continuous w.r.t.* π.

Proof. (i) Since 1 is an eigenvalue of U and $|U| = 1$, it follows that 1 is also an eigenvalue of the adjoint V of U. Hence there exists a non-zero $\mu \in \operatorname{ca}(W)$ such that

$$\mu(A) = \int_W \mu(\mathrm{d}w) Q(w, A), \quad A \in \mathscr{W}. \tag{3.3.7}$$

There is no loss of generality in assuming μ to be real-valued i.e. a signed measure. Then, by the Hahn decomposition (see Section A1.4), we have $\mu = \mu^+ - \mu^-$, where μ^+ and μ^- are finite measures which put mass only on the sets W^+ and W^- respectively, and $W^+ \cup W^- = W$, $W^+ \cap W^- = \varnothing$. If $\mu^+(W^+) = 0$, then μ^- verifies (3.3.7) and π is obtained by conveniently norming μ^-. Now, assume that $\mu^+(W^+) > 0$. Then we have

$$\begin{aligned}\int_W f(w)\mu(\mathrm{d}w) &= \int_W f(w)\mu^+(\mathrm{d}w) - \int_W f(w)\mu^-(\mathrm{d}w)\\ &\leqslant \int_W f(w)\mu^+(\mathrm{d}w) = \int_{W^+} f(w)\mu^+(\mathrm{d}w)\end{aligned}$$

for any non-negative function f which is integrable w.r.t. μ. Set $A^+ = \operatorname{supp}\mu^+$. Putting $A = A^+$ in (3.3.7) and taking into account that the above inequality holds for $f(w) = Q(w, A^+)$, we get

$$\mu(A^+) \leqslant \int_{W^+} \mu^+(\mathrm{d}w) Q(w, A^+).$$

But since $\mu(A^+) = \mu^+(A^+)$, we have

$$\mu^+(A^+) \leqslant \int_{A^+} \mu^+(\mathrm{d}w) Q(w, A^+).$$

Applying the preceding lemma for $v = \mu^+$ and $\Sigma = A^+$, we conclude that $A^+ = W$ and hence $W^+ = W$. Therefore, by conveniently norming μ^+, we get a stationary probability, which we shall denote π. By the same lemma we obtain $\operatorname{supp}\pi = W$.

(ii) The stationarity of π implies that

$$\pi(A) = \int_W \pi(\mathrm{d}w) Q^n(w, A)$$

for all $n \in \mathbb{N}^*$ and $A \in \mathscr{W}$. Let $A \in \mathscr{W}$ be such that $\pi(A) = 0$; then we deduce from the above equation that there exists $N \in \mathscr{W}$ with $\pi(N) = 0$ such that $Q^n(w, A) = 0$ for all $w \notin N$. Assume that there exists $w_0 \in N$ such that

$Q^n(w_0, A) = a > 0$. From the continuity of Q^n we deduce that there exists an open neighbourhood V of w_0 such that $Q_n(w, A) \geqslant a/2 > 0$ for all $w \in V$. But, by a well-known characterization of the support of a probability on a separable metric space (see Section A1.16), it follows that $\pi(V) > 0$. Then $\pi(A) \geqslant a\pi(V)/2 > 0$, which leads to a contradiction. □

Lemma 3.3.13. *The stationary probability π in Lemma 3.3.12 is unique and*

$$\lim_{n \to \infty} Q_n(w, A) = \pi(A) \tag{3.3.8}$$

for all $w \in W$ and $A \in \mathscr{W}$.

Proof. The uniqueness of π will follow from the validity of equation (3.3.8). Let π be a stationary probability. As $Q(w, \cdot)$ is absolutely continuous w.r.t. π, Birkhoff's individual ergodic Theorem A2.6 ensures the existence of the limit in (3.3.8) for all $w \notin N$ with $\pi(N) = 0$. Now, by the preceding lemma, the limit should exist for an arbitrary $\bar{w} \in W$, since we can write

$$\begin{aligned}\lim_{n \to \infty} Q_n(\bar{w}, A) &= \lim_{n \to \infty} \frac{1}{n} \sum_{k=1}^{n} Q^{k+1}(\bar{w}, A) = \lim_{n \to \infty} \frac{1}{n} \sum_{k=1}^{n} \int_W Q^k(w, A) Q(\bar{w}, \mathrm{d}w) \\ &= \int_W \left[\lim_{n \to \infty} Q_n(w, A) \right] Q(\bar{w}, \mathrm{d}w).\end{aligned}$$

Denote this limit by $S(\bar{w}, A)$. Since, for a fixed $A \in \mathscr{W}$, $(Q_n(\cdot, A))_{n \in \mathbb{N}^*}$ is a sequence of bounded functions, it follows that $S(\cdot, A) \in B(W)$. Letting $n \to \infty$ in the equation

$$\int_W Q_n(w, A) Q(\bar{w}, \mathrm{d}w) = \frac{1}{n} \sum_{k=1}^{n} Q^{k+1}(\bar{w}, A),$$

we conclude that

$$S(\bar{w}, A) = \int_W Q(\bar{w}, \mathrm{d}w) S(w, A) = U[S(\cdot, A)](\bar{w}).$$

Hence, by Lemma 3.1.30, $S(\cdot, A)$ is a continuous function for any $A \in \mathscr{W}$. The last equation shows also that the hypotheses in Lemma 3.3.7 are fulfilled; therefore we have $S(w, A) \equiv S(A)$ for any $A \in \mathscr{W}$. On the other hand, the equation

$$\pi(A) = \int_W \pi(\mathrm{d}w) Q_n(w, A) \tag{3.3.9}$$

which holds for all $n \in \mathbb{N}^*$ and $A \in \mathscr{W}$, implies that $S(A) = \pi(A)$ for all $A \in \mathscr{W}$. □

Remark. The existence of the limit in (3.3.8) has been proved by Šidák

(1967) under less stringent conditions; however, the stronger assumptions here allowed us to obtain the value of that limit and to shorten the proof considerably.

Corollary 3.3.14. *We have*

$$\lim_{n\to\infty} |U_n f - b| = 0 \tag{3.3.10}$$

for all $f \in C(W)$, *where* $b = \int_W \pi(\mathrm{d}w) f(w)$.

Proof. The validity of (3.3.8) implies that

$$\lim_{n\to\infty} U_n f(w) = b \tag{3.3.11}$$

for all $f \in C(W)$ and $w \in W$. Since $|U_n| = 1$ for all $n \in \mathbb{N}^*$, it follows that the sequence $(U_n f)_{n \in \mathbb{N}^*}$ converges in $C(W)$ to b if it is weakly sequentially compact (see Dunford & Schwartz (1958, p. 345)). But weak convergence in $C(W)$ is pointwise convergence (see Section A2.5), so that (3.3.10) is an immediate consequence of (3.3.11). □

Definition 3.3.15. *Let* $(W, \mathscr{W})$ *be an arbitrary measurable space and let* φ *be a* σ*-finite measure on* $\mathscr{W}$. *An MC with state space* $(W, \mathscr{W})$ *is said to be* φ-irreducible *iff* $L(w, A) > 0$ *for all* $w \in W$ *and* $A \in \mathscr{W}$ *with* $\varphi(A) > 0$. *It is said to be* φ-recurrent *iff* $L(w, A) = 1$ *for all* $w \in W$ *and* $A \in \mathscr{W}$ *with* $\varphi(A) > 0$. *It is said to be* uniformly φ-recurrent *iff*

$$\sum_{m=1}^{n} K^m(w, A) \to 1 \tag{3.3.12}$$

uniformly w.r.t. $w \in W$ *as* $n \to \infty$ *for all* $A \in \mathscr{W}$ *with* $\varphi(A) > 0$. *A set* $B \in \mathscr{W}$ *is said to be* φ-uniform *iff, for any* $A \in \mathscr{W}$ *with* $\varphi(A) > 0$,

$$\sup_{w \in B} \sum_{m=n}^{\infty} K^m(w, A) \to 0 \tag{3.3.13}$$

as $n \to \infty$. *If* (3.3.13) *holds with* φ *replaced by a stationary probability* π, *the set* B *will simply be said to be* uniform.

Lemma 3.3.16. (i) *The function* $L(\cdot, A)$ *is continuous for all* $A \in \mathscr{W}$.

(ii) *We have* $\inf_{w \in W} L(w, A) > 0$ *for all* $A \in \mathscr{W}$ *with* $\pi(A) > 0$.

Proof. (i) Obviously $L(\cdot, A) \in B(W)$. Since

$$L(w, A) = Q(w, A) + \int_{A^c} Q(w, \mathrm{d}w') L(w', A) = Q(w, A) + U[\chi_{A^c}(\cdot) L(\cdot, A)](w),$$

the first statement is a consequence of Lemma 3.1.30.

(ii) Let us suppose that there exists a set $A \in \mathscr{W}$ with $\pi(A) > 0$ such that $\inf_{w \in W} L(w, A) = 0$. By the continuity of $L(\cdot, A)$, there is a $w_0 \in W$ such that $L(w_0, A) = 0$. As $\sup_{n \in \mathbb{N}^*} Q^n(w_0, A) \leqslant L(w_0, A)$, it follows that $Q^n(w_0, A) = 0$ for all $n \in \mathbb{N}^*$, which, by Lemma 3.3.13 leads to a contradiction. □

Proposition 3.3.17. *The chain is π-recurrent.*

To prove this proposition we need the following result, due to Chung (1964), which is valid for an arbitrary MC.

Lemma 3.3.18. *If $A, B \in \mathscr{W}$ and $\inf_{w \in A} L(w, B) > 0$, then*

$$M(w, A) = M(w, A, B) \tag{3.3.14}$$

for all $w \in W$. Here $M(w, A) = \mathbf{P}_w(\limsup_{n \to \infty} \{\zeta_n \in A\})$ and $M(w, A, B) = \mathbf{P}_w(\limsup_{n \to \infty} \{\zeta_n \in A\} \cap \limsup_{n \to \infty} \{\zeta_n \in B\})$.

Proof. Let us set $\limsup_{n \to \infty} \{\zeta_n \in A\} = \Lambda(A)$, $\limsup_{n \to \infty} \{\zeta_n \in B\} = \Lambda(B)$, and $B_k = \bigcup_{n \geqslant k} \{\zeta_n \in B\}$, $k \in \mathbb{N}^*$. Obviously, $\Lambda(B) = \bigcap_{k \in \mathbb{N}^*} B_k$; moreover,

$$\mathbf{P}_w(\Lambda(B) | \zeta_1, \ldots, \zeta_n) \leqslant \mathbf{P}_w(B_{n+1} | \zeta_1, \ldots, \zeta_n) \leqslant \mathbf{P}_w(B_k | \zeta_1, \ldots, \zeta_n), \ \mathbf{P}_w\text{-a.s.}$$

for all $n \geqslant k$. By Theorem A1.11, letting first $n \to \infty$ and then $k \to \infty$, we get

$$\chi_{\Lambda(B)} \leqslant \lim_{n \to \infty} \mathbf{P}_w(B_{n+1} | \zeta_1, \ldots, \zeta_n) \leqslant \lim_{k \to \infty} \chi_{B_k} = \chi_{\Lambda(B)}.$$

Since

$$L(\zeta_n, B) = \mathbf{P}_w(B_{n+1} | \zeta_n) = \mathbf{P}_w(B_{n+1} | \zeta_1, \ldots, \zeta_n), \ \mathbf{P}_w\text{-a.s.},$$

we deduce that

$$\lim_{n \to \infty} L(\zeta_n, B) = \chi_{\Lambda(B)}, \ \mathbf{P}_w\text{-a.s.} \tag{3.3.15}$$

If $\omega \in \Lambda(A)$, then $\zeta_n(\omega) \in A$ for infinitely many values of n and hence

$$L(\zeta_n(\omega), B) \geqslant \inf_{w \in A} L(w, B) > 0$$

for such values. This last inequality, together with equation (3.3.15), implies that $\mathbf{P}_w(\Lambda(A)) = \mathbf{P}_w(\Lambda(A) \cap \Lambda(B))$ for all $w \in W$. □

Proof of Proposition 3.3.17. Obviously, we have $M(w, W) = 1$ for all $w \in W$. For a set $C \in \mathscr{W}$ with $\pi(C) > 0$, by Lemma 3.3.16 (ii) we have $\inf_{w \in W} L(w, C) > 0$. Applying Lemma 3.3.18, with $A = W$ and $B = C$, we get $M(w, W, C) = M(w, W)$ or, equivalently, $M(w, C) = 1$ for all $w \in W$. The proof is completed by noticing that $L(w, C) \geqslant M(w, C)$. □

Lemma 3.3.19. (i) *Any compact set is uniform.*
(ii) *The chain is uniformly π-recurrent.*

Proof. For $A\in\mathscr{W}$ with $\pi(A)>0$, let us define

$$U_A f(w) = U(\chi_{A^c} f)(w) = \int_{A^c} Q(w, dw') f(w'), \quad f\in B(W).$$

It is clear that U_A is a linear operator on $B(W)$, and that $U_A f\in C(W)$ for any $f\in B(W)$. Denote by τ_A the first entrance time of the chain into the set A. By Proposition 3.3.17, we have

$$\lim_{n\to\infty} \mathbf{P}_w(\tau_A > n) = \lim_{n\to\infty} \sum_{m\geqslant n+1} K^m(w, A) = 0.$$

But

$$\mathbf{P}_w(\tau_A > n) = \mathbf{P}_w(\zeta_k\in A^c, 1\leqslant k\leqslant n)$$
$$= \int_{A^c} Q(w, dw_1) \int_{A^c} Q(w_1, dw_2) \cdots \int_{A^c} Q(w_{n-1}, dw_n) = U_A^n 1(w),$$

which shows that $\mathbf{P}_{\cdot}(\tau_A > n)\in C(W)$ for all $n\in\mathbb{N}^*$. The first statement follows by using Dini's Theorem A1.15, while the second statement is an immediate consequence of the first one. □

In the remainder of this subsection we shall need the result below, due to Cogburn (1975).

Proposition 3.3.20. *The following two statements are equivalent:*
(i) *B is a uniform set;*
(ii) *we have*

$$\lim_{n\to\infty} \sup_{w\in B} \| Q_n(w, \cdot) - \pi(\cdot) \| = 0. \tag{3.3.16}$$

Proposition 3.3.21. *The chain is ergodic w.r.t. $C(W)$.*

Proof. Put $U^\infty f = \int_W \pi(dw) f(w)$ for any $f\in C(W)$. Then using Lemma 2.1.1, we obtain

$$| U_n - U^\infty | \leqslant \frac{1}{2} \sup_{\substack{f\in C(W)\\ |f|=1}} (\operatorname{osc} f) \sup_{w\in W} \| Q_n(w, \cdot) - \pi(\cdot) \|.$$

By Lemma 3.3.19(i), W is a uniform set, so that equation (3.3.16) holds if we replace B by W. This equation, combined with the above inequality, shows that the chain is ergodic w.r.t. $C(W)$. □

3.3.4 *The general case (several ergodic kernels)*

By Lemma 3.3.6, there exists a finite number of ergodic kernels. The restriction of the chain to one of these ergodic kernels is still a continuous

MC and hence has all the properties proved in Subsection 3.3.3. Let π_j be the unique stationary distribution corresponding to the restriction of the chain to the kernel $\ _j$. We shall extend this probability to the entire space by putting

$$\pi_j(A) = \pi_j(A \cap \ _j), \quad A \in \mathscr{W}.$$

Let us denote

$$Q^{\infty}(w, A) = \sum_{j=1}^{l} g_j(w)\pi_j(A) \tag{3.3.17}$$

for all $w \in W$ and $A \in \mathscr{W}$, where the π_j are those defined above and the g_j are given by (3.3.5). Obviously, $Q^{\infty}(w, \cdot)$ is a probability for each $w \in W$. The measurability of $Q^{\infty}(\cdot, A)$ for each $A \in \mathscr{W}$ is a consequence of its definition. Hence Q^{∞} is a t.p.f.

Lemma 3.3.22. *The set* $\{g_1, \ldots, g_l\}$ *is a basis for* $E(1)$.

Proof. The functions $g_j^n, 1 \leqslant j \leqslant l$, introduced by (3.3.4) satisfy the equation $Ug_j^n = g_j^{n+1}$, whence we obtain $g_j = Ug_j$ by the Lebesgue dominated convergence Theorem A1.8. As $g_j \in B(W)$, it must be in $C(W)$ by Lemma 3.1.30 (ii).

The functions g_j are linearly independent by their definition. For an arbitrary g in $E(1)$, put $k = g - \sum_{j=1}^{l} \hat{g}_j g_j$, where $\hat{g}_j$ is the value of g_j on E_j. Then $k \in E(1)$ and $k \equiv 0$ by Lemma 3.3.7; hence the g_j, $1 \leqslant j \leqslant l$ span $E(1)$. □

Lemma 3.3.23. *We have*

$$Q_n(w, \cdot) \Rightarrow Q^{\infty}(w, \cdot), \tag{3.3.18}$$

the convergence being uniform w.r.t. $w \in W$. *Moreover, the convergence is also uniform w.r.t.* $f \in C(W)$ *with* $|f| = 1$, *i.e.*

$$\lim_{n \to \infty} \sup_{\substack{f \in C(W) \\ |f| = 1}} \sup_{w \in W} \left| \int_W Q_n(w, \mathrm{d}w') f(w') - \int_W Q^{\infty}(w, \mathrm{d}w') f(w') \right| = 0.$$

Proof. For any $f \in C(W)$ we have

$$\int_W Q_n(w, \mathrm{d}w') f(w') = \frac{1}{n} \sum_{k=1}^{n} \int_{\mathrm{T}} Q^k(w, \mathrm{d}w') f(w')$$
$$+ \sum_{j=1}^{l} \frac{1}{n} \sum_{k=1}^{n} \int_{\mathrm{E}_j} Q^k(w, \mathrm{d}w') f(w'), \quad w \in W.$$

Since, by Lemma 3.3.8, there exist positive constants $\rho < 1$ and a such that

$$\sup_{w \in W} \left| \frac{1}{n} \sum_{k=1}^{n} \int_{\mathrm{T}} Q^k(w, \mathrm{d}w') f(w') \right| \leqslant \frac{1}{n} \sum_{k=1}^{n} |f| \sup_{w \in W} Q^k(w, \mathrm{T}) \leqslant |f| a \frac{1}{n} \sum_{k=1}^{n} \rho^k,$$

it follows that the first term in the decomposition of $\int_W Q_n(w, dw')f(w')$ above tends to 0 as $n \to \infty$, uniformly w.r.t. $w \in W$ and $f \in C(W)$ with $|f| = 1$. It is then sufficient to prove that

$$\lim_{n\to\infty} \int_{E_j} Q_n(w, dw')f(w') = g_j(w) \int_{E_j} \pi_j(dw')f(w'), \quad 1 \leqslant j \leqslant l, \tag{3.3.19}$$

uniformly w.r.t. $w \in W$ and $f \in C(W)$ with $|f| = 1$. Set

$$h_n^j(w) = \int_{E_j} Q_n(w, dw')f(w'), \quad b_j = \int_{E_j} \pi(dw')f(w'), \quad 1 \leqslant j \leqslant l.$$

Using Lemma 3.3.10, by straightforward computations, (3.3.3) leads to

$$h_n^j(w) = \frac{1}{n}\sum_{r=1}^{n} \int_{E_j} K^r(w, dw')f(w')$$
$$+ \frac{1}{n}\sum_{r=1}^{n-1}(n-r)\int_{E_j} K^r(w, dw')h_{n-r}^j(w'), \quad 1 \leqslant j \leqslant l.$$

But

$$\left|\frac{1}{n}\sum_{r=1}^{n}\int_{E_j} K^r(w, dw')f(w')\right| \leqslant \frac{1}{n}|f|g_j^n(w) \leqslant \frac{1}{n}|f|, \quad 1 \leqslant j \leqslant l,$$

which shows that this term tends to 0 as $n \to \infty$, uniformly w.r.t. $w \in W$ and $f \in C(W)$ with $|f| = 1$. On the other hand, we have

$$\left|\frac{1}{n}\sum_{r=1}^{n-1} r \int_{E_j} K^r(w, dw')h_{n-r}^j(w')\right| \leqslant |f|\frac{1}{n}\sum_{r=1}^{n-1} rK^r(w, E_j), \quad 1 \leqslant j \leqslant l,$$

which implies, by Lemma 3.3.9, that this term also tends to 0, as $n \to \infty$, uniformly w.r.t. $w \in W$ and $f \in C(W)$ with $|f| = 1$. So we should eventually prove that

$$\lim_{n\to\infty} \sum_{r=1}^{n-1} \int_{E_j} K^r(w, dw')h_{n-r}^j(w') = b_j g_j(w), \quad 1 \leqslant j \leqslant l, \tag{3.3.20}$$

uniformly w.r.t. $w \in W$ and $f \in C(W)$ with $|f| = 1$.

Take an arbitrary $\varepsilon > 0$ and choose $N = N(\varepsilon)$ such that

$$\sup_{w \in E_j} |h_p^j(w) - b_j| = |h_p^j - b_j|_{E_j} < \varepsilon, \quad 1 \leqslant j \leqslant l,$$

for all $p \geqslant N$. By Proposition 3.3.21, this is actually possible. Then, for $n > N$, we have

$$\left|\frac{\sum_{r=1}^{n-1}\int_{E_j} K^r(w, dw')h_{n-r}^j(w')}{\sum_{r=1}^{n-1} K^r(w, E_j)} - b_j\right|$$

$$\leqslant \frac{\sum_{r=1}^{n-N} K^r(w, E_j)}{\sum_{r=1}^{n-1} K^r(w, E_j)} \varepsilon + \max_{1 \leqslant p \leqslant N} |h_p^j - b_j|_{E_j} \frac{\sum_{r=n-N+1}^{n-1} K^r(w, E_j)}{\sum_{r=1}^{n-1} K^r(w, E_j)}, \quad 1 \leqslant j \leqslant l.$$

But, by Lemma 3.3.19, the second term tends to 0 as $n \to \infty$, uniformly w.r.t. $w \in W$. Let us note that the convergence is also uniform w.r.t. $f \in C(W)$ with $|f| = 1$, since $\max_{1 \leqslant p \leqslant N} |h_p^j - b_j|_{E_j} \leqslant 2|f|$. It then follows that

$$\limsup_{n \to \infty} \left| \frac{\sum_{r=1}^{n-1} \int_{E_j} K^r(w, dw') h_{n-r}^j(w')}{\sum_{r=1}^{n-1} K^r(w, E_j)} - b_j \right| \leqslant \varepsilon, \quad 1 \leqslant j \leqslant l,$$

which concludes the proof. □

Proof of Theorem 3.3.2. Statement (i) was proved in Lemma 3.3.23. The representation formula in (ii) is a consequence of the definition formula (3.3.17).

Statement (ii)(a) is an immediate consequence of Lemma 3.3.22, while (ii)(b) follows from (3.3.17) and from the stationarity of the π_j, $1 \leqslant j \leqslant l$. □

Proof of Theorem 3.3.3. Lemma 3.3.6 ensures the validity of (i). Note also that Q^∞ was constructed in such a manner that the representation formula in (ii) holds.

Statement (ii)(a) was proved in Lemma 3.3.22. It only remains to prove (ii)(b). To this end let $\mu \in \mathrm{ca}(W)$ for which $V\mu = \mu$. For any $f \in C(W)$ we can write

$$(V_n \mu, f) = (\mu, U_n f) \to (\mu, U^\infty f) = (V^\infty \mu, f)$$

as $n \to \infty$; here V^∞ is the adjoint of U^∞. But, since $V_n \mu = \mu$ for all $n \in \mathbb{N}^*$, we get

$$\mu(\cdot) = V^\infty \mu(\cdot) = \int_W \mu(dw) Q^\infty(w, \cdot) = \sum_{j=1}^{l} \left[\int_W \mu(dw) g_j(w) \right] \pi_j(\cdot),$$

which completes the proof, as the π_j, $1 \leqslant j \leqslant l$, are obviously linearly independent. □

Corollary 3.3.24. *We have*

$$Q^\infty(w, A) = \int_W Q^\infty(w', A) Q^k(w, dw')$$

for all $w \in W$, $A \in \mathscr{W}$, and $k \in \mathbb{N}^*$.

Proof. Since $Ug_j = g_j$, it follows that $U^k g_j = g_j$ for all $k \in \mathbb{N}^*$. Then (3.3.17) leads to

$$\int_W Q^\infty(w', A) Q^k(w, dw') = \sum_{j=1}^{l} \pi_j(A) \int_W g_j(w') Q^k(w, dw')$$
$$= \sum_{j=1}^{l} \pi_j(A) U^k g_j(w) = Q^\infty(w, A). \qquad \square$$

3.3.5 Subergodic decomposition, ergodicity, regularity

We shall assume, as in Subsection 3.3.3, that the MC considered has only one ergodic kernel, which coincides with the entire space W. (In fact, this amounts to considering the restriction of the chain to one of its ergodic kernels – see Remark 2 after Theorem 3.2.8.)

To get the subergodic decomposition of the chain we need the following result, due to Orey (1971, p. 30).

Proposition 3.3.25. (i) *For a φ-recurrent chain there exists a maximal cycle $\{C_1, \ldots, C_p\}$ such that $W - \bigcup_{j=1}^{p} C_j$ is a φ-null set.*

(ii) *If the chain is uniformly φ-recurrent then it has a unique stationary probability π, and*

$$\lim_{n\to\infty} \sup_{w\in W} \left\| \frac{1}{p} \sum_{k=0}^{p-1} Q^{n+k}(w, .) - \pi(.) \right\| = 0. \tag{3.3.21}$$

This proposition, whose proof is omitted, enables us to obtain, for a continuous MC, the stronger result stated in Theorem 3.3.26.

Theorem 3.3.26. *There exists a maximal cycle $\{D_1, \ldots, D_p\}$ such that*

(i) *the sets D_j, $1 \leqslant j \leqslant p$, are topologically closed.*

(ii) $W = \bigcup_{j=1}^{p} D_j$.

Proof. (i) By Lemma 3.3.19, the chain is uniformly π-recurrent, hence Proposition 3.3.25 applies. Let $\{C_1, \ldots, C_p\}$ be the maximal cycle appearing in that proposition. Let us define

$$D_j = \{w : Q(w, C_{j+1}) = 1\}, \quad 1 \leqslant j \leqslant p-1,$$
$$D_p = \{w : Q(w, C_1) = 1\}.$$

By their construction, the sets D_j, $1 \leqslant j \leqslant p$, are topologically closed. Moreover, $D_j \supset C_j$, $1 \leqslant j \leqslant p$, and $Q(w, D_{j+1}) \geqslant Q(w, C_{j+1}) = 1$ for any $w \in D_j$, $1 \leqslant j \leqslant p-1$, $Q(w, D_1) \geqslant Q(w, C_1) = 1$ for any $w \in D_p$. The D_j are also disjoint, for if not, there is $w_0 \in D_a \cap D_b$, $a \neq b$, and $Q(w_0, C_{a+1}) =$

$Q(w_0, C_{b+1}) = 1$, which yields the false conclusion $Q(w_0, C_{a+1} \cap C_{b+1}) = 1$. So $\{D_1, \ldots, D_p\}$ is a maximal cycle, verifying (i).

(ii) The set $W - \bigcup_{j=1}^{p} D_j$ must be empty. For, if not, as a subset of $W - \bigcup_{j=1}^{p} C_j$, it is a π-null open set, which is a contradiction. □

Now, we are able to give some necessary and sufficient conditions for a continuous MC to be ergodic or regular w.r.t. $C(W)$.

Theorem 3.3.27. *A continuous MC is ergodic w.r.t. $C(W)$ iff it has just one ergodic kernel or, equivalently, iff it possesses a unique stationary probability.*

The simple proof is left to the reader.

To give another necessary and sufficient criterion for a continuous MC to be ergodic, we need a further definition.

Definition 3.3.28. *A point $w_0 \in W$ is said to be an* accumulation point *of the chain iff*

$$L(w, S_\varepsilon(w_0)) > 0 \tag{3.3.22}$$

for every $w \in W$ and $\varepsilon > 0$; here $S_\varepsilon(w_0)$ is the open sphere of centre w_0 and radius ε.

Theorem 3.3.29. *A continuous MC is ergodic w.r.t. $C(W)$ iff it has an accumulation point.*

Proof. If the chain is ergodic, there exists a unique stationary probability π, the support of which is the unique ergodic kernel E of the chain. Then $\pi(S_\varepsilon(w_0)) > 0$ for any $w_0 \in$ E and $\varepsilon > 0$, whence $L(w, S_\varepsilon(w_0)) = 1$ for all $w \in W$ and $\varepsilon > 0$.

To prove sufficiency, take an arbitrary $B \in \mathscr{W}$ and denote by w_M and w_m the points at which $Q^\infty(\cdot, B)$ attains its maximum and minimum, respectively. Let $\varepsilon > 0$ be arbitrary. Then, by (3.3.22), there exists $k \in \mathbb{N}^*$ such that $Q^k(w_M, S_\varepsilon(w_0)) > 0$.

We shall first prove that there exists a $w_\varepsilon \in S_\varepsilon(w_0)$ such that $Q^\infty(w_\varepsilon, B) = Q^\infty(w_M, B)$. Assume that this is not true. Then we have $Q^\infty(w, B) < Q^\infty(w_M, B)$ for all $w \in S_\varepsilon(w_0)$, which, using Corollary 3.3.24, yields

$$\begin{aligned} Q^\infty(w_M, B) &= \int_{S_\varepsilon(w_0)} Q^\infty(w, B) Q^k(w_M, dw) + \int_{S_\varepsilon^c(w_0)} Q^\infty(w, B) Q^k(w_M, dw) \\ &< Q^\infty(w_M, B) Q^k(w_M, S_\varepsilon(w_0)) + Q^\infty(w_M, B) Q^k(w_M, S_\varepsilon^c(w_0)) \\ &= Q^\infty(w_M, B) \end{aligned}$$

which is, clearly, a contradiction. Take $(\varepsilon_n)_{n\in\mathbb{N}^*}$ such that $\lim_{n\to\infty}\varepsilon_n=0$ and denote by $w_n=w_{\varepsilon_n}$ the corresponding sequence of points, the existence of which has just been proved. The sequence $(w_n)_{n\in\mathbb{N}^*}$ converges to w_0 and, by the continuity of $Q^\infty(\cdot,B)$, we get $\lim_{n\to\infty}Q^\infty(w_n,B)=Q^\infty(w_0,B)$, from which $Q^\infty(w_M,B)=Q^\infty(w_0,B)$. A similar argument involving w_m shows that $Q^\infty(w_m,B)=Q^\infty(w_0,B)$. Hence $Q^\infty(w,B)=Q^\infty(w_0,B)$ for all $w\in W$, i.e. $Q^\infty(\cdot,B)$ only depends on $B\in\mathscr{W}$. As the set B was taken arbitrarily, it follows that the chain is ergodic w.r.t. $C(W)$. □

Remark. The sufficiency part in the above theorem was proved under similar assumptions by Isaac (1962).

Theorem 3.3.30. *A continuous MC is regular w.r.t. $C(W)$ iff it has just one ergodic kernel and this kernel is of period* 1.

Proof. The sufficiency part is immediate.

Conversely, if the chain is regular, it is also ergodic, so there is only one ergodic kernel. Suppose the period p of this kernel is greater than 1. Then let D_{a+1} be an arbitrary element of a cycle and take $w_a\in\mathrm{D}_a$. By the definition of a cycle, equation (3.3.21) yields

$$\pi(\mathrm{D}_{a+1})=\frac{1}{p}. \tag{3.3.23}$$

On the other hand, since the chain is regular, the sequence $Q^n(w_a,\cdot)$ converges weakly to π, and so do any of its subsequences. Applying Theorem A1.20, we obtain

$$\limsup_{n\to\infty} Q^{np+1}(w_a,\mathrm{D}_{a+1})\leqslant\pi(\mathrm{D}_{a+1}),$$

since D_{a+1} is topologically closed. This inequality, together with (3.3.23), implies $1\leqslant 1/p$, so that p must equal 1. □

Theorem 3.3.30 ensures the possibility of getting another characterization of the regularity of a continuous MC; this characterization is analogous to that for compact MCs, given in Theorem 3.2.13.

Theorem 3.3.31. *A continuous MC is regular w.r.t. $C(W)$ iff there exists $w_0\in W$ such that*

$$\lim_{n\to\infty} d(\Sigma_n(w),w_0)=0 \tag{3.3.24}$$

for all $w\in W$. Here $\Sigma_n(w)=\operatorname{supp}Q^n(w,\cdot)$, $n\in\mathbb{N}^*$.

Proof. To prove sufficiency, let E_j^m be a set in the maximal cycle of the ergodic kernel E_j, and p_j be the period of the cycle. Take $w \in E_j^m$. Then, for any $n \in \mathbb{N}^*$, $n = kp_j$, $k \in \mathbb{N}^*$, we have $Q^n(w, E_j^m) = 1$ and hence $\Sigma_n(w) \subset E_j^m$. But $d(w_0, E_j^m) \leqslant d(w_0, \Sigma_n(w))$ and, letting $n \to \infty$, we deduce that $d(w_0, E_j^m) = 0$, so that $w_0 \in E_j^m$. It follows that w_0 belongs to all the sets E_j^m, $1 \leqslant m \leqslant p_j$, $1 \leqslant j \leqslant l$, which is a contradiction, unless there exists but one such set. Therefore the chain is regular w.r.t. $C(W)$.

To prove necessity, let E be the unique ergodic kernel, and take an arbitrary $w_0 \in E$. Let $S_\varepsilon(w_0)$ be the sphere of centre w_0 and radius ε. Using the regularity of the chain and applying Theorem A1.20, we obtain

$$\liminf_{n \to \infty} Q^n(w, S_\varepsilon(w_0)) \geqslant \pi(S_\varepsilon(w_0)) > 0,$$

the last inequality being a result of Lemma 3.3.13. Therefore $Q^n(w, S_\varepsilon(w_0)) > 0$ for a sufficiently large $n \in \mathbb{N}^*$, whence $\Sigma_n(w) \cap S_\varepsilon(w_0) \neq \varnothing$, which implies the validity of (3.3.24). □

The lemma below specifies the method by which the supports of the probabilities $Q^n(w, \cdot)$, $n \in \mathbb{N}^*$, $w \in W$, can be iteratively computed.

Lemma 3.3.32. *For a continuous MC we have*

$$\Sigma_{m+n}(w) = \overline{\bigcup_{w' \in \Sigma_m(w)} \Sigma_n(w')}.$$

for all $m, n \in \mathbb{N}^*$ *and* $w \in W$.

Proof. Putting

$$\Sigma = \bigcup_{w' \in \Sigma_m(w)} \Sigma_n(w'),$$

we have

$$Q^{m+n}(w, \Sigma) = \int_{\Sigma_m(w)} Q^m(w, dw') Q^n(w', \Sigma) = Q^m(w, \Sigma_m(w)) = 1,$$

since $Q^n(w', \Sigma) \geqslant Q^n(w', \Sigma_m(w')) = 1$ for all $w' \in \Sigma_m(w)$. This implies that $\bar{\Sigma} \supset \Sigma_{m+n}(w)$.

To prove the converse, let us first remark that the equation

$$1 = Q^{m+n}(w, \Sigma_{m+n}(w)) = \int_W Q^m(w, dw') Q^n(w', \Sigma_{m+n}(w))$$

implies that $Q^m(w, \Sigma^*) = 1$, where $\Sigma^* = \{w' : Q^n(w', \Sigma_{m+n}(w)) = 1\}$. By the continuity of Q^n, it follows that Σ^* is a topologically closed set. Hence $\Sigma^* \supset \Sigma_m(w)$ and, if $w' \in \Sigma_m(w)$, we also have $\Sigma_{m+n}(w) \supset \Sigma_n(w')$. In other words, $\Sigma_{m+n}(w) \supset \Sigma$ and, since $\Sigma_{m+n}(w)$ is a closed set, we also have $\Sigma_{m+n}(w) \supset \bar{\Sigma}$. □

Remarks. 1. If $\Sigma_1(w)$ is a finite set for any $w \in W$, then, using Lemma 3.3.32 and an induction argument, we conclude that so is $\Sigma_n(w)$, $n \in \mathbb{N}^*$, for any $w \in W$. If such is the case, we simply have

$$\Sigma_{m+n}(w) = \bigcup_{w' \in \Sigma_m(w)} \Sigma_n(w').$$

2. The proof of the preceding lemma is completely analogous to that of Lemma 3.2.14.

3.3.6 The structure of the tail σ-algebra

Necessary and sufficient conditions are known for the tail σ-algebra of a φ-irreducible MC with a general state space, to be trivial. Such conditions are set forth in the theorem below due to Orey (1971, p. 19).

Theorem 3.3.33. *For a φ-irreducible MC with an arbitrary state space $(W, \mathscr{W})$, the following statements are equivalent:*

(i) *the tail σ-algebra of the chain is $\mathbf{P}_\mu$-trivial, regardless of the initial distribution μ of the chain;*

(ii) *we have*

$$\lim_{n\to\infty} \| \mu Q^n - \nu Q^n \| = 0 \tag{3.3.25}$$

for any probabilities μ and ν on $\mathscr{W}$.

(iii) *there are no bounded harmonic space-time functions*† *for the chain but the constants.*

By Lemma 3.3.16, a continuous MC, which is ergodic w.r.t. $C(W)$ is π-irreducible. If, in addition, the chain is regular w.r.t. $C(W)$, then (3.3.25) is fulfilled as a consequence of Proposition 3.3.25(ii). We conclude that the regularity w.r.t. $C(W)$ of a continuous MC is a sufficient condition for its tail σ-algebra to be $\mathbf{P}_\mu$-trivial, regardless of the initial distribution μ of the chain. Checking that a continuous MC is regular w.r.t. $C(W)$ amounts to proving that there exists a probability $\mathbf{Q}^\infty$ on $\mathscr{W}$ such that

$$\lim_{n\to\infty} \sup_{\substack{f\in C(W)\\ |f|=1}} \sup_{w\in W} \left| \int_W Q^n(w, dw') f(w') - \int_W \mathbf{Q}^\infty(dw') f(w') \right| = 0. \tag{3.3.26}$$

Of course, such a check cannot easily be done in concrete cases. Hence the following natural question arises: could we relax (3.3.26), at the same

† A function $h: W \times \mathbb{N}^* \to \mathbb{R}$ is said to be a *harmonic space-time function* of the chain iff

$$h(w, n) = \int_W h(w', n+1) Q(w, dw'), \quad n \in \mathbb{N}^*, \quad w \in W.$$

time strengthening some other assumptions which are more easily verified? The rest of this subsection is essentially devoted to the proof of Theorem 3.3.34, which provides an answer to this question.

Theorem 3.3.34. *Assume that for a continuous MC the t.p.f. Q is strongly continuous. Then a sufficient condition for its tail σ-algebra to be $\mathbf{P}_\mu$-trivial, regardless of the initial distribution μ, is the existence of a probability $\mathbf{Q}^\infty$ on $\mathscr{W}$ such that*

$$Q^n(w,\cdot) \Rightarrow \mathbf{Q}^\infty(.), \quad w \in W. \tag{3.3.27}$$

Remark. It is not known whether this condition is also necessary. Nor is it known whether it is still sufficient in the case where the t.p.f. Q is only continuous. However, (3.3.27) fails to be sufficient in the case where the t.p.f. Q is only weakly continuous, as illustrated in Problem 26.

To prove Theorem 3.3.34 we need the next two lemmas.

Lemma 3.3.35. *Let (W, d) be a totally bounded metric space, and $(h_n)_{n\in\mathbb{N}^*}$ an equicontinuous sequence of real-valued functions defined on W. If the limit*

$$h(w) = \lim_{n\to\infty} h_n(w) \tag{3.3.28}$$

exists for any $w \in W$, then h is a continuous function and the convergence in (3.3.28) *is uniform.*

Proof. It follows from the equicontinuity of the sequence $(h_n)_{n\in\mathbb{N}^*}$ that, for any $\varepsilon > 0$, there exists $\eta = \eta(\varepsilon) > 0$ such that

$$|h_n(w_1) - h_n(w_2)| < \varepsilon \tag{3.3.29}$$

for any $w_1, w_2 \in W$ with $d(w_1, w_2) < \eta$ and all $n \in \mathbb{N}^*$. Letting $n \to \infty$, we deduce that h is a continuous function.

Since W is totally bounded, for any $\delta > 0$, there exists a finite number of open spheres $S_\delta(w_1), \ldots, S_\delta(w_p)$ of radius δ which cover the entire space. Choose δ such that

$$|h(w) - h(w_j)| < \tfrac{1}{3}\varepsilon \tag{3.3.30}$$

for any $w \in S_\delta(w_j)$, $1 \leqslant j \leqslant p$, and

$$|h_n(w) - h_n(w_j)| < \tfrac{1}{3}\varepsilon \tag{3.3.31}$$

for any $w \in S_\delta(w_j)$, $1 \leqslant j \leqslant p$, and all $n \in \mathbb{N}^*$. Such a choice is possible by (3.3.29) and the continuity of h. But

$$|h_n(w_j) - h(w_j)| < \tfrac{1}{3}\varepsilon \tag{3.3.32}$$

for all $n \geqslant N_j$, $1 \leqslant j \leqslant p$, where $N_j = N(w_j, \varepsilon)$. Taking $N_0 = N_0(\varepsilon) = \max_{1\leqslant j\leqslant p} N_j$, equations (3.3.30)–(3.3.32) yield

$$|h_n(w) - h(w)| < \varepsilon$$

for all $n \geqslant N_0$ and $w \in W$. □

Lemma 3.3.36. *Let (W, d) be a separable complete metric space, and let $(\mu_n)_{n\in\mathbb{N}^*}$ and μ be probabilities on $\mathscr{W}$ such that $\mu_n \Rightarrow \mu$. Let $(f_n)_{n\in\mathbb{N}^*}$ be a sequence of equally bounded continuous real-valued functions defined on W. If f_n converges to a function f as $n \to \infty$, uniformly on compact sets, then*

$$\lim_{n\to\infty} \int_W f_n(w')\mu_n(\mathrm{d}w') = \int_W f(w')\mu(\mathrm{d}w'). \tag{3.3.33}$$

Proof. By the hypotheses there exists $M > 0$ such that $|f_n| < M$ for all $n \in \mathbb{N}^*$, and, for any $\varepsilon > 0$, there exists a compact set K such that, for any $n \in \mathbb{N}^*$, $\mu_n(W - K) < \varepsilon/(2M)$.[†] Then the first two terms on the right of the inequality

$$\left|\int_W f_n \,\mathrm{d}\mu_n - \int_W f\,\mathrm{d}\mu\right| \leqslant \left|\int_W f\,\mathrm{d}(\mu_n - \mu)\right| + \left|\int_K (f_n - f)\,\mathrm{d}\mu_n\right| + \left|\int_{W-K} (f_n - f)\,\mathrm{d}\mu_n\right|$$

tend to 0 as $n \to \infty$, while the last term is less than ε. Since $\varepsilon > 0$ is arbitrary, the proof is complete. □

Proof of Theorem 3.3.34. By Theorem 3.3.33, it is sufficient to prove that the constants are the only functions $h: W \times \mathbb{N}^* \to \mathbb{R}$ satisfying the conditions:

$$|h(w, n)| \leqslant M, \quad w \in W, \quad n \in \mathbb{N}^* \tag{3.3.34}$$

for some $M > 0$, and

$$h(w, n) = \int_W Q(w, \mathrm{d}w')h(w', n+1), \quad w \in W, \quad n \in \mathbb{N}^*. \tag{3.3.35}$$

Applying Lemma 2.1.1, (3.3.34) and (3.3.35) yield

$$|h(w_1, n) - h(w_2, n)| \leqslant M \,\| Q(w_1, \cdot) - Q(w_2, \cdot) \|. \tag{3.3.36}$$

[†] The second statement follows from Prohorov's theorem: A family $(\mu_\theta)_{\theta\in\Theta}$ of probabilities on a separable complete metric space (W, d) is relatively compact iff it is tight, i.e. iff for any $\varepsilon > 0$ there exists a compact set $K = K_\varepsilon$ such that $\mu_\theta(W - K) < \varepsilon$ for all $\theta \in \Theta$ (see Billingsley (1968, p. 37)).

Since the t.p.f. Q is strongly continuous, we deduce that $h(\cdot, n)$ is continuous for any fixed $n \in \mathbb{N}^*$.

The space (W, d) is compact, thus separable. Let $V = (v_j)_{j \in \mathbb{N}^*}$ be a dense set in W. Clearly, $(h(v_1, n))_{n \in \mathbb{N}^*}$ is a bounded sequence of real numbers, hence it contains a convergent subsequence. Let $\mathbb{N}_1 \subset \mathbb{N}^*$ be the set for which the sequence $(h(v_1, n))_{n \in \mathbb{N}_1}$ is convergent, and denote by $g(v_1)$ its limit. By an iterative procedure, we obtain a sequence of sets $\mathbb{N}^* \supset \mathbb{N}_1 \supset \cdots \supset \mathbb{N}_k \supset \mathbb{N}_{k+1} \supset \cdots$ of natural integers such that the limit

$$g(v_j) = \lim_{\substack{n \to \infty \\ n \in \mathbb{N}_j}} h(v_j, n)$$

exists for each $j \in \mathbb{N}^*$. Let n_k^k be the kth term in $\mathbb{N}_k, k \in \mathbb{N}^*$. Then we obtain

$$g(v_j) = \lim_{k \to \infty} h(v_j, n_k^k) \tag{3.3.37}$$

for any $j \in \mathbb{N}^*$. By (3.3.36), for any $\varepsilon > 0$, there exists $\delta = \delta(\varepsilon) > 0$ such that

$$|h(v_r, n_k^k) - h(v_s, n_k^k)| < \varepsilon$$

for any $v_r, v_s \in V$ with $d(v_r, v_s) < \delta$, and all $k \in \mathbb{N}^*$. Letting $k \to \infty$, we obtain

$$|g(v_r) - g(v_s)| < \varepsilon$$

for any $v_r, v_s \in V$ with $d(v_r, v_s) < \delta$, so that the function $g: V \to \mathbb{R}$ defined by (3.3.37) is uniformly continuous. Then there exists a uniformly continuous function $h: W \to \mathbb{R}$ with $h|_V = g$.† The equation

$$h(w) = \lim_{k \to \infty} h(w, n_k^k) \tag{3.3.38}$$

for all $w \in W$ is a consequence of the inequality

$$|h(w) - h(w, n_k^k)| \leqslant |h(w) - h(v_j)| + |h(v_j) - h(v_j, n_k^k)| + |h(v_j, n_k^k) - h(w, n_k^k)|$$

with a suitable choice of v_j. Using Lemma 3.3.35 (note that (W, d) is compact, thus totally bounded), we conclude that the convergence in (3.3.38) is uniform.

Iterating (3.3.35), we get

$$h(w, m) = \int_W Q^{n_k^k - m}(w, dw') h(w', n_k^k)$$

for all $k \in \mathbb{N}^*$ for which $n_k^k > m$. Letting $k \to \infty$ in this equation, and using Lemma 3.3.36, we conclude that $h(w, m) = \text{constant}$. □

† Here we used the following result: If $g: V \to Y$ is a uniformly continuous function, where V is a subset of a metric space W, and Y is a complete metric space, then there exists a uniformly continuous function $h: \bar{V} \to Y$ which coincides with g on V (see Royden (1968, p. 136)).

3.4 Applications to ergodicity

In this section we shall prove that a homogeneous RSCC $\{(W,\mathscr{W}),(X,\mathscr{X}),u,P\}$ or a GRSCC $\{(W,\mathscr{W}),(X,\mathscr{X}),\Pi,P\}$, whose associated MC is regular w.r.t. a certain Banach space of functions defined on $(W,\mathscr{W})$, is uniformly ergodic. Moreover, we shall show that the sequence $(\varepsilon_n)_{n\in\mathbb{N}^*}$ tends exponentially to 0; here

$$\varepsilon_n = \sup|P_j^n(w,A) - \mathbf{P}_j^\infty(A)|,$$

the supremum being taken over all $w\in W, A\in\mathscr{X}^j$, and $j\in\mathbb{N}^*$, while $\mathbf{P}_j^\infty$ is the probability on $\mathscr{X}^j$, whose existence is ensured by uniform ergodicity.

Let us first consider an RSCC, whose associated MC is regular w.r.t. $B(W,\mathscr{W})$. Then, by Lemma 3.1.19, there exist positive constants $\alpha<1$ and c such that

$$|U^n - U^\infty| \leqslant c\alpha^n, \quad n\in\mathbb{N}^*. \tag{3.4.1}$$

Hence

$$|U^n f - U^\infty f| \leqslant c|f|\alpha^n, \quad n\in\mathbb{N}^*, \quad f\in B(W,\mathscr{W}). \tag{3.4.2}$$

But, for any $j\in\mathbb{N}^*$, we have $P_j(\cdot,A)\in B(W,\mathscr{W})$ for all $A\in\mathscr{X}^j$. Put

$$\mathbf{P}_j^\infty(A) = U^\infty P_j(\cdot,A) = \int_W P_j(w,A)\mathbf{Q}^\infty(\mathrm{d}w).$$

(The existence of a probability $\mathbf{Q}^\infty$ on $\mathscr{W}$ such that the above equation holds is an immediate consequence of the regularity of the chain w.r.t. $B(W,\mathscr{W})$.) Then we obtain

$$|P_j^n(\cdot,A) - \mathbf{P}_j^\infty(A)| \leqslant c|P_j(\cdot,A)|\alpha^n, \quad n\in\mathbb{N}^*, \tag{3.4.3}$$

and, since $\sup_{A\in\mathscr{X}^j}|P_j(\cdot,A)|\leqslant 1$, this shows that we have proved the following theorem.

Theorem 3.4.1. *An RSCC, whose associated MC is regular w.r.t. $B(W,\mathscr{W})$, is uniformly ergodic. We have*

$$\mathbf{P}_j^\infty(A) = U^\infty P_j(\cdot,A) = \int_W P_j(w,A)\mathbf{Q}^\infty(\mathrm{d}w), \quad j\in\mathbb{N}^*, \quad A\in\mathscr{X}^j,$$

and there exist positive constants $\alpha<1$ and c such that

$$\varepsilon_n \leqslant c\alpha^n, \quad n\in\mathbb{N}^*.$$

Our aim now is to reach similar conclusions for both a GRSCC, whose associated MC is continuous and regular w.r.t. $C(W)$, and an RSCC whose associated MC is compact and regular w.r.t. $L(W)$.

In the first case, inequality (3.4.1) is valid by the same Lemma 3.1.19, and (3.4.2) holds for any $f\in C(W)$. For any $j\in\mathbb{N}^*$, by Lemma 3.1.30, we have $P_j^2(\cdot,A) = UP_j(\cdot,A)\in C(W)$ for all $A\in\mathscr{X}^j$. Since $U^\infty U = U^\infty$ and $\sup_{j\in\mathbb{N}^*}\sup_{A\in\mathscr{X}^j}|P_j(\cdot,A)|\leqslant 1$, applying (3.4.2) to the function $f(\cdot) = P_j^2(\cdot,A)$, we get

$$|P_j^{n+1}(\cdot, A) - \mathbf{P}_j^\infty(A)| \leqslant c\alpha^n, \quad n \in \mathbb{N}^*, \quad A \in \mathscr{X}^j,$$

which in this context plays the part of (3.4.3) in the preceding argument. Hence we have proved Theorem 3.4.2.

Theorem 3.4.2. *A GRSCC, whose associated MC is continuous and regular w.r.t. $C(W)$, is uniformly ergodic. We have*

$$\mathbf{P}_j^\infty(A) = U^\infty P_j(\cdot, A) = \int_W P_j(w, A)\mathbf{Q}^\infty(\mathrm{d}w), \quad j \in \mathbb{N}^*, \quad A \in \mathscr{X}^j,$$

and there exist positive constants $\alpha < 1$ and c such that

$$\varepsilon_n \leqslant c\alpha^n, \quad n \in \mathbb{N}^*.$$

In the case of an RSCC with contraction, whose associated MC is compact and regular w.r.t. $L(W)$, since $|\cdot| \leqslant \|\cdot\|_L$, the same argument is still valid if we are able to prove that $P_j(\cdot, A) \in L(W), j \in \mathbb{N}^*, A \in \mathscr{X}^j$, and $\sup_{j \in \mathbb{N}^*} \sup_{A \in \mathscr{X}^j} \|P_j(\cdot, A)\|_L < \infty$. These facts are consequences of Lemma 3.4.3.

Lemma 3.4.3. *For an RSCC with contraction we have*

$$\sup_{j \in \mathbb{N}^*} \sup_{A \in \mathscr{X}^j} \mathrm{s}(P_j(\cdot, A)) < \infty. \tag{3.4.4}$$

To prove this lemma we need the auxiliary result below, where we use the notation in (3.1.15) and (3.1.16).

Lemma 3.4.4. *For an RSCC with contraction we have*

$$R_{i+j} \leqslant r_i R_j + R_i, \quad i, j \in \mathbb{N}^*, \tag{3.4.5}$$

$$r_{i+j} \leqslant r_i r_j, \quad i, j \in \mathbb{N}^*. \tag{3.4.6}$$

Proof. By (1.1.3), we have

$$\int_{X^{i+j}} P_{i+j}(w', \mathrm{d}x^{(i+j)}) d(w'x^{(i+j)}, w''x^{(i+j)})$$

$$= \int_{X^i} P_i(w', \mathrm{d}x^{(i)}) \int_{X^j} P_j(w_1, \mathrm{d}y^{(j)}) d(w_1 y^{(j)}, w_2 y^{(j)})$$

for all $w', w'' \in W$ and $i, j \in \mathbb{N}^*$; here $w_1 = w'x^{(i)}$ and $w_2 = w''x^{(i)}$. Since the second integral on the right is dominated by $r_j d(w_1, w_2)$ and

$$\int_{X^i} P_i(w', \mathrm{d}x^{(i)}) d(w_1, w_2) \leqslant r_i d(w', w''),$$

the above equation immediately leads to (3.4.6).

By the same equation (1.1.3) we have

$$P_{i+j}(w', A) = \int_{X^i} P_i(w', \mathrm{d}x^{(i)}) h(w', x^{(i)})$$

for all $w' \in W, i, j \in \mathbb{N}^*$, and $A \in \mathscr{X}^{i+j}$, where

$$h(w', x^{(i)}) = \int_{X^j} P_j(w'x^{(i)}, \mathrm{d}y^{(j)}) \chi_A(x^{(i)}, y^{(j)}).$$

For any $i \in \mathbb{N}^*$, $w', w'' \in W$, and $B \in \mathscr{X}^i$, put $\Lambda_i(w', w'', B) = P_i(w', B) - P_i(w'', B)$; then

$$\Lambda_{i+j}(w', w'', A) = \int_{X^i} P_i(w', \mathrm{d}x^{(i)})[h(w', x^{(i)}) - h(w'', x^{(i)})] + \int_{X^i} \Lambda_i(w', w'', \mathrm{d}x^{(i)}) h(w'', x^{(i)}). \tag{3.4.7}$$

Using Lemma 2.1.1, we get

$$|h(w', x^{(i)}) - h(w'', x^{(i)})| \leqslant R_j d(w'x^{(i)}, w''x^{(i)}),$$

$$\left| \int_{X^i} \Lambda_i(w', w'', \mathrm{d}x^{(i)}) h(w'', x^{(i)}) \right| \leqslant R_i d(w', w'').$$

These two inequalities, together with (3.4.7), lead to

$$|\Lambda_{i+j}(w', w'', A)| \leqslant R_j \int_{X^i} P_i(w', \mathrm{d}x^{(i)}) d(w'x^{(i)}, w''x^{(i)}) + R_i d(w', w'') \leqslant (r_i R_j + R_i) d(w', w''),$$

from which we obtain (3.4.5). □

Remark. Since for an RSCC with contraction, $R_1 < \infty$ and $r_1 < \infty$, it is sufficient to take $i = 1$ in (3.4.5) and (3.4.6) to conclude that $R_j < \infty$ and $r_j < \infty$ for all $j \in \mathbb{N}^*$.

Proof of Lemma 3.4.3. Inequality (3.4.4) can also be written as $\sup_{j \in \mathbb{N}^*} R_j < \infty$ and, since $R_{j+1} \geqslant R_j$, $j \in \mathbb{N}^*$, we only have to prove that

$$\lim_{j \to \infty} R_j < \infty. \tag{3.4.8}$$

But for an RSCC with contraction there exists $k \in \mathbb{N}^*$ such that $r_k < 1$. Taking $i = k$ in (3.4.5) and noting that $R_j \leqslant R_{j+k}$, $j \in \mathbb{N}^*$, yields

$$R_{j+k} \leqslant \frac{R_k}{1 - r_k} < \infty, \quad j \in \mathbb{N}^*,$$

and letting $j \to \infty$, we obtain (3.4.8). □

Remark. Putting

$$R_\infty = \sup_{A \in \mathscr{X}} \sup_{w' \neq w''} \frac{|\mathbf{P}_{w'}(A) - \mathbf{P}_{w''}(A)|}{d(w', w'')},$$

it is obvious that $R_\infty \geqslant \sup_{j \in \mathbb{N}^*} R_j$. In fact, since $\mathscr{X}$ is the σ-algebra generated by $\bigcup_{j \in \mathbb{N}^*} \mathscr{X}^j$, it is possible to prove that $R_\infty = \sup_{j \in \mathbb{N}^*} R_j$. Hence by Lemma 3.4.3, this shows that, for an RSCC with contraction, $\mathbf{P}_{\cdot}(A) \in L(W)$ for all $A \in \mathscr{X}$.

The results proved above enable us to state the following theorem.

Theorem 3.4.5. *An RSCC with contraction, whose associated MC is compact and regular w.r.t. $L(W)$, is uniformly ergodic. We have*

$$\mathbf{P}_j^\infty(A) = U^\infty P_j(\cdot, A) = \int_W P_j(w, A) \mathbf{Q}^\infty(dw), \quad j \in \mathbb{N}^*, \quad A \in \mathscr{X}^j,$$

and there exist positive constants $\alpha < 1$ and c such that

$$\varepsilon_n \leqslant c\alpha^n, \quad n \in \mathbb{N}^*.$$

The hypothesis concerning the regularity of the associated MC can be replaced by weaker hypotheses; however, if so, the conclusions are also weaker. As an illustration we give Theorem 3.4.6.

Theorem 3.4.6. *Consider an absorbing RSCC with contraction. Then, for any $j \in \mathbb{N}^*$, there exist a t.p.f. P_j^∞ from $(W, \mathscr{W})$ to $(X^j, \mathscr{X}^j)$ and two positive constants $\alpha < 1$ and c such that*

$$\| P_j^n(\cdot, A) - P_j^\infty(\cdot, A) \|_L \leqslant c\alpha^n$$

for all $n \in \mathbb{N}^*$ and $A \in \mathscr{X}^j$.

Proof. Since, by Lemma 3.4.3, $f(\cdot) = P_j(\cdot, A) \in L(W)$, $j \in \mathbb{N}^*, A \in \mathscr{X}^j$, and $\sup_{j \in \mathbb{N}^*} \sup_{A \in \mathscr{X}^j} \| P_j(\cdot, A) \|_L < \infty$, Theorem 3.2.11 applies. Putting

$$P_j^\infty(\cdot, A) = U^\infty P_j(\cdot, A) = \int_W P_j(w, A) \mathbf{Q}^\infty(\cdot, dw), \quad j \in \mathbb{N}^*,$$

we obtain the inequality in the theorem. □

Remark. Theorems 3.4.1, 3.4.2, and 3.4.5 give the form of the probabilities $\mathbf{P}_j^\infty$, $j \in \mathbb{N}^*$, in terms of $\mathbf{Q}^\infty$. However, to determine $\mathbf{Q}^\infty$ effectively is a difficult problem even in the simplest special cases. For special cases when this is possible, the reader should consult Dubins & Freedman (1966, 1967), Halfin (1975), Karlin (1953), Marliss & McGregor (1968, 1971), McGregor & Hui (1962), McGregor & Zidek (1965a, b), and Usačev (1971). A case where $\mathbf{Q}^\infty$ can be effectively determined will also be seen in Section 5.2.

Problems

1. Prove that any finite MC statisfies Condition (D).

2. Prove that any of Conditions 1, 2, and 3 in Subsection 3.1.2 implies Condition (D).

3. Consider the function $u:[0,1]\to[0,1]$ defined as
$$u(w)=\begin{cases} w+1/2, & \text{if}\, 0\leqslant w\leqslant \frac{1}{2} \\ 1/4, & \text{if}\, \frac{1}{2}<w\leqslant 1 \end{cases}$$
and the MC with state space $[0,1]$ and t.p.f. $Q(w,A)=\delta(u(w),A)$, $w\in[0,1]$, $A\in\mathscr{B}_{[0,1]}$. Prove that this MC satisfies Condition (D), but is not a Doeblin–Fortet chain, and $U(C(W))\not\subset C(W)$. (Herkenrath (1977))
(*Hint*: Apply Proposition 3.1.9.)

4. Show that the converse of statement (ii) in Theorem 3.1.28 does not hold without additional assumptions.
(*Hint*: Use the MC in the preceding problem as a counter-example.)

5. Let $\{(W,\mathscr{W}),(X,\mathscr{X}),\Pi,P\}$ be a GRSCC for which:
   (i) there exists a finite measure ν on $\mathscr{W}$ such that $\Pi(w,x,\cdot)$ is absolutely continuous w.r.t. ν for all $w\in W$ and $x\in X$;
   (ii) there exists $K>0$ such that
$$\sup_{\substack{w,t\in W\\ x\in X}} |\pi(w,x,t)|\leqslant K,$$
   where $\pi(w,x,t)=\mathrm{d}\Pi(w,x,\cdot)/\mathrm{d}\nu(\cdot)$.
Prove that its associated MC satisfies Condition (D). (Herkenrath (1977))

6. Let $\{(W,\mathscr{W}),(X,\mathscr{X}),\Pi,P\}$ be a GRSCC for which there exists a set $D_0\in\mathscr{X}$ such that
   (i) $D_0\subset\{x\in X$: the t.p.f. $\Pi(\cdot,x,\cdot)$ satisfies Condition (D)$\}$;
   (ii) there exists $\delta>0$ such that $P(w,D_0)\geqslant\delta$ for all $w\in W$.
Prove that its associated MC satisfies Condition (D). (Herkenrath (1977)

7. Consider the RSCC $\{(W,\mathscr{W}),(X,\mathscr{X}),u,P\}$ for which
$$W=[0,1],\mathscr{W}=\mathscr{B}_{[0,1]},X=\{0\},\mathscr{X}=\mathscr{P}(X),$$
$$u(w,x)=u(w)=\begin{cases}(1/2)w, & \text{if}\, 0\leqslant w\leqslant\frac{1}{2}\\ (-1/2)w+1/2, & \text{if}\,\frac{1}{2}<w\leqslant 1,\end{cases}$$
$$P(w,\{0\})=1, w\in W, x\in X.$$

Prove that its associated MC is a Doeblin–Fortet chain, but it does not satisfy Condition (D). Draw the conclusion that the associated Markov operator is quasi-compact on $L(W)$, but is neither quasi-compact on $B(W)$ nor quasi-compact on $C(W)$, in spite of the fact that $U(C(W)) \subset C(W)$. (Herkenrath (1977))
(*Hint*: Prove first that this RSCC is with contraction. Then show that its associated MC does not satisfy the necessary and sufficient condition in Proposition 3.1.9.)

8. Let W be a compact path-connected subset of $\mathbb{R}^n$ and define the t.p.f. Q by $Q(w, A) = \delta(u(w), A)$, $w \in W$, $A \in W \cap \mathscr{B}^n$, where the function $u: W \to W$ is $(\mathscr{W}, \mathscr{W})$-measurable. Prove that the Markov operator associated with Q is compact on $C(W)$ iff u is a constant function. (Herkenrath (1977))

9. Prove that if a Markov operator U is compact on $C(W)$, where (W, d) is a compact metric space, then the sequence $(U^n f)_{n \in \mathbb{N}^*}$ is equicontinuous for any $f \in C(W)$. (In this case, the operator U is said to be *uniformly stable*.)

10. Prove that if a Markov operator U is quasi-compact on $C(W)$, where (W, d) is a compact metric space, then the sequence $((1/n)\sum_{k=1}^{n} U^k f)_{n \in \mathbb{N}^*}$ is equicontinuous for any $f \in C(W)$. (In this case, the operator U is said to be *uniformly stable in mean*.) (Jamison (1965))

11. Prove that if U is a Doeblin–Fortet operator on a compact metric space, then U is uniformly stable. (Norman (1972, p. 51))

12. Prove that if a Markov operator U is quasi-compact on $L(W)$, where (W, d) is a compact metric space, then U is uniformly stable in mean.

13. Prove that any continuous MC is uniformly stable.

14. Prove that if a Markov operator U is orderly on $L(W)$, where (W, d) is a complete separable metric space, then there exists a t.p.f. Q^∞ on $(W, \mathscr{W})$ for which equations (3.1.27) and (3.1.29) hold.

15. Under the conditions of the preceding problem, assume that U is ergodic w.r.t. $L(W)$. Then prove that the t.p.f. $Q^\infty(w, \cdot)$ does not depend on w, and is the unique fixed point of the adjoint V of U.

16. Prove that the statements in the preceding problem still hold for a Markov operator on $B(W, \mathscr{W})$, which is ergodic.

17. Consider the RSCC $\{(W, \mathscr{W}), (X, \mathscr{X}), u, P\}$ for which the sets $W \subset \mathbb{R}^n$

and $X \subset \mathbb{R}^n$ have λ^n-null frontiers, $\mathscr{W} = W \cap \mathscr{B}^n$, and $\mathscr{X} = X \cap \mathscr{B}^n$. (Here λ^n is the Lebesgue measure on $\mathscr{B}^n$.) Assume also that W is a compact set, and:

(i) there exists a function $p: W \times X \to \mathbb{R}_+$, which is $\mathscr{W} \otimes \mathscr{X}$-measurable, such that

$$P(w, B) = \int_B p(w, x)\lambda^n(\mathrm{d}x), \quad w \in W, \quad B \in \mathscr{X};$$

(ii) denoting by $u_w^j(x), 1 \leqslant j \leqslant n$, the components of the vector $u_w(x) = u(w, x)$, the partial derivatives $\partial u_w^i/\partial x^j, 1 \leqslant i, j \leqslant n$, do exist in the interior X^0 of X and are continuous. Moreover, the Jacobian $Ju_w(x)$ satisfies the inequalities $0 < |Ju_w(x)| < \infty$ for all $w \in W$ and $x \in X^0$;

(iii) there exists an inverse map $u_w^{-1}: W \to X$ of the map $u_w: X \to W$;

(iv) the functions p and u_w^{-1} are continuous for all $w \in W^0$, while the function $\varphi_{w'}(w) = |Ju_w^{-1}(w')|$ is continuous for all $w' \in W^0$.

Prove that the t.p.f. of the associated MC is strongly continuous. (Herkenrath (1979))

18. Prove that any finite MC is compact as well as continuous, regardless of the metric defined on its state space.

19. Consider the RSCC $\{(W, \mathscr{W}), (X, \mathscr{X}), u, P\}$ for which

(i) (W, d) is a compact metric space and $\mathscr{W}$ is the collection of the Borel subsets of W;

(ii) we have $s(P(\cdot, A)) \leqslant a < \infty$ for all $A \in \mathscr{X}$;

(iii) we have $\mu_d(u(\cdot, x)) \leqslant 1$ for all $x \in X$, where

$$\mu_d(u(\cdot, x)) = \sup_{w' \neq w''} \frac{d(w'x, w''x)}{d(w', w'')}.$$

(iv) there exist $n_0 \in \mathbb{N}^*$ and $A_0 \in \mathscr{X}^{no}$ such that

$$A_0 \subset \{x^{(no)}: \mu_d(u(\cdot, x^{(no)})) < 1\},$$

$$\sup_{x^{(no)} \in A_0} \mu_d(u(\cdot, x^{(no)})) < 1,$$

$$\inf_{w \in W} P_{no}(w, A_0) > 0.$$

Prove that this RSCC is with contraction. (Iosifescu & Theodorescu (1969, p. 112))

20. Consider an RSCC $\{(W, \mathscr{W}), (X, \mathscr{X}), u, P\}$ for which the space X is finite. Assume that (i) and (iii) in the preceding problem hold, and that

(ii') we have $s(P(\cdot, \{x\})) < \infty$ for all $x \in X$;

(iv') for any $w \in W$, there exist $k = k(w) \in \mathbb{N}^*$ and $x^{(k)} = x^{(k)}(w) \in X^k$

such that

$$\mu_d(u(\cdot, x^{(k)})) < 1, \quad P_k(w, \{x^{(k)}\}) > 0.$$

Prove that this RSCC is with contraction. (Norman (1968))

21. (*The five operators model*) Consider the RSCC $\{(W, \mathscr{W}), (X, \mathscr{X}), u, P\}$ for which

$$W = [0,1], \quad \mathscr{W} = \mathscr{B}_{[0,1]}, \quad X = \{0,1\} \times \{0,1\} \times \{0,1\} \quad \mathscr{X} = \mathscr{P}(X),$$
$$u(w, x) = w + k\theta_{ij}(j - w),$$

where $x = (i, j, k) \in X$ and

$$P(w, \{x\}) = \begin{cases} w\pi_{1j}c_{1j} & \text{if } i = 1, k = 1 \\ w\pi_{1j}c'_{1j} & \text{if } i = 1, k = 0 \\ w'\pi_{0j}c_{0j} & \text{if } i = 0, k = 1 \\ w'\pi_{0j}c'_{0j} & \text{if } i = 0, k = 0, \end{cases}$$

using the convention $\alpha' = 1 - \alpha$ for any $0 \leqslant \alpha \leqslant 1$. Here $0 \leqslant \theta_{ij}, \pi_{ij}, c_{ij} \leqslant 1, \pi_{i0} + \pi_{i1} = 1, i, j = 0, 1$. Put $\omega_i = \theta_{ii'}\pi_{ii'}c_{ii'}, i = 0, 1$. Prove that, if ω_0 and ω_1 are both positive and at least one of them is less than 1, then the associated MC is regular w.r.t. $L([0,1])$. (Norman (1972, p. 177))

22. Let $f: [0,1] \to (0,1]$ be continuous with

$$\lim_{w \nearrow 1} \frac{f(w) - f(1)}{w - 1} = +\infty.$$

Define the t.p.f. Q on $([0,1], \mathscr{B}_{[0,1]})$ by

$$Q(w, A) = \frac{1}{f(w)} \int_{A \cap [0, f(w)]} d\lambda, \quad w \in [0,1], \quad A \in \mathscr{B}_{[0,1]}.$$

Prove that the MC with state space $[0,1]$ and t.p.f. Q is continuous but not compact.

23. Consider the RSCC $\{(W, \mathscr{W}), (X, \mathscr{X}), u, P\}$ for which

$$W = [0,1], \quad \mathscr{W} = \mathscr{B}_{[0,1]}, \quad X = \{0,1\}, \quad \mathscr{X} = \mathscr{P}(X),$$
$$P(w, \{0\}) = P(w, \{1\}) = 1/2,$$
$$u(w, \{0\}) = f(w), \quad u(w, \{1\}) = g(w),$$

where the functions $f, g: [0,1] \to [0,1]$ satisfy

(i) $f(0) = 0$ and $f(w) < w, \quad w \in (0,1]$;

(ii) $g(1) = 1$ and $g(w) > w, \quad w \in [0,1)$;

(iii) f and g are differentiable and $|f'(w)| < 1, |g'(w)| < 1, w \in [0,1]$.

Prove that the associated MC is compact (Kaijser (1972)), but that it is not continuous.

24. Prove that an MC whose state space is a totally bounded metric space and whose t.p.f. is strongly continuous, has a finite number of ergodic kernels.

25. Assume that the functions $f, g: [0,1] \to [0,1]$ are non-decreasing, everywhere differentiable and that they satisfy
    (i) $f(w) < g(w)$, $w \in [0,1]$;
    (ii) $f'(w) < 1$ and $g'(w) < 1$, $w \in [0,1]$.
    Consider the MC with state space $[0,1]$ and t.p.f. Q given by
$$Q(w, A) = \frac{1}{g(w) - f(w)} \int_{A \cap [f(w), g(w)]} d\lambda, \quad w \in [0,1], \quad A \in \mathscr{B}_{[0,1]}.$$
    Prove that this MC is continuous and regular w.r.t. $C(W)$.

26. Consider the MC with state space $[0,1]$ and t.p.f. Q given by
$$Q(w, A) = \chi_A(aw), \quad w \in [0,1], \quad A \in \mathscr{B}_{[0,1]},$$
    where $a \in (0,1)$ is a rational number. Prove that this MC is compact and regular w.r.t. $L(W)$. Prove also that its t.p.f. is weakly continuous and that there exists a probability $\mathbf{Q}^\infty$ on $\mathscr{B}_{[0,1]}$ such that $Q^n(w, \cdot) \Rightarrow \mathbf{Q}^\infty(\cdot)$ for all $w \in [0,1]$. Show that, nevertheless, its tail σ-algebra is not trivial. (Grigorescu 1976b))
    (*Hint*: The functions
$$h(w,n) = h(w) = \begin{cases} 1, & \text{if } w \text{ is rational,} \\ 0, & \text{if } w \text{ is irrational,} \end{cases} \qquad n \in \mathbb{N}^*,$$
    are space–time harmonic functions for the chain.)

27. Consider an RSCC with contraction, whose associated MC $(\zeta_n)_{n \in \mathbb{N}^*}$ is compact and regular w.r.t. $L(W)$. Prove that the MC $(\zeta'_n)_{n \in \mathbb{N}^*}$, where $\zeta'_n = (\zeta_n, \xi_n)$, $n \in \mathbb{N}^*$, is regular w.r.t. $L' = L'(W \times X, \mathscr{W} \otimes \mathscr{X})$, where $L' = \{f: W \times X \to \mathbb{R}: \|f\|'_L < \infty\}$ and $\|f\|'_L = |f|' + s'(f)$, $|f|' = \sup_{w \in W, x \in X} |f(w,x)|$, $s'(f) = \sup_{x \in X} s(f(\cdot, x))$.

28. Let (W, d) be a metric space and $W = \bigcup_{i=1}^m W_i$, $m \in \mathbb{N}^*$, a partition of W, where the W_i are Borel sets in W. Let $L(W_1, \ldots, W_m)$ be the collection of all bounded complex-valued functions f on W for which $s(f_i) < \infty$, where f_i is the restriction of f to W_i, $1 \leqslant i \leqslant m$. Let also $C(W_1, \ldots, W_m)$ be the collection of all bounded complex-valued functions f on W for which f_i is continuous on W_i, $1 \leqslant i \leqslant m$. Show that $C(W_1, \ldots, W_m)$ is a Banach space under the supremum norm $|f| = \sup_{w \in W} |f(w)|$, while $L(W_1, \ldots, W_m)$ is a Banach space under the norm $\|f\|_L = |f| + s(f)$, where $s(f) = \max_{1 \leqslant i \leqslant m} s(f_i)$. Show that this pair of Banach spaces satisfies Condition (ITM_1) in Section A2.9.

29. (Continuation of Problem 28) Consider an RSCC $\{(W, d), (X, \mathscr{X}), u, P\}$, where (W, d) is a separable metric space, and assume that there exists a partition $W = \bigcup_{i=1}^{m} W_i$, $m \in \mathbb{N}^*$, where the W_i are Borel sets in W, such that for any $x \in X$ and $1 \leqslant i \leqslant m$ there exists $j = j(x, i)$ for which $W_j \supset u(W_i, x)$. Generalize the concept of an RSCC with contraction to the present setting (note that the usual concept corresponds to the special case $m = 1$), and also the concept of a Doeblin–Fortet operator, by replacing the space $L(W)$ by the space $L(W_1, \ldots, W_m)$ in such a manner that the results in Sections 3.1 and 3.2 keep their validity. (For a special case see Kalpazidou (1985a))

4
Asymptotic behaviour

4.1 Limit theorems for the infinite order chain

4.1.1 Preliminaries

Throughout this section we shall consider a uniformly strong ergodic (in general, non-homogeneous) RSCC, and we shall put

$$\varepsilon_n = \sup |{}^mP_r^n(w, A) - \mathbf{P}_r^\infty(A)|, \quad n \in \mathbb{N}^*,$$

the supremum being taken over all $m \in \mathbb{N}$, $w \in W$, $A \in \mathscr{X}^r$, and $r \in \mathbb{N}^*$. Thus $\varepsilon_n \downarrow 0$ as $n \to \infty$, and, by Theorem 2.1.5 and the remark concluding Subsection 2.3.2, the sequence $(\xi_n)_{n \in \mathbb{N}^*}$ constructed in Theorem 1.1.6 is φ-mixing under both probabilities $\mathbf{P}_\infty$ and $\mathbf{P}_w$, for any $w \in W$.

Let F be a real r.v. on $X^{\mathbb{N}^*}$, and put

$$f_n = F(\xi_n, \xi_{n+1}, \ldots), \quad n \in \mathbb{N}^*.$$

It is obvious that, under $\mathbf{P}_\infty$, the sequence $(f_n)_{n \in \mathbb{N}^*}$, is strictly stationary (but, in general, it is *not* φ-mixing). In the theorems we shall prove we make some assumptions on the moments of f_1 under $\mathbf{P}_\infty$. Therefore it is important to know what assumptions should be made on the moments under $\mathbf{P}_w$ to ensure that the necessary assumptions on the moments under $\mathbf{P}_\infty$ do hold.

Proposition 4.1.1. *If for arbitrarily fixed* $w \in W$ *and* $\delta > 0$ *we have* $\liminf_{n \to \infty} \mathbf{E}_w(|f_n|^\delta) < \infty$, *then* $\mathbf{E}_\infty(|f_1|^\delta) < \infty$.

Proof. Uniform strong ergodicity implies that

$$|(\mathbf{P}_w f_n^{-1})(A) - (\mathbf{P}_\infty f_1^{-1})(A)| \leqslant \varepsilon_n$$

for any $n \in \mathbb{N}^*$ and $A \in \mathscr{B}$. Therefore, for any arbitrary $c > 0$, by Lemma 2.1.1, we have

$$\mathbf{E}_w(|f_n|^\delta) = \int_{\mathbb{R}} |u|^\delta (\mathbf{P}_w f_n^{-1})(du) \geqslant \int_{-c}^{+c} |u|^\delta (\mathbf{P}_w f_n^{-1})(du)$$

$$\left(\xrightarrow[n \to \infty]{} \int_{-c}^{+c} |u|^\delta (\mathbf{P}_\infty f_1^{-1})(du) \xrightarrow[c \to \infty]{} \mathbf{E}_\infty(|f_1|^\delta) \right).$$

Hence, letting first $n \to \infty$ and then $c \to \infty$,

$$\mathsf{E}_\infty(|f_1|^\delta) \leqslant \liminf_{n\to\infty} \mathsf{E}_w(|f_n|^\delta). \qquad \square$$

4.1.2 The law of large numbers

We shall now prove a generalization of Kolmogorov's law of large numbers.

Theorem 4.1.2. (Law of large numbers) *If $\mathsf{E}_\infty(|f_1|) < \infty$, then, for any $w \in W$, the r.v. $(\sum_{i=1}^n f_i)/n$ converges P_w-a.s. to $\mathsf{E}_\infty(f_1)$ as $n \to \infty$.*

Proof. For any $a \in \mathbb{R}$ the random event

$$A_a = \left\{ \lim_{n\to\infty} \frac{\sum_{i=1}^n f_i}{n} = a \right\}$$

belongs to the tail σ-algebra of the sequence $(\xi_n)_{n\in\mathbb{N}^*}$. By Theorem 2.1.5, for any $w \in W$ and $a \in \mathbb{R}$, we have $\mathsf{P}_w(A_a) = \mathsf{P}_\infty(A_a)$. By the same theorem and Proposition A3.3 we also have $\mathsf{P}_\infty(A_a) = 1$ for $a = \mathsf{E}_\infty(f_1)$.

$\square$

An important special case is obtained by taking $F(x_1, x_2, \ldots) = \chi_A(x^{(k)})$ for arbitrarily fixed $k \in \mathbb{N}^*$ and $A \in \mathcal{X}^k$. Clearly, in this case, $\mathsf{E}_\infty(f_1) = \mathsf{P}_k^\infty(A)$, while $\sum_{i=1}^n f_i = \sum_{i=1}^n \chi_A(\xi_i, \ldots, \xi_{i+k-1}) = \nu_{n,k}(A)$ is the absolute frequency of the elements of A among the n k-tuples of consecutive terms of the sequence $\xi_1, \ldots, \xi_{n+k-1}$. We can therefore state the following generalization of Borel's law of large numbers.

Corollary 4.1.3. *For any $w \in W$, the relative frequency $\nu_{n,k}(A)/n$ converges P_w-a.s. to $\mathsf{P}_k^\infty(A)$ as $n \to \infty$, for all $k \in \mathbb{N}^*$ and $A \in \mathcal{X}^k$.*

The asymptotic behaviour as $n \to \infty$ of the r.v. $\nu_{n,k}(A)$ for an absorbing RSCC, is radically different from that just described for a uniformly strong ergodic RSCC. It is in fact reminiscent of the behaviour of finite absorbing MCs. Thus we have Proposition 4.1.4.

Proposition 4.1.4. *Consider an absorbing RSCC with contraction, whose absorbing points are $a_1, \ldots, a_l$. For an arbitrarily fixed $k \in \mathbb{N}^*$, for any set $A \in \mathcal{X}^k$ for which $P_k(a_j, A) = 0$, $1 \leqslant j \leqslant l$, the limit $\nu_{\infty,k}(A) = \lim_{n\to\infty} \nu_{n,k}(A)$ exists and is finite P_w-a.s. for any $w \in W$. Moreover, there exist constants*

$c>0$ and $0<\alpha<1$ such that

$$\| \mathbf{E}(\nu_{\infty,k}(A)) \|_L \leqslant \frac{c\alpha}{1-\alpha}. \tag{4.1.1}$$

The function $\gamma(w)=\mathbf{E}_w(\nu_{\infty,k}(A))$ is the unique continuous solution of the equation $\gamma(w)=P_k(w,A)+U\gamma(w)$ for which $\gamma(a_j)=0$, $1\leqslant j\leqslant l$.

Proof. Since $P_k^{\infty}(w,A)=U^{\infty}(P_k(\cdot,A))(w)=\sum_{j=1}^{l} g_j(w)P_k(a_j,A)=0$, by Theorem 3.4.6, we have $\| P_k^n(\cdot,A)\|_L\leqslant c\alpha^n$, $n\in\mathbb{N}^*$. It follows that the series $\sum_{n\in\mathbb{N}^*} P_k^n(\cdot,A)$ converges in $L(W)$ and

$$\| \sum_{n\in\mathbb{N}^*} P_k^n(\cdot,A)\|_L \leqslant \sum_{n\in\mathbb{N}^*} \| P_k^n(\cdot,A)\|_L \leqslant \frac{c\alpha}{1-\alpha}. \tag{4.1.2}$$

The fact that for any $w\in A$ the r.v. $\nu_{\infty,k}(A)$ is $\mathbf{P}_w$-a.s. finite follows from the convergence in $L(W)$ of the series $\sum_{n\in\mathbb{N}^*} P_k^n(\cdot,A)$ and the equation

$$\mathbf{E}_w(\chi_A(\xi_i,\ldots,\xi_{i+k-1}))=P_k^i(w,A), \quad i\in\mathbb{N}^*.$$

Inequality (4.1.1) is a consequence of the equation

$$\mathbf{E}_w(\nu_{\infty,k}(A))=\sum_{n\in\mathbb{N}^*} P_k^n(w,A)$$

and of (4.1.2).

It is easy to see that

$$\gamma(w)=\mathbf{E}_w(\chi_A(\xi_1,\ldots,\xi_k))+\mathbf{E}_w\Big(\sum_{i\geqslant 2}\chi_A(\xi_i,\ldots,\xi_{i+k-1})\Big)=P_k(w,A)+U\gamma(w)$$

and $\gamma(a_j)=0$, $1\leqslant j\leqslant l$. If γ' is another continuous solution of the above equation for which $\gamma'(a_j)=0$, $1\leqslant j\leqslant l$, then $\gamma-\gamma'\in C(W)$, $U(\gamma-\gamma')=\gamma-\gamma'$, and $(\gamma-\gamma')(a_j)=0$, $1\leqslant j\leqslant l$. By Lemma 3.3.7, the difference $\gamma-\gamma'$ attains its maximum on the set $\{a_1,\ldots,a_l\}$, hence $\gamma\equiv\gamma'$. □

4.1.3 *The functional central limit theorem*

This subsection is devoted to the proof of an invariance principle for the central limit theorem. The reader should refer to Appendix 3.

Assume, without any loss of generality, that $\mathbf{E}_{\infty}(f_1)=0$. The general case is obtained by considering $F-\mathbf{E}_{\infty}(f_1)$ instead of F. Putting $S_0=0$, $S_n=\sum_{i=1}^{n} f_i$, $n\in\mathbb{N}^*$, let us define the stochastic processes

$$\boldsymbol{\eta}_n^C:\boldsymbol{\eta}_n^C(t)=\frac{1}{\sigma\sqrt{n}}(S_{[nt]}+(nt-[nt])f_{[nt]+1}), \quad t\in[0,1], \quad n\in\mathbb{N}^*,$$

and

$$\boldsymbol{\eta}_n^D:\boldsymbol{\eta}_n^D(t)=\frac{1}{\sigma\sqrt{n}}S_{[nt]}, \quad t\in[0,1], \quad n\in\mathbb{N}^*,$$

where σ is a positive number, which is defined in the theorem below.

Theorem 4.1.5. *Assume that* $\mathbf{E}_\infty(|f_1|^{2+\delta})<\infty$ *for some* $\delta\geqslant 0$ (or $|F|\leqslant c<\infty$), $\sum_{n\in\mathbb{N}^*}\varepsilon_n^{(1+\delta)/(2+\delta)}<\infty$ (*or* $\sum_{n\in\mathbb{N}^*}\varepsilon_n<\infty$), *and that* $\sum_{n\in\mathbb{N}^*}\mathbf{E}_\infty^{1/2}([f_1-\mathbf{E}_\infty(f_1|(\xi_i)_{1\leqslant i\leqslant n})]^2)<\infty$. *Then according to Theorem* A3.7 *the series*

$$\sigma^2=\mathbf{E}_\infty(f_1^2)+2\sum_{n\in\mathbb{N}^*}\mathbf{E}_\infty(f_1f_{n+1})$$

converges absolutely and $\sigma^2\geqslant 0$. *If* $\sigma>0$, *then* $\mathbf{P}_w\boldsymbol{\eta}_n^{-1}\Rightarrow\mathbf{B}$ *for any* $w\in W$, *where* $\boldsymbol{\eta}_n$ *stands for either* $\boldsymbol{\eta}_n^C$ *or* $\boldsymbol{\eta}_n^D$.

Proof. By Theorems 2.1.5 and A3.7, we have

$$\mathbf{P}_\infty\boldsymbol{\eta}_n^{-1}\Rightarrow\mathbf{B}. \tag{4.1.3}$$

Let us construct a new process $\bar{\boldsymbol{\eta}}_n$ by suppressing an initial segment of the process $\boldsymbol{\eta}_n$. More precisely, let $(p_n)_{n\in\mathbb{N}^*}$ be a sequence of natural integers converging to ∞ and for which $1>p_n/\sqrt{n}\to 0$ as $n\to\infty$. Put

$$\bar{\boldsymbol{\eta}}_n^C:\bar{\boldsymbol{\eta}}_n^C(t)=\begin{cases}0, & \text{if } 0\leqslant t<p_n/\sqrt{n}\\ \dfrac{1}{\sigma\sqrt{n}}\displaystyle\sum_{p_n\leqslant i\leqslant nt}f_i+(nt-[nt])f_{[nt]+1}, & \text{if } p_n/\sqrt{n}\leqslant t\leqslant 1,\end{cases}$$

$$\bar{\boldsymbol{\eta}}_n^D:\bar{\boldsymbol{\eta}}_n^D(t)=\begin{cases}0, & \text{if } 0\leqslant t<p_n/\sqrt{n}\\ \dfrac{1}{\sigma\sqrt{n}}\displaystyle\sum_{p_n\leqslant i\leqslant nt}f_i, & \text{if } p_n/\sqrt{n}\leqslant t\leqslant 1.\end{cases}$$

We have

$$\left.\begin{aligned}d(\bar{\boldsymbol{\eta}}_n^C,\boldsymbol{\eta}_n^C)&=\sup_{0\leqslant t\leqslant 1}|\bar{\boldsymbol{\eta}}_n^C(t)-\boldsymbol{\eta}_n^C(t)|\\ d_0(\bar{\boldsymbol{\eta}}_n^D,\boldsymbol{\eta}_n^D)&\leqslant\sup_{0\leqslant t\leqslant 1}|\bar{\boldsymbol{\eta}}_n^D(t)-\boldsymbol{\eta}_n^D(t)|\end{aligned}\right\}\leqslant\frac{1}{\sigma\sqrt{n}}\sum_{i=1}^{p_n}|f_i|=\frac{p_n}{\sigma\sqrt{n}}\frac{\sum_{i=1}^{p_n}|f_i|}{p_n}.$$

By the choice of p_n, as a result of Proposition A3.3 and Theorem 4.1.2, the distances $d(\bar{\boldsymbol{\eta}}_n^C,\boldsymbol{\eta}_n^C)$ and $d_0(\bar{\boldsymbol{\eta}}_n^D,\boldsymbol{\eta}_n^D)$ converge both $\mathbf{P}_\infty$-a.s. and $\mathbf{P}_w$-a.s. to 0 as $n\to\infty$ for any $w\in W$. Then (4.1.3) and Proposition A1.22 yield

$$\mathbf{P}_\infty\bar{\boldsymbol{\eta}}_n^{-1}\Rightarrow\mathbf{B}.$$

Note now that for any $A\in\mathscr{B}_C(A\in\mathscr{B}_D)$ the random event $\{\bar{\boldsymbol{\eta}}_n\in A\}$ belongs to the σ-algebra $\sigma((\xi_i)_{i\geqslant p_n})$, so that Theorem 2.1.5 implies that

$$|\mathbf{P}_w(\bar{\boldsymbol{\eta}}_n\in A)-\mathbf{P}_\infty(\bar{\boldsymbol{\eta}}_n\in A)|\leqslant\varepsilon_{p_n}.$$

Letting $n\to\infty$ we get

$$\lim_{n\to\infty}\mathbf{P}_w(\bar{\boldsymbol{\eta}}_n\in A)=\mathbf{B}(A)$$

for all $w\in W$ and $A\in\mathscr{B}_C(A\in\mathscr{B}_D)$ with $\mathbf{B}(\partial A)=0$, i.e. $\mathbf{P}_w\bar{\boldsymbol{\eta}}_n^{-1}\Rightarrow\mathbf{B}$ for all $w\in W$. Using Proposition A1.22 again, we finally get $\mathbf{P}_w\boldsymbol{\eta}_n^{-1}\Rightarrow\mathbf{B}$ for all $w\in W$.

□

Theorem 4.1.5 has several important corollaries.

Corollary 4.1.6. (Central limit theorem) *For any $w \in W$ we have*

$$\lim_{n\to\infty} \mathbf{P}_w\left(\frac{S_n}{\sigma\sqrt{n}} < a\right) = \frac{1}{\sqrt{2\pi}} \int_{-\infty}^{a} e^{-u^2/2}\, du$$

for all $a \in \mathbb{R}$.

Proof. Consider the r.v. h on C defined as $h(\mathbf{x}) = \mathrm{pr}_1(\mathbf{x}) = \mathbf{x}(1)$, $\mathbf{x} \in C$. It is obvious that h is continuous. By Proposition A1.21 we can write

$$\mathbf{P}_w(h(\boldsymbol{\eta}_n^C))^{-1} = (\mathbf{P}_w(\boldsymbol{\eta}_n^C)^{-1})h^{-1} \Rightarrow \mathbf{B}h^{-1}.$$

As $h(\boldsymbol{\eta}_n^C) = S_n/\sigma\sqrt{n}$ and

$$\mathbf{B}h^{-1}((-\infty, a)) = \mathbf{B}(\mathbf{x}: \mathbf{x} \in C,\quad \mathbf{x}(1) < a) = \frac{1}{\sqrt{2\pi}} \int_{-\infty}^{a} e^{-u^2/2}\, du,$$

the proof is complete. □

For any $n \in \mathbb{N}^*$ put

$$m_n = \min_{0 \leqslant i \leqslant n} S_i, \quad M_n = \max_{0 \leqslant i \leqslant n} S_i.$$

Corollary 4.1.7. (Limit distribution of extrema of partial sums) *For any $w \in W$ the limit distribution of the random vector $(\sigma\sqrt{n})^{-1}(m_n, M_n, S_n)$ and its limit marginal distributions are given by*

$$\lim_{n\to\infty} \mathbf{P}_w(a\sigma\sqrt{n} < m_n \leqslant M_n < b\sigma\sqrt{n}, c\sigma\sqrt{n} < S_n < d\sigma\sqrt{n})$$
$$= \frac{1}{\sqrt{2\pi}}\left(\sum_{k\in\mathbb{Z}} \int_{c+2k(b-a)}^{d+2k(b-a)} e^{-u^2/2}\, du - \sum_{k\in\mathbb{Z}} \int_{2b-d+2k(b-a)}^{2b-c+2k(b-a)} e^{-u^2/2}\, du\right)$$

for all $a \leqslant 0 \leqslant b$ and $a \leqslant c < d \leqslant b$;

$$\lim_{n\to\infty} \mathbf{P}_w(a\sigma\sqrt{n} < m_n \leqslant M_n < b\sigma\sqrt{n}) = \frac{1}{\sqrt{2\pi}} \sum_{k\in\mathbb{Z}} (-1)^k \int_{a+k(b-a)}^{b+k(b-a)} e^{-u^2/2}\, du$$

for all $a \leqslant 0 \leqslant b$;

$$\lim_{n\to\infty} \mathbf{P}_w(a\sigma\sqrt{n} < m_n, c\sigma\sqrt{n} < S_n < d\sigma\sqrt{n})$$
$$= \frac{1}{\sqrt{2\pi}}\left[\int_{c}^{d} e^{-u^2/2}\, du - \int_{2a-d}^{2a-c} e^{-u^2/2}\, du\right]$$

for all $a \leqslant c < d$, $a \leqslant 0$;

$$\lim_{n\to\infty} \mathbf{P}_w(M_n < b\sigma\sqrt{n}, c\sigma\sqrt{n} < S_n < d\sigma\sqrt{n})$$
$$= \frac{1}{\sqrt{2\pi}}\left[\int_{c}^{d} e^{-u^2/2}\, du - \int_{2b-d}^{2b-c} e^{-u^2/2}\, du\right]$$

for all $c < d \leqslant b$, $b \geqslant 0$;

$$\lim_{n\to\infty} \mathbf{P}_w(a\sigma\sqrt{n} < m_n) = \sqrt{\frac{2}{\pi}} \int_a^0 e^{-u^2/2}\, du$$

for all $a \leqslant 0$;

$$\lim_{n\to\infty} \mathbf{P}_w(M_n < b\sigma\sqrt{n}) = \sqrt{\frac{2}{\pi}} \int_0^b e^{-u^2/2}\, du$$

for all $b \geqslant 0$.

For any $w \in W$ *the limit distribution of the r.v.* $(\sigma\sqrt{n})^{-1} \max_{0\leqslant i\leqslant n}|S_i|$ *is given by*

$$\begin{aligned}
&\lim_{n\to\infty} \mathbf{P}_w(\max_{0\leqslant i\leqslant n} |S_i| < b\sigma\sqrt{n}) \\
&= \frac{1}{\sqrt{2\pi}} \sum_{k\in\mathbb{Z}} (-1)^k \int_{(2k-1)b}^{(2k+1)/b} e^{-u^2/2}\, du \\
&= \frac{4}{\pi} \sum_{k\in\mathbb{N}} \frac{(-1)^k}{2k+1} e^{-\pi^2(2k+1)^2/8b^2}
\end{aligned}$$

for all $b \geqslant 0$.

By continuity, all strict inequalities above can be weakened.

Proof. Consider the r.v.s h_1, h_2, h_3 on C defined as $h_1(\mathbf{x}) = \inf_{0\leqslant t\leqslant 1} \mathbf{x}(t)$, $h_2(\mathbf{x}) = \sup_{0\leqslant t\leqslant 1} \mathbf{x}(t)$, $h_3(x) = \mathbf{x}(1)$, $\mathbf{x} \in C$. Clearly, all of them are continuous, hence $h = (h_1, h_2, h_3)$ is a continuous map from C into $\mathbb{R}^3$. By Proposition A1.21 we can write

$$\mathbf{P}_w(h\boldsymbol{\eta}_n^C))^{-1} = (\mathbf{P}_w(\boldsymbol{\eta}_n^C)^{-1})h^{-1} \Rightarrow \mathbf{B}h^{-1}.$$

But $h(\boldsymbol{\eta}_n^C) = (\sigma\sqrt{n})^{-1}(m_n, M_n, S_n)$ and for any $a \leqslant 0 \leqslant b$ and $a \leqslant c < d \leqslant b$ we have (see Billingsley (1968, §11))

$$\begin{aligned}
&\mathbf{B}h^{-1}((a,0),(b,\infty),(c,d)) \\
&= \mathbf{B}(\mathbf{x}: \mathbf{x}\in C,\ \inf_{0\leqslant t\leqslant 1} \mathbf{x}(t) > a,\ \sup_{0\leqslant t\leqslant 1} \mathbf{x}(t) < b, c < \mathbf{x}(1) < d) \\
&= \frac{1}{\sqrt{2\pi}}\left(\sum_{k\in\mathbb{Z}} \int_{c+2k(b-a)}^{d+2k(b-a)} e^{-u^2/2}\, du - \sum_{k\in\mathbb{Z}} \int_{2b-d+2k(b-a)}^{2b-c+2k(b-a)} e^{-u^2/2}\, du\right).
\end{aligned}$$

To obtain the limit marginal distributions we should successively take $c = a$ and $d = b$; $b = \infty$; $a = -\infty$; $c = a$ and $d = b = \infty$; $c = a = -\infty$ and $d = b$.

To obtain the last limit distribution we should take $a = -b$ in the equation giving the limit marginal distribution for the r.v.s m_n and M_n. (For the transformation of the series of normal integrals into a series of exponentials see Feller (1966, pp. 330 and 594).) □

For any $n \in \mathbb{N}^*$ put

$$T_n = \max\{1 \leqslant i \leqslant n: \{S_i = 0\} \cup \{S_{i-1} > 0 > S_i\} \cup \{S_{i-1} < 0 < S_i\}\}$$
$$= \text{last moment when } S_1, \ldots, S_n \text{ cross level } 0;$$
$$U_n = \text{number of indices } 1 \leqslant i \leqslant n \text{ for which } S_i > 0.$$

Corollary 4.1.8. (Arc sine law) *For any* $w \in W$ *we have*

$$\lim_{n\to\infty} \mathbf{P}_w\left(\frac{T_n}{n} < a\right) = \lim_{n\to\infty} \mathbf{P}_w\left(\frac{U_n}{n} < a\right) = \frac{2}{\pi}\arcsin\sqrt{a}$$

for all $0 \leqslant a \leqslant 1$.

Proof. Consider the r.v.s h_1 and h_2 on D defined as

$$h_1(\mathbf{x}) = \begin{cases} \inf\{0 \leqslant t \leqslant 1: \mathbf{x}(s)\mathbf{x}(1) > 0, t \leqslant s \leqslant 1\}, & \text{if } \mathbf{x}(1) \neq 0 \\ 1 & \text{if } \mathbf{x}(1) = 0, \end{cases}$$

$$h_2(\mathbf{x}) = \text{Lebesgue measure of } \{0 \leqslant t \leqslant 1: \mathbf{x}(t) > 0\},\ \mathbf{x} \in D.$$

It can be proved (see Billingsley (1968, Appendix 2)) that both h_1 and h_2 are $\mathbf{B}$-a.s. continuous on D. By Proposition A1.21 we can write

$$\mathbf{P}_w(h_i(\boldsymbol{\eta}_n^D))^{-1} = (\mathbf{P}_w(\boldsymbol{\eta}_n^D)^{-1})h_i^{-1} \Rightarrow \mathbf{B}h_i^{-1}, \quad i = 1, 2.$$

As

$$h_1(\boldsymbol{\eta}_n^D) = \begin{cases} \dfrac{T_n}{n}, & \text{if } S_{T_n} \neq 0 \\ \dfrac{T_n + 1}{n}, & \text{if } S_{T_n} = 0, \end{cases} \qquad h_2(\boldsymbol{\eta}_n^D) = \frac{U_n}{n},$$

and (see Billingsley (1968, §11))

$$\mathbf{B}h_1^{-1}((0, a)) = \mathbf{B}(\mathbf{x}: \mathbf{x} \in C, 0 < h_1(\mathbf{x}) < a) = \frac{2}{\pi}\arcsin\sqrt{a},$$

$$\mathbf{B}h_2^{-1}((0, a)) = \mathbf{B}(\mathbf{x}: \mathbf{x} \in C, 0 < h_2(\mathbf{x}) < a) = \frac{2}{\pi}\arcsin\sqrt{a},$$

the proof is complete. □

Remark. Putting V_n = number of indices $1 \leqslant i \leqslant T_n$ for which $S_i > 0$, it can be shown (*cf.* Billingsley (1968, §11)) that for any $w \in W$ we have

$$\lim_{n\to\infty} \mathbf{P}_w\left(\frac{T_n}{n} < t, \frac{V_n}{n} < v, \frac{S_n}{\sigma\sqrt{n}} < a\right)$$
$$= \frac{1}{2\pi}\int_0^v \mathrm{d}z \int_z^t \mathrm{d}y \int_{-\infty}^a \frac{|x|}{(y(1-y))^{3/2}} \exp\left(-\frac{x^2}{2(1-y)}\right)\mathrm{d}x$$

for all $0 \leqslant v < t \leqslant 1$, $a \in \mathbb{R}$. Hence, for any $w \in W$,

$$\lim_{n\to\infty} \mathbf{P}_w\left(\frac{V_n}{n} < v\right) = \frac{2}{\pi}(\arcsin\sqrt{v} + \sqrt{v(1-v)})$$

for all $0 \leqslant v \leqslant 1$.

Theorem 4.1.5 also leads to a functional central limit theorem for the renewal function.

In the context of Subsection 4.1.1 assume that F is non-negative and that $a = \mathbf{E}_\infty(f_1) > 0$. If $\mathbf{E}_\infty(|f_1|^{2+\delta}) < \infty$ for some $\delta \geqslant 0$ (or $F \leqslant c < \infty$), $\sum_{n\in\mathbb{N}^*} \varepsilon_n^{(1+\delta)/(2+\delta)} < \infty$ (or $\sum_{n\in\mathbb{N}^*} \varepsilon_n < \infty$), and

$$\sum_{n\in\mathbb{N}^*} \mathbf{E}_\infty^{1/2}([f_1 - \mathbf{E}_\infty(f_1|(\xi_i)_{1\leqslant i\leqslant n})]^2) < \infty,$$

then by Theorem 4.1.5 the series

$$\sigma_1^2 = \mathbf{E}_\infty(f_1^2) - a^2 + 2\sum_{n\in\mathbb{N}^*}(\mathbf{E}_\infty(f_1 f_{n+1}) - a^2)$$

is absolutely convergent and $\sigma_1^2 \geqslant 0$. Consider the renewal function

$$v(t) = \max\{k: k\in\mathbb{N}, S_k \leqslant t\}, \quad t \geqslant 0,$$

and, if $\sigma_1 > 0$, for any $n \in \mathbb{N}^*$ define the stochastic process

$$\mathbf{v}_n: \mathbf{v}_n(t) = \frac{v(nt) - nt/a}{\sigma_1 a^{-3/2}\sqrt{n}}, \quad 0 \leqslant t \leqslant 1.$$

Clearly, $\mathbf{v}_n$ is a D-valued r.v.

Corollary 4.1.9. *If $\sigma_1 > 0$, then for any $w \in W$ we have $\mathbf{P}_w \mathbf{v}_n^{-1} \Rightarrow \mathbf{B}$.*

Proof. This is an immediate consequence of Theorem 4.1.5 and Billingsley (1968, Theorem 17.3). □

The next result shows that in the statement of Theorem 4.1.5 the probabilities $\mathbf{P}_w$ can conveniently be replaced.

Theorem 4.1.10. *Let $w \in W$ be arbitrarily fixed. Then Theorem 4.1.5 is still true when $\mathbf{P}_w$ is replaced by any probability $\mathbf{Q} \ll \mathbf{P}_w$.*

Proof. We have to prove that, under the assumptions of Theorem 4.1.5, $\mathbf{Q}\boldsymbol{\eta}_n^{-1} \Rightarrow \mathbf{B}$ if $\mathbf{Q} \ll \mathbf{P}_w$ for a $w \in W$ arbitrarily fixed. Let $A \in \mathscr{B}_C$ ($A \in \mathscr{B}_D$) for which $\mathbf{B}(\partial A) = 0$ and $E \in \sigma((\xi_i)_{1\leqslant i\leqslant l})$, $l \in \mathbb{N}^*$. With $\bar{\boldsymbol{\eta}}_n^C$ and $\bar{\boldsymbol{\eta}}_n^D$ defined as in the proof of Theorem 4.1.5, we have $\bar{\boldsymbol{\eta}}_n^{-1}(A) \in \sigma((\xi_i)_{p_n\leqslant i\leqslant n})$. Therefore, by Theorem 2.1.5, for n large enough,

$$|\mathbf{P}_w(\{\bar{\boldsymbol{\eta}}_n \in A\} \cap E) - \mathbf{P}_w(\bar{\boldsymbol{\eta}}_n \in A)\mathbf{P}_w(E)| \leqslant 2\varepsilon_{p_n - l + 1}.$$

Hence

$$\lim_{n\to\infty} \mathbf{P}_w(\{\bar{\boldsymbol{\eta}}_n \in A\} \cap E) = \mathbf{B}(A)\mathbf{P}_w(E). \tag{4.1.4}$$

Now, remark that the collection of the sets E of the type considered is an algebra which generates the σ-algebra $\sigma((\xi_i)_{i\in\mathbb{N}^*})$. Then (4.1.4) implies that

$$\lim_{n\to\infty} \mathbf{Q}(\bar{\boldsymbol{\eta}}_n \in A) = \mathbf{B}(A),$$

i.e. $\mathbf{Q}\bar{\boldsymbol{\eta}}_n^{-1} \Rightarrow \mathbf{B}$.[†]

As we have seen in the proof of Theorem 4.1.5, both $d(\bar{\boldsymbol{\eta}}_n^C, \boldsymbol{\eta}_n^C)$ and $d_0(\bar{\boldsymbol{\eta}}_n^D, \boldsymbol{\eta}_n^D)$ converge $\mathbf{P}_w$-a.s. to 0 as $n \to \infty$. Therefore, since $\mathbf{Q} \ll \mathbf{P}_w$ there will also be $\mathbf{Q}$-a.s. convergence to 0 of the two distances. To conclude the proof it remains to apply Proposition A1.22. □

The result below shows that convergence to Wiener measure still holds when the partial sums are selected at random.

Theorem 4.1.11. *Let ν_n, $n\in\mathbb{N}^*$, be positive integer valued r.v.s on Ω such that for any $w\in W$ we have $\nu_n/a_n \xrightarrow{\mathbf{P}_w} \theta$, where θ is a positive r.v., and $(a_n)_{n\in\mathbb{N}^*}$ is a sequence of positive numbers converging to ∞. Under the assumptions of Theorem 4.1.5, we have $\mathbf{P}_w\boldsymbol{\rho}_n^{-1} \Rightarrow \mathbf{B}$, where $\boldsymbol{\rho}_n$ stands for either of the processes*

$$\boldsymbol{\rho}_n^C : \boldsymbol{\rho}_n^C(t,\omega) = \frac{1}{\sigma\sqrt{\nu_n(\omega)}}(S_{[\nu_n(\omega)t]}(\omega) + (\nu_n(\omega)t - [\nu_n(\omega)t])f_{[\nu_n(\omega)t]+1}),$$

$$t\in[0,1], \quad \omega\in\Omega, \quad n\in\mathbb{N}^*,$$

and

$$\boldsymbol{\rho}_n^D : \boldsymbol{\rho}_n^D(t,\omega) = \frac{1}{\sigma\sqrt{\nu_n(\omega)}} S_{[\nu_n(\omega)t]}(\omega), \quad t\in[0,1], \quad \omega\in\Omega, \quad n\in\mathbb{N}^*.$$

The proof is completely similar to that of Theorem 17.2 in Billingsley (1968), using equation (4.1.4) and the fact that both $d(\bar{\boldsymbol{\eta}}_n^C, \boldsymbol{\eta}_n^C)$ and $d_0(\bar{\boldsymbol{\eta}}_n^D, \boldsymbol{\eta}_n^D)$ converge $\mathbf{P}_w$-a.s. to 0 as $n \to \infty$ (see the proof of Theorem 4.1.5).

Note that all the corollaries of Theorem 4.1.5 can be transposed to the context of Theorem 4.1.11, yielding the corresponding results for partial sums selected at random.

[†] Here we use the following result (Billingsley (1968, Theorem 16.2)). Let $\mathscr{K}_0$ be an algebra of random events in the probability space $(\Omega, \mathscr{K}, \mathbf{P})$. If $E_n \in \sigma(\mathscr{K}_0)$, $n\in\mathbb{N}^*$, and $\lim_{n\to\infty} \mathbf{P}(E_n \cap E) = \alpha \mathbf{P}(E)$ for all $E\in\mathscr{K}_0$, then for any probability $\mathbf{Q} \ll \mathbf{P}$, $\lim_{n\to\infty} \mathbf{Q}(E_n) = \alpha$.

4.1.4 The central limit theorem with remainder

This subsection is devoted to the estimation of the rate of convergence to the normal distribution in the central limit theorem (Corollary 4.1.6). We shall assume that the r.v. F only depends on finitely many coordinates of a current point of $X^{\mathbb{N}^*}$, i.e. F is a real r.v. on X^k, for a given $k \in \mathbb{N}^*$. Therefore, in this subsection, we have $f_n = F(\xi_n, \ldots, \xi_{n+k-1})$, $n \in \mathbb{N}^*$, and, as it is easy to see, the sequence $(f_n)_{n\in\mathbb{N}^*}$ is φ-mixing under both $\mathbf{P}_\infty$ and $\mathbf{P}_w$, for any $w \in W$. Put $S_{m,n} = \sum_{i=m}^{m+n-1} f_i$, $m, n \in \mathbb{N}^*$, $S_n = S_{1,n}$.

Theorem 4.1.12. *Assume that $\mathbf{E}_w(|f_n|^{2+\delta}) \leqslant c < \infty$, $n \in \mathbb{N}^*$, for $w \in W$ and $\delta > 0$ arbitrarily fixed (or $|F| \leqslant c < \infty$), and that $\sum_{n\in\mathbb{N}^*} \varepsilon_n^{(1+\delta)/(2+\delta)} < \infty$ (or $\sum_{n\in\mathbb{N}^*} \varepsilon_n < \infty$). Then the series*

$$\sigma^2 = \mathbf{E}_\infty(f_1^2) - \mathbf{E}_\infty^2(f_1) + 2 \sum_{n\in\mathbb{N}^*} (\mathbf{E}_\infty(f_1 f_{n+1}) - \mathbf{E}_\infty^2(f_1))$$

converges absolutely and $\sigma^2 \geqslant 0$. If $\sigma > 0$, then there exist two positive constants $v < 1$ and c_0 (depending on F, δ, and c) such that

$$\left| \mathbf{P}_w\left(\frac{S_{m,n} - n\mathbf{E}_\infty(f_1)}{\sigma\sqrt{n}} < a \right) - \Phi(a) \right| < c_0 n^{-v}$$

for all $a \in \mathbb{R}$ and $m, n \in N^*$.

Proof. The properties of σ^2 follow from Proposition 4.1.1 and Theorem 4.1.5.

Without any loss of generality we can assume that $\mathbf{E}_\infty(f_1) = 0$. Let $r \in \mathbb{N}$, $r < n$. Since $S_{m+r,n-r}$ is $\sigma((\xi_i)_{m+r\leqslant i\leqslant m+n+k-2})$-measurable, by uniform ergodicity, we have

$$\left| \mathbf{P}_{w'}\left(\frac{S_{m+r,n-r}}{\sigma\sqrt{n}} < a \right) - \mathbf{P}_\infty\left(\frac{S_{n-r}}{\sigma\sqrt{n}} < a \right) \right| \leqslant \varepsilon_{r+1} \tag{4.1.5}$$

for all $w' \in W$, $m, n \in \mathbb{N}^*$, and $a \in \mathbb{R}$. Consider the random events

$$E_1 = \{S_{m+r,n-r} < a\sigma\sqrt{n} - \tau\}, \quad E_2 = \{-\tau < S_{m,r} < \tau\},$$
$$E_3 = \{S_{m,n} < a\sigma\sqrt{n}\}, \quad E_4 = \{S_{m+r,n-r} < a\sigma\sqrt{n} + \tau\},$$

where $\tau > 0$ is arbitrary but fixed. We have $E_1 \cap E_2 \subset E_3$ and $E_2 \cap E_3 \subset E_4$. Hence

$$\mathbf{P}_w(E_3) \geqslant \mathbf{P}_w(E_1 \cap E_2) \geqslant \mathbf{P}_w(E_1) + \mathbf{P}_w(E_2) - 1,$$
$$\mathbf{P}_w(E_4) \geqslant \mathbf{P}_w(E_2 \cap E_3) \geqslant \mathbf{P}_w(E_2) + \mathbf{P}_w(E_3) - 1,$$

and, since by Chebyshev's inequality

$$\mathbf{P}_w(E_2) \geqslant 1 - \frac{\mathbf{E}_w(S_{m,r}^2)}{\tau^2},$$

we can write

$$\mathsf{P}_w(E_1) - \frac{\mathsf{E}_w(S_{m,r}^2)}{\tau^2} \leqslant \mathsf{P}_w(E_3) \leqslant \mathsf{P}_w(E_4) + \frac{\mathsf{E}_w(S_{m,r}^2)}{\tau^2}.$$

By (4.1.5), we have

$$\left|\mathsf{P}_w(E_1) - \mathsf{P}_\infty\left(\frac{S_{n-r}}{\sigma\sqrt{n}} < a - \frac{\tau}{\sigma\sqrt{n}}\right)\right| \leqslant \varepsilon_{r+1},$$

$$\left|\mathsf{P}_w(E_4) - \mathsf{P}_\infty\left(\frac{S_{n-r}}{\sigma\sqrt{n}} < a + \frac{\tau}{\sigma\sqrt{n}}\right)\right| \leqslant \varepsilon_{r+1}.$$

On the other hand, Theorem A3.5 implies that

$$\left|\mathsf{P}_\infty\left(\frac{S_{n-r}}{\sigma\sqrt{n}} < a \pm \frac{\tau}{\sigma\sqrt{n}}\right) - \Phi\left(\sqrt{\frac{n}{n-r}}a \pm \frac{\tau}{\sigma\sqrt{n-r}}\right)\right| \leqslant c_1(n-r)^{-v_0}$$

with suitable positive constants $v_0 < 1$ and c_1.

Remark that

$$\Phi\left(\sqrt{\frac{n}{n-r}}a + \frac{\tau}{\sigma\sqrt{n-r}}\right) \leqslant \Phi\left(\sqrt{\frac{n}{n-r}}a\right) + \frac{\tau}{\sigma\sqrt{n-r}},$$

$$\Phi\left(\sqrt{\frac{n}{n-r}}a - \frac{\tau}{\sigma\sqrt{n-r}}\right) \geqslant \Phi\left(\sqrt{\frac{n}{n-r}}a\right) - \frac{\tau}{\sigma\sqrt{n-r}},$$

and that†

$$\left|\Phi\left(\sqrt{\frac{n}{n-r}}a\right) - \Phi(a)\right| \leqslant \frac{1}{\sqrt{2\pi e}}\left|1 - \sqrt{\frac{n}{n-r}}\right|.$$

Finally, note that

$$\mathsf{E}_w(S_{m,r}^2) \leqslant r\sum_{n=m}^{m+r-1} \mathsf{E}_w(f_n^2) \leqslant r^2 \max(c, 1).$$

† For $k > 0$ we have

$$|\Phi(a) - \Phi(ka)| \leqslant \frac{1}{\sqrt{2\pi e}}|k-1| \max\left(1, \frac{1}{k}\right), \quad a \in \mathbb{R}.$$

Indeed

$$\Phi(a) - \Phi(ka) = \frac{1}{\sqrt{2\pi}}\int_{ka}^{a} e^{-x^2/2}\,dx$$

and, obviously,

$$\left|\int_{ka}^{a} e^{-x^2/2}\,dx\right| \leqslant |k-1||a| \max(e^{-a^2/2}, e^{-k^2a^2/2}).$$

It remains to note that $|a|e^{-a^2/2} \leqslant 1/\sqrt{e}$ for all $a \in \mathbb{R}$.

We therefore have

$$\left|\mathbf{P}_w\left(\frac{S_{m,n}}{\sigma\sqrt{n}}<a\right)-\Phi(a)\right|\leqslant \varepsilon_{r+1}+\frac{1}{\sqrt{2\pi e}}\left|1-\sqrt{\frac{n}{n-r}}\right|$$

$$+c_1(n-r)^{-v_0}+\frac{\tau}{\sigma\sqrt{n-r}}+\max(c,1)\frac{r^2}{\tau^2}.$$

To conclude the proof we should conveniently choose r and τ. (We can, for example, take $r=[n^{1/4}]$ and $\tau=[n^{1/3}]$.) □

4.1.5 The law of the iterated logarithm

In this subsection, using results from Section A3.1, we shall prove Strassen's law of the iterated logarithm. As in the preceding subsection, we assume that F is a real r.v. on X^k for a given $k\in\mathbb{N}^*$, thus $f_n=F(\xi_n,\ldots,\xi_{n+k-1})$, $n\in\mathbb{N}^*$.

Theorem 4.1.13. *Assume, without any loss of generality, that $\mathbf{E}_\infty(f_1)=0$. If $\mathbf{E}_\infty(|f_1|^{2+\delta})<\infty$ for some $\delta\geqslant 0$ (or $|F|\leqslant c<\infty$) and $\sum_{n\in\mathbb{N}^*}\varepsilon_n^{(1+\delta)/(2+\delta)}<\infty$ (or $\sum_{n\in\mathbb{N}^*}\varepsilon_n<\infty$), then the series*

$$\sigma^2=\mathbf{E}_\infty(f_1^2)+2\sum_{n\in\mathbb{N}^*}\mathbf{E}_\infty(f_1f_{n+1})$$

converges absolutely and $\sigma^2\geqslant 0$. If $\sigma>0$, then for any $w\in W$ the sequence

$$\left(\frac{\boldsymbol{\eta}_n^C}{\sqrt{2\log\log n}}\right)_{n\geqslant 3}$$

viewed as a subset of C, is a relatively compact set, whose derived set coincides $\mathbf{P}_w$-a.s. with K, the collection of all absolutely continuous functions $\mathbf{x}\in C$ for which $\mathbf{x}(0)=0$ and $\int_0^1[\mathbf{x}'(t)]^2\,dt\leqslant 1$.

Proof. If we denote by E the random event occurring in the statement of the theorem, then by Theorem A3.9 we have $\mathbf{P}_\infty(E)=1$. On the other hand, E belongs to the tail σ-algebra $\bigcap_{n\in\mathbb{N}^*}\sigma((\xi_i)_{i\geqslant n})$. By Theorem 2.1.5 we have $\mathbf{P}_w(E)=\mathbf{P}_\infty(E)$ for all $w\in W$. □

Theorem 4.1.13 has many important consequences. Apart from the classical law of the iterated logarithm (Corollary A3.10), which asserts that for any $w\in W$ the set of accumulation points of the sequence $((S_n-n\mathbf{E}_\infty(f_1))/\sigma\sqrt{2n\log\log n})_{n\geqslant 3}$ coincides $\mathbf{P}_w$-a.s. with the segment $[-1,1]$, we also note the two results below.

Corollary 4.1.14. *For any $a \geqslant 1$, $m \in \mathbb{N}^*$, and $w \in W$ we have*

$$\mathbf{P}_w\left(\limsup_{n\to\infty}\frac{\sum_{s=1}^n |S_{m,s}-s\mathbf{E}_\infty(f_1)|^a}{n^{1+a/2}(2\sigma^2\log\log n)^{a/2}}=\frac{2(a+2)^{a/2-1}}{\left(\int_0^1\frac{dt}{\sqrt{1-t^a}}\right)^a a^{a/2}}\right)=1.$$

(For $a=1$ and $a=2$ the values of lim sup *are* $\frac{1}{3}$ *and* $4/\pi^2$, *respectively.)*

Corollary 4.1.15. Let h be a real Riemann integrable function on $[0,1]$. *For any $m \in \mathbb{N}^*$ and $w \in W$ we have*

$$\mathbf{P}_w\left(\limsup_{n\to\infty}\frac{\sum_{s=1}^n h\left(\frac{s}{n}\right)(S_{m,s}-s\mathbf{E}_\infty(f_1))}{\sigma n\sqrt{2n\log\log n}}=\sqrt{\int_0^1\left(\int_t^1 h(u)du\right)^2 dt}\right)=1.$$

4.1.6 Some non-parametric statistics

In this subsection we shall prove the fundamental theorem of mathematical statistics (Glivenko–Cantelli) as well as an invariance principle for empirical distribution functions.

To begin with, we shall assume that F is a real s-dimensional random vector on X^k for a given $k \in \mathbb{N}^*$, thus $f_n = F(\xi_n,\ldots,\xi_{n+k-1})$ is a real s-dimensional random vector for any $n \in \mathbb{N}^*$. For $u^{(s)}=(u_1,\ldots,u_s)\in\mathbb{R}^s$, $a\in\mathbb{R}$, let us put

$$F_\infty(a|u^{(s)})=\mathbf{P}_k^\infty\left(\left\{x^{(k)}:\sum_{i=1}^s u_i \mathrm{pr}_i F(x^{(k)})\leqslant a\right\}\right)$$

and (see Subsection 4.1.2)

$$F_n^*(a|u^{(s)})=\frac{1}{n}\nu_{n,k}\left(\left\{x^{(k)}:\sum_{i=1}^s u_i \mathrm{pr}_i F(x^{(k)})\leqslant a\right\}\right).$$

Theorem 4.1.16. *For any $w \in W$ we have*

$$\mathbf{P}_w\left(\lim_{n\to\infty}\sup_{u^{(s)}\in\mathbb{R}^s}\sup_{a\in\mathbb{R}}|F_n^*(a|u^{(s)})-F_\infty(a|u^{(s)})|=0\right)=1.$$

Proof. By Corollary 4.1.3, for any $u^{(s)}\in\mathbb{R}^s$ and $w\in W$, the empirical distribution function $F_n^*(a|u^{(s)})$ converges $\mathbf{P}_w$-a.s. to $F_\infty(a|u^{(s)})$ as $n\to\infty$. Now the result stated follows as in the case of independent identically distributed r.v.s (see Wolfowitz (1954)). □

Further, for the sake of clarity, we shall limit ourselves to the case $s=1$, i.e. we return to the framework of the previous two subsections. In addition, we shall assume that $0\leqslant F\leqslant 1$, which in the present context is not a restriction. For $0 \leqslant t \leqslant 1$ let us put

$$F_\infty(t) = F_\infty(t|1) = \mathbf{P}_k^\infty(\{x^{(k)}: F(x^{(k)}) \leqslant t\}) = \mathbf{P}_\infty(f_1 \leqslant t),$$

$$F_n^*(t) = F_n^*(t|1) = \frac{1}{n} v_{n,k}(\{x^{(k)}: F(x^{(k)}) \leqslant t\}).$$

Therefore, $F_n^*(t)$ is the relative frequency of the values f_i, $1 \leqslant i \leqslant n$, which do not exceed t.

For any $n \in \mathbb{N}^*$ consider the stochastic process

$$\mathbf{Y}_n: \mathbf{Y}_n(t) = \sqrt{n}(F_n^*(t) - F_\infty(t)), \quad 0 \leqslant t \leqslant 1.$$

Theorem 4.1.17. *For any $w \in W$, if $\sum_{n \in \mathbb{N}^*} \varepsilon_n < \infty$ and F_∞ is a continuous function, then $\mathbf{P}_w \mathbf{Y}_n^{-1} \Rightarrow \mathbf{G}$ (in D), where $\mathbf{G}$ is a probability on $\mathscr{B}_D$, whose support is C. Under $\mathbf{G}$ the stochastic process $(\mathrm{pr}_t)_{0 \leqslant t \leqslant 1}$ is a Gaussian process, whose trajectories are a.s. continuous, of mean 0 and covariance $r(s,t) = \mathbf{E}_\infty(g_s(f_1) g_t(f_1)) + \sum_{n \in \mathbb{N}^*} \mathbf{E}_\infty(g_s(f_1) g_t(f_{n+1})) + \sum_{n \in \mathbb{N}^*} \mathbf{E}_\infty(g_t(f_1) g_s(f_{n+1}))$, $0 \leqslant s$, $t \leqslant 1$, both series being absolutely convergent, where*

$$g_t(u) = \begin{cases} 1 - F_\infty(t), & \text{if } 0 \leqslant u \leqslant t \\ -F_\infty(t), & \text{if } t < u \leqslant 1. \end{cases}$$

The conclusion still holds when, for w arbitrarily fixed, $\mathbf{P}_w$ is replaced by any probability $\mathbf{Q} \ll \mathbf{P}_w$.

Proof. We can write

$$\mathbf{Y}_n(t) = \frac{1}{\sqrt{n}} \sum_{k=1}^n g_t(f_k).$$

Let us define

$$\bar{\mathbf{Y}}_n(t) = \frac{1}{\sqrt{n}} \sum_{k=p_n}^n g_t(f_k), \quad 0 \leqslant t \leqslant 1,$$

where $(p_n)_{n \in \mathbb{N}^*}$ is a sequence of natural integers converging to ∞ and for which $1 > p_n/\sqrt{n} \to 0$ as $n \to \infty$. We have

$$d_0(\bar{\mathbf{Y}}_n, \mathbf{Y}_n) \leqslant \sup_{0 \leqslant t \leqslant 1} |\bar{\mathbf{Y}}_n(t) - \mathbf{Y}_n(t)| \leqslant \frac{1}{\sqrt{n}} \sup_{0 \leqslant t \leqslant 1} \sum_{k=1}^{p_n - 1} |g_t(f_k)|.$$

Since $|g_t(f_k)| \leqslant 1$, $0 \leqslant t \leqslant 1$, $k \in \mathbb{N}^*$, we deduce that $d_0(\bar{\mathbf{Y}}_n, \mathbf{Y}_n)$ converges to 0 as $n \to \infty$ both $\mathbf{P}_\infty$-a.s. and $\mathbf{P}_w$-a.s., for any $w \in W$. On the other hand, by Theorems 2.1.5 and A3.8, we have $\mathbf{P}_\infty \bar{\mathbf{Y}}_n^{-1} \Rightarrow \mathbf{G}$. From now on the proof is completely similar to that of Theorem 4.1.5, and is, therefore, left to the reader. The possibility of replacing $\mathbf{P}_w$ by $\mathbf{Q} \ll \mathbf{P}_w$ can be proved along the lines of the proof of Theorem 4.1.10. □

From Theorem 4.1.17 we can obtain results corresponding to the classical non-parametric Kolmogorov–Simirnov criteria. Consider the real r.v.s h_1 and h_2 on D defined as

$$h_1(\mathbf{x}) = \sup_{0 \leqslant t \leqslant 1} |\mathbf{x}(t)|, \quad h_2(\mathbf{x}) = \sup_{0 \leqslant t \leqslant 1} \mathbf{x}(t), \quad \mathbf{x} \in D.$$

Since the support of $\mathbf{G}$ is the whole of C, h_1 and h_2 are $\mathbf{G}$-a.s. continuous and Proposition A1.21 allows us to state Corollary 4.1.18.

Corollary 4.1.18. *Under the assumptions of Theorem* 4.1.17, *for any* $w \in W$ *we have*

$$\lim_{n\to\infty} \mathbf{P}_w(\sqrt{n} \sup_{0 \leqslant t \leqslant 1} |F_n^*(t) - F_\infty(t)| \leqslant \theta) = \mathbf{G} h_1^{-1}([0,\theta])$$

$$\lim_{n\to\infty} \mathbf{P}_w(\sqrt{n} \sup_{0 \leqslant t \leqslant 1} (F_n^*(t) - F_\infty(t)) \leqslant \theta) = \mathbf{G} h_2^{-1}([0,\theta])$$

for all $\theta > 0$.

Now, a natural problem is to find conditions under which the values of the limits in Corollary 4.1.18 can be expressed in simple closed forms. Until now the only case where this appears to be possible, is that where the terms of the two series occurring in Theorem 4.1.17 are 0, i.e.

$$r(s,t) = \mathbf{E}_\infty(g_s(f_1)g_t(f_1))(= F_\infty(s)(1 - F_\infty(t)), s \leqslant t,$$

as in the classical case where the f_n, $n \in \mathbb{N}^*$, are independent under $\mathbf{P}_\infty$. In this case the values of the two limits are $k(\theta) = 1 + 2\sum_{n\in\mathbb{N}^*}(-1)^n e^{-2n^2\theta^2}$, and $1 - e^{-2\theta^2}$, respectively (see Billingsley (1968, §11)).

Note that $k(\theta)$ can also be written as $k(\theta) = \prod_{n\in\mathbb{N}^*}(1 - e^{-2n\theta^2})/(1 + e^{-2n\theta^2})$ and that $k(\theta)$ is the distribution function of the r.v. $(\sum_{n\in\mathbb{N}^*} e_n^2/2n^2)^{1/2}$, where the e_n, $n \in \mathbb{N}^*$, are independent identically distributed exponential r.v.s of mean 1 (see Chung (1982)).

4.2 Limit theorems for the associated Markov chain

4.2.1 Preliminaries

Throughout this section we consider the MC $(\zeta_n)_{n\in\mathbb{N}}$ associated with an RSCC with contraction for which the space (W, d) is compact. We assume that the chain is regular w.r.t. $L(W)$. We saw (see Theorems 3.1.24 and 3.2.4, and Lemma 3.1.19) that under these assumptions there exists a probability $\mathbf{Q}^\infty$ on $\mathscr{W}$ such that

$$|U^n f - U^\infty f| \leqslant \varepsilon_n \|f\|_L, \quad f \in L(W), \tag{4.2.1}$$

where

$$U^\infty f = \int_W f(w)\mathbf{Q}^\infty(\mathrm{d}w)$$

with $\varepsilon_n = O(\alpha^n)$ for some $0 \leqslant \alpha < 1$ as $n \to \infty$.

Let us remark (see Theorem 3.4.5) that the equation

$$\mathbf{P}_l^\infty(A) = \int_W \mathbf{Q}^\infty(\mathrm{d}w) P_l(w, A), \quad l \in \mathbb{N}^*, \quad A \in \mathscr{X}^l, \tag{4.2.2}$$

enables us to conclude that

$$\mathbf{P}_\infty(E) = \int_W \mathbf{Q}^\infty(\mathrm{d}w)\mathbf{P}_w(E), \quad E \in \mathscr{K}. \tag{4.2.3}$$

Indeed, (4.2.2) shows that (4.2.3) holds for any $E \in \sigma((\xi_i)_{1 \leqslant i \leqslant l})$ since for such an E we have

$$\mathbf{P}_\infty(E) = \mathbf{P}_l^\infty(\mathrm{pr}_{\{1,\ldots,l\}} E), \quad \mathbf{P}_w(E) = P_l(w, \mathrm{pr}_{\{1,\ldots,l\}} E)$$

(see the proofs of Theorems 1.1.2 and 2.1.5). Therefore (4.2.3) holds for any set E belonging to the algebra $\bigcup_{l \in \mathbb{N}^*} \sigma((\xi_i)_{1 \leqslant i \leqslant l}) \subset \mathscr{K}$. If $(E_n)_{n \in \mathbb{N}^*}$ is a non-decreasing sequence of sets in this algebra, then

$$\lim_{n\to\infty} \mathbf{P}_\infty(E_n) = \mathbf{P}_\infty\left(\bigcup_{n \in \mathbb{N}^*} E_n\right), \quad \lim_{n\to\infty} \mathbf{P}_w(E_n) = \mathbf{P}_w\left(\bigcup_{n \in \mathbb{N}^*} E_n\right)$$

and, by dominated convergence (see Section A1.9), equation (4.2.3) also holds for $E = \bigcup_{n \in \mathbb{N}^*} E_n$. Thus (4.2.3) holds for any set E belonging to the σ-algebra generated by $\bigcup_{l \in \mathbb{N}^*} \sigma((\xi_i)_{1 \leqslant i \leqslant l})$, which is identical to the σ-algebra $\mathscr{K}$.

4.2.2 *The law of large numbers*

In this subsection we show that for the MC considered, a certain law of large numbers holds.

It was shown in the proof of Lemma 3.1.19 (under less stringent assumptions) that $U^\infty U f = U^\infty f$, $f \in L(W)$. This amounts to

$$\int_W Uf(w)\mathbf{Q}^\infty(\mathrm{d}w) = \int_W f(w)\mathbf{Q}^\infty(\mathrm{d}w),$$

or

$$\int_W \left(\int_W Q(w, \mathrm{d}w') f(w')\right)\mathbf{Q}^\infty(\mathrm{d}w) = \int_W f(w)\mathbf{Q}^\infty(\mathrm{d}w)$$

for any $f \in L(W)$. Hence, by Theorem 3.1.22(i),

$$\int_W \mathbf{Q}^\infty(\mathrm{d}w) Q(w, B) = \mathbf{Q}^\infty(B), \quad B \in \mathscr{W}, \tag{4.2.4}$$

which is also a consequence of Theorem 3.1.24(iv). Moreover, this asserts that $\mathbf{Q}^\infty$ is the only probability on $\mathscr{W}$ which satisfies (4.2.4). Here is a direct proof of this assertion. If

$$\int_W \mu(dw)Q(w,B) = \mu(B), \quad B\in\mathscr{W},$$

then

$$\int_W Uf(w)\mu(dw) = \int_W f(w)\mu(dw), \quad f\in L(W),$$

whence

$$\int_W U^n f(w)\mu(dw) = \int_W f(w)\mu(dw), \quad n\in\mathbb{N}^*.$$

Letting $n\to\infty$, by (4.2.1), we get

$$U^\infty f = \int_W f(w)\mu(dw), \quad f\in L(W),$$

which implies that $\mu = \mathbf{Q}^\infty$.

Equation (4.2.4) and the above result show that $\mathbf{Q}^\infty$ is the only stationary probability of the MC $(\zeta_n)_{n\in\mathbb{N}}$. Using (4.2.3) we can assert that $(\zeta_n)_{n\in\mathbb{N}}$ is also an MC under $\mathbf{P}_\infty$, when the initial distribution of the chain coincides with $\mathbf{Q}^\infty$, i.e. $\mathbf{P}_\infty = \mathbf{P}_{\mathbf{Q}^\infty}$ (see Section A3.2).

Theorem 4.2.1. *For any $w\in W$ and $f\in C(W)$ we have*

$$\lim_{n\to\infty}\frac{1}{n}\sum_{i=1}^{n} f(\zeta_i) = \int_W f(w)\mathbf{Q}^\infty(dw), \quad \mathbf{P}_w\text{-a.s.}$$

Proof. We have already seen that $\mathbf{Q}^\infty$ is the only stationary probability of $(\zeta_n)_{n\in\mathbb{N}}$. The continuity of Uf for $f\in C(W)$ follows immediately from the fact that $r_1<\infty$ and $R_1<\infty$ (see Definition 3.1.18).† Therefore Theorem A3.11 applies, yielding the stated result. □

4.2.3 *The central limit theorem with remainder*

This subsection is devoted to estimating the convergence rate to normality of suitably normed partial sums associated with the MC $(\zeta_n)_{n\in\mathbb{N}}$. It is instructive to compare the method we use here with that employed in Subsection 4.1.4.

Let $f\in L_r(W)$ be fixed throughout this subsection and put $f_n = f(\zeta_n)$ $S_n = \sum_{j=1}^{n} f_j$, $n\in\mathbb{N}^*$. We shall assume that $U^\infty f = \mathbf{E}_\infty(f_1) = 0$. (The general case reduces to this one by considering the function $f - U^\infty f$.)

† For more general results see Problems 7 and 8 at the end of Chapter 3.

For any $w \in W$ we first have $\mathsf{E}_w(f_j) = U^j f(w) = T^j f(w)$, $j \in \mathbb{N}^*$, since $U^\infty f = 0$. Here $T = U - U^\infty$, hence $T^n = U^n - U^\infty$, and $\| T^n \|_L \leqslant \varepsilon_n$, $n \in \mathbb{N}^*$. Therefore $|\mathsf{E}_w(f_j)| \leqslant |T^j f| \leqslant \| f \|_L \varepsilon_j$, i.e. $\mathsf{E}_w(f_j) = O(\alpha^j)$, $j \in \mathbb{N}^*$, $w \in W$, the constant implied in O being independent of $w \in W$. Next, considering the second moments, for $j < k$, we have

$$\begin{aligned} \mathsf{E}_w(f_j f_k) &= \mathsf{E}_w(f_j \mathsf{E}_w(f_k | \zeta_j)) = \mathsf{E}_w(f_j T^{k-j} f(\zeta_j)) \\ &= U^j(f T^{k-j} f)(w) = \rho_{k-j} + T^j(f T^{k-j} f)(w), \end{aligned}$$

where

$$\rho_h = U^\infty(f T^h f) = \mathsf{E}_\infty(f_1 f_{h+1}) = O(\alpha^h), \quad h \in \mathbb{N}. \tag{4.2.5}$$

Noting that

$$|T^j(f T^{k-j} f)| \leqslant \| f \|_L \| T^{k-j} f \|_L \varepsilon_j \leqslant \| f \|_L^2 \varepsilon_j \varepsilon_{k-j} = O(\alpha^k),$$

we have

$$\mathsf{E}_w(f_j f_k) = \rho_{k-j} + O(\alpha^k), \quad 1 \leqslant j < k. \tag{4.2.6}$$

Put $\sigma^2 = \mathsf{E}_\infty(f_1^2) + 2\sum_{n \in \mathbb{N}^*} \mathsf{E}_\infty(f_1 f_{n+1})$, and note that the absolute convergence of the series is ensured by (4.2.5). Let us show that $\mathsf{E}_w(S_n^2) = n\sigma^2 + O(1)$ as $n \to \infty$. Indeed, (4.2.6) enables us to write the equation

$$\mathsf{E}_w(S_n^2) = n\rho_0 + 2\sum_{u=0}^{n-1} (n-u)\rho_u + O(1) = \mathsf{E}_\infty(S_n^2) + O(1) \tag{4.2.7}$$

as $n \to \infty$. It remains to remark that

$$n\sigma^2 - \left(n\rho_0 + 2\sum_{u=0}^{n-1} (n-u)\rho_u \right) = 2\left(\sum_{u \geqslant n} n\rho_u + \sum_{u=1}^{n-1} u\rho_u \right) \leqslant 2\sum_{u \in \mathbb{N}^*} u\rho_u = O(1)$$

as $n \to \infty$.

Clearly, (4.2.7) implies that

$$\lim_{n \to \infty} \mathsf{E}_\infty(S_n^2/n) = \sigma^2 \geqslant 0. \tag{4.2.8}$$

To begin with, we investigate a perturbation of the operator U, which will allow us to obtain the optimal convergence rate $O(n^{-1/2})$ in the central limit theorem. Define the perturbed operator U_θ as $U_\theta g = U(g \exp(i\theta f))$, $g \in L(W)$, $\theta \in \mathbb{R}$. Clearly, $U_0 = U$.

Proposition 4.2.2 *For any $\theta \in \mathbb{R}$, the operator U_θ is a bounded linear operator on $L(W)$. The map $\theta \to U_\theta$ from $\mathbb{R}$ into $\mathscr{L}(L(W))$ is analytic.*

Proof. For any $g \in L(W)$ and $\theta \in \mathbb{R}$ we have $g \exp(i\theta f) \in L(W)$, and

$$\| U_\theta g \|_L = \| U(g \exp(i\theta f)) \|_L \leqslant \| U \|_L \| g \|_L \| \exp(i\theta f) \|_L.$$

But

$$\begin{aligned} \| \exp(i\theta f) \|_L &= |\exp(i\theta f)| + \mathrm{s}(\exp(i\theta f)) \leqslant 1 + \mathrm{s}(\cos\theta f) + \mathrm{s}(\sin\theta f) \\ &\leqslant 1 + 2|\theta| \mathrm{s}(f). \end{aligned}$$

Therefore $\| U_\theta \|_L \leqslant (1 + 2|\theta| s(f)) \| U \|_L$, i.e. U_θ is a bounded linear operator on $L(W)$.

Next, we have $\| U(gf^n) \|_L \leqslant \| U \|_L \| g \|_L \| f \|_L^n$, $g \in L(W)$, $n \in \mathbb{N}^*$. Hence

$$\sum_{n \in \mathbb{N}} \| ((i\theta)^n/n!) U(gf^n) \|_L \leqslant \| U \|_L \| g \|_L \sum_{n \in \mathbb{N}} (|\theta|^n/n!) \| f \|_L^n$$
$$= \| U \|_L \| g \|_L \exp(|\theta| \, \| f \|_L).$$

This shows that for any $g \in L(W)$ and $\theta \in \mathbb{R}$ the series $\sum_{n \in \mathbb{N}} ((i\theta)^n/n!) U(gf^n)$ converges in $L(W)$ to $U_\theta g$, which implies that the map $\theta \to U_\theta$ is analytic. □

The result below, though elementary, has a key importance in what follows.

Proposition 4.2.3. *For any $\theta \in \mathbb{R}$, $g \in L(W)$, and $n \in \mathbb{N}^*$ we have*

$$U_\theta^n g(w) = \mathsf{E}_w(g(\zeta_n) \exp(i\theta S_n)), \quad w \in W.$$

Proof. For $n = 1$ the equation above is obvious. Assuming it to be true for some $n \in \mathbb{N}^*$ we can write

$$U_\theta^{n+1} g(w) = U_\theta(U_\theta^n g)(w) = U(U_\theta^n(g) \exp(i\theta f))(w)$$
$$= \int_W Q(w, dw') \exp(i\theta f(w')) \mathsf{E}_{w'}(g(\zeta_n) \exp(i\theta S_n))$$
$$= \mathsf{E}_w\left(\exp(i\theta f(\zeta_1)) \mathsf{E}_{\zeta_1}\left(g(\zeta_{n+1}) \exp\left(i\theta \sum_{j=2}^{n+1} f(\zeta_j) \right) \right) \right)$$
$$= \mathsf{E}_w(\mathsf{E}_{\zeta_1}(g(\zeta_{n+1}) \exp(i\theta S_{n+1}))) = \mathsf{E}_w(g(\zeta_{n+1}) \exp(i\theta S_{n+1})),$$

which completes the proof. □

Theorem 4.2.4. *There exists $a > 0$ such that for $|\theta| < a$ we have*

(i) $U_\theta^n = \sigma_1^n(\theta) U_{1,\theta} + T_\theta^n$, $n \in \mathbb{N}^*$,

where $\sigma_1(\theta)$ is the only eigenvalue of U_θ of maximum absolute value, which satisfies $|\sigma_1(\theta)| \geqslant (2 + \alpha)/3$; $U_{1,\theta}$ is a projection (i.e. $U_{1,\theta}^2 = U_{1,\theta}$) on the one-dimensional eigenspace E_θ associated with $\sigma_1(\theta)$; and T_θ is a bounded linear operator on $L(W)$ of spectral radius $\leqslant (1 + 2\alpha)/3$, with $T_\theta E_\theta = 0$;

(ii) *the maps $\theta \to \sigma_1(\theta)$ from $(-a, a)$ into $\mathbb{C}$, and $\theta \to U_{1,\theta}$ and $\theta \to T_\theta$ from $(-a, a)$ into $\mathscr{L}(L(W))$ are analytic;*

(iii) $\| T_\theta^n 1 \|_L \leqslant c |\theta| ((1 + 2\alpha)/3)^{n+1}$, $n \in \mathbb{N}^*$, *for some $c > 0$.*

Proof. The first two assertions are special instances of Rellich's analytic perturbation theory of linear operators (see Baumgärtel (1985, Ch. 6), Dunford & Schwartz (1958, VII.6) or Kato (1976, Ch. 7)). Here we simply outline the construction of the operators $U_{1,\theta}$ and T_θ (cf. Le Page (1982) and Nagaev (1957)).

Let $R(z)$ be the resolvent of $U = U^\infty + T$ defined as

$$R(z) = (zI - U)^{-1} = \frac{U^\infty}{z-1} + \frac{I - U^\infty}{z} + \sum_{n \in \mathbb{N}^*} \frac{T^n}{z^{n+1}}, \quad z \in \mathbb{C},$$

for $|z| > \alpha$, $z \neq 1$. Put

$$R_\theta(z) = R(z) \sum_{n \in \mathbb{N}} [(U_\theta - U)R(z)]^n.$$

If $\| U_\theta - U \|_L < 1/\| R(z) \|_L$, then the series above converges in $L(W)$ and defines the resolvent $R_\theta(z)$ of U_θ, the map $\theta \to R_\theta(z)$ being analytic. Let $I_1 = \{z : |z-1| = (1-\alpha)/3\}$ and $I_2 = \{z : |z| = (1+2\alpha)/3\}$. Choose $\delta > 0$ such that $\delta < (1-\alpha)/3$, which implies that $\alpha + \delta < (1+2\alpha)/3$. Put $M_\delta = \sup \| R(z) \|_L$, where the supremum is taken over all z for which $|z| > \alpha + \delta$ and $|z-1| > \delta$. Clearly, $M_\delta < \infty$. Then, if $\| U_\theta - U \|_L < 1/M_\delta$, the circles I_1 and I_2 are contained in the resolvent set of U_θ. Consider the projections

$$U_{1,\theta} = \frac{1}{2\pi i} \int_{I_1} R_\theta(z) \mathrm{d}z, \quad T_\theta^{(0)} = \frac{1}{2\pi i} \int_{I_2} R_\theta(z) \mathrm{d}z.$$

We have $U_{1,\theta} + T_\theta^{(0)} = I$. Clearly, $U_{1,0} = U^\infty$, $T_0^{(0)} = I - U^\infty$. Since $\theta \to R_\theta(z)$ is analytic, the same is true of the maps $\theta \to U_{1,\theta}$ and $\theta \to T_\theta^{(0)}$. Hence, if θ is small enough, we have $\| U_{1,\theta} - U^\infty \|_L < 1$, which implies that $\dim U_{1,\theta}(L(W)) = \dim U^\infty(L(W)) = 1$. Indeed, if there were two linearly independent $h_1, h_2 \in U_{1,\theta}(L(W))$, then $U^\infty h_1 = cU^\infty h_2$ for some $c \in \mathbb{C}$, and $(U_{1,\theta} - U^\infty)(h_1 - ch_2) = h_1 - ch_2$, which would imply that $\| U_{1,\theta} - U^\infty \|_L \geq 1$. Let $g_\theta \in L(W)$ generate $E_\theta = U_{1,\theta}(L(W))$ so that $g_\theta = U_{1,\theta} g_\theta$. Then, since $U_\theta R_\theta(z) = R_\theta(z) U_\theta$, we have

$$U_\theta U_{1,\theta} g_\theta = U_{1,\theta} U_\theta g_\theta = \sigma_1(\theta) g_\theta = \sigma_1(\theta) U_{1,\theta} g_\theta$$

for some $\sigma_1(\theta) \in \mathbb{C}$. Hence $U_\theta U_{1,\theta} g = \sigma_1(\theta) U_{1,\theta} g$ for any $g \in L(W)$. Then, clearly, $U_\theta^n U_{1,\theta} = \sigma_1^n(\theta) U_{1,\theta}$, $n \in \mathbb{N}^*$. By the analytic perturbation theory of linear operators, for θ small enough, $\sigma_1(\theta)$ is the only eigenvalue of U_θ of maximum absolute value, and it lies in the disk bordered by I_1. Hence $|\sigma_1(\theta)| \geq 1 - (1-\alpha)/3 = (2+\alpha)/3$, and the map $\theta \to \sigma_1(\theta)$ is analytic in $(-a, a)$, where a is chosen such that all restrictions above on θ are fulfilled.

Next, we have

$$U_\theta^n = U_\theta^n (U_{1,\theta} + T_\theta^{(0)}) = \sigma_1^n(\theta) U_{1,\theta} + T_\theta^n, \quad n \in \mathbb{N}^*,$$

where

$$T_\theta^n = U_\theta^n T_\theta^{(0)} = \frac{1}{2\pi i} \int_{I_2} z^n R_\theta(z) \, \mathrm{d}z, \quad n \in \mathbb{N}^*.$$

(Note that $U_\theta R_\theta(z) = zR_\theta(z) - I$.) Since the map $\theta \to R_\theta(z)$ is analytic we can write $R_\theta(z) = R(z) + \theta R_\theta^{(1)}(z)$ for some bounded linear operator $R_\theta^{(1)}(z)$ on $L(W)$. Hence

$$T_\theta^n = \frac{1}{2\pi i}\int_{I_2} z^n R_\theta(z)\,dz = T^n + \frac{\theta}{2\pi i}\int_{I_2} z^n R_\theta^{(1)}(z)\,dz, \quad n\in\mathbb{N}^*. \quad (4.2.9)$$

Thus, setting $c = \sup \| R_\theta^{(1)}(z)\|_L < \infty$, where the supremum is taken over all $|\theta| < a$ and $z\in I_2$, we get

$$\| T_\theta^n \|_L \leqslant \| T^n \|_L + c|\theta|\left(\frac{1+2\alpha}{3}\right)^{n+1}, \quad n\in\mathbb{N}^*.$$

Since $\| T^n \|_L = O(\alpha^n)$, $n\in\mathbb{N}^*$, we deduce that the spectral radius of T_θ is $\leqslant (1+2\alpha)/3$.

Finally, the inequality in (iii) follows from (4.2.9) because $T1 = 0$. □

Recalling (4.2.8) we can now state Proposition 4.2.5.

Proposition 4.2.5. *We have*

(i) $\sigma_1'(0) = i\mathsf{E}_\infty(f_1) = 0$;

(ii) $\sigma_1''(0) = -\lim_{n\to\infty} \mathsf{E}_\infty(S_n^2/n) = -\sigma^2$.

Proof. (i) By Proposition 4.2.3 we have

$$\mathsf{E}_\infty(\exp[(i\theta/n)S_n]) = \int_W (U_{\theta/n}^n 1)(w)\mathbf{Q}^\infty(dw), \quad \theta\in\mathbb{R}, \quad n\in\mathbb{N}^*.$$

Hence, by Theorem 4.2.4, for n large enough ($|\theta/n| < a$), we have

$$\mathsf{E}_\infty(\exp[(i\theta/n)S_n]) = \sigma_1^n(\theta/n)\int_W (U_{1,\theta/n}1)(w)\mathbf{Q}^\infty(dw) + \int_W (T_{\theta/n}^n 1)(w)\mathbf{Q}^\infty(dw),$$

and

$$\left|\int_W (T_{\theta/n}^n 1)(w)\mathbf{Q}^\infty(dw)\right| \leqslant \| T_{\theta/n}^n 1 \|_L \leqslant \frac{c|\theta|}{n}\left(\frac{1+2\alpha}{3}\right)^{n+1}.$$

On the other hand,

$$U_{1,\theta/n} = U^\infty + \frac{\theta}{n}U_1^{(1)} + \frac{\theta^2}{2n^2}U_1^{(2)} + \frac{\theta^2}{n^2}\tilde{U}_{1,\theta/n},$$

where $U_1^{(1)}$, $U_1^{(2)}$, and $\tilde{U}_{1,\varepsilon}$ are bounded linear operators on $L(W)$, with $\lim_{\varepsilon\to 0}\| \tilde{U}_{1,\varepsilon}\|_L = 0$. Hence

$$\lim_{n\to\infty}\int_W (U_{1,\theta/n}1)(w)\mathbf{Q}^\infty(dw) = 1.$$

Also

$$\sigma_1(\theta/n) = 1 + \frac{\theta}{n}\sigma_1'(0) + \frac{\theta^2}{2n^2}\sigma_1''(0) + \frac{\theta^2}{n^2}\tilde{\sigma}_1(\theta/n),$$

with $\lim_{\varepsilon\to 0}\tilde{\sigma}_1(\varepsilon)=0$, whence

$$\lim_{n\to\infty}\sigma_1^n(\theta/n)=\exp(\theta\sigma_1'(0)).$$

We have therefore proved that

$$\lim_{n\to\infty}\mathbf{E}_\infty(\exp[(i\theta/n)S_n])=\exp(\theta\sigma_1'(0)).$$

But, by Theorem 4.2.1, the ratio S_n/n converges $\mathbf{Q}^\infty$-a.s. to $\mathbf{E}_\infty(f_1)=0$ as $n\to\infty$. Hence $\exp(i\theta\mathbf{E}_\infty(f_1))=\exp(\theta\sigma_1'(0))$, i.e. $\sigma_1'(0)=i\mathbf{E}_\infty(f_1)=0$.

(ii) Note first that

$$\frac{\partial^2}{\partial\theta^2}\mathbf{E}_\infty(\exp[(i\theta/\sqrt{n})S_n])|_{\theta=0}=-\mathbf{E}_\infty(S_n^2/n)$$

for all $n\in\mathbb{N}^*$. By Theorem 4.2.4, for n large enough ($|\theta/\sqrt{n}|<a$), we have

$$\mathbf{E}_\infty(\exp[(i\theta/\sqrt{n})S_n])=\sigma_1^n(\theta/\sqrt{n})\int_W(U_{1,\theta/\sqrt{n}}1)(w)\mathbf{Q}^\infty(\mathrm{d}w)$$
$$+\int_W(T^n_{\theta/\sqrt{n}}1)(w)\mathbf{Q}^\infty(\mathrm{d}w).$$

Equation (4.2.9) yields

$$T^n_{\theta/\sqrt{n}}1=\frac{1}{2\pi i}\int_{I_2}z^n(R_{\theta/\sqrt{n}}(z)1)\,\mathrm{d}z.$$

Since the map $\theta\to R_\theta(z)$ is analytic we can write

$$R_{\theta/\sqrt{n}}(z)=R(z)+\frac{\theta}{\sqrt{n}}R^{(1)}(z)+\frac{\theta^2}{2n}R^{(2)}(z)+\frac{\theta^2}{n}\tilde{R}_{\theta/\sqrt{n}}(z),$$

where $R^{(1)}(z)$, $R^{(2)}(z)$, and $\tilde{R}_\varepsilon(z)$ are bounded linear operators on $L(W)$, with $\lim_{\varepsilon\to 0}\|\tilde{R}_\varepsilon(z)\|_L=0$. Therefore

$$T^n_{\theta/\sqrt{n}}1=\frac{\theta}{2\sqrt{n}\pi i}\int_{I_2}z^n(R^{(1)}(z)1)\mathrm{d}z+\frac{\theta^2}{4n\pi i}\int_{I_2}z^n(R^{(2)}(z)1)\,\mathrm{d}z$$
$$+\frac{\theta^2}{2n\pi i}\int_{I_2}z^n(\tilde{R}_{\theta/\sqrt{n}}(z)1)\mathrm{d}z,$$

whence

$$\lim_{n\to\infty}\frac{\partial^2}{\partial\theta^2}\left(\int_W(T^n_{\theta/\sqrt{n}}1)(w)\mathbf{Q}^\infty(\mathrm{d}w)\right)\bigg|_{\theta=0}=0.$$

Using the expansions of $\sigma_1(\theta)$ and $U_{1,\theta}$, it is easy to deduce that

$$\lim_{n\to\infty}\frac{\partial^2}{\partial\theta^2}\left(\sigma_1^n(\theta/\sqrt{n})\int_W(U_{1,\theta/\sqrt{n}}1)(w)\mathbf{Q}^\infty(\mathrm{d}w)\right)=\sigma_1''(0),$$

which completes the proof. □

Proposition 4.2.6. Assume that $\sigma > 0$. Then there exists $a > 0$ such that for $|u| < a\sqrt{n}$, $u \in \mathbb{R}$, $n \in \mathbb{N}^*$, we have

$$|U^n_{u/\sigma\sqrt{n}}1 - \exp(-u^2/2)| \leqslant (c_1|u|^3/\sigma^3\sqrt{n} + c_2|u|/\sigma\sqrt{n})\exp(-u^2/4) + (c|u|/\sigma\sqrt{n})((1+2\alpha)/3)^{n+1},$$

where c_1, c_2, and c are positive constants.

Proof. By Theorem 4.2.4 for θ small enough ($|\theta| < a/\sigma$) we can write

$$\begin{aligned} U^n_\theta &= \sigma^n_1(\theta)U_{1,\theta} + T^n_\theta \\ &= (1 + \theta\sigma'_1(0) + (\theta^2/2)\sigma''_1(0) + (\theta^3/6)\sigma'''_1(0) + \theta^3\tilde{\sigma}_1(\theta))^n \\ &\quad \times (U^\infty + \theta U^{(1)}_1 + (\theta^2/2)U^{(2)}_1 + \theta^2\tilde{U}_{1,\theta}) + T^n_\theta, \quad n \in \mathbb{N}^*. \end{aligned}$$

Here $\lim_{\theta\to 0}\tilde{\sigma}_1(\theta) = 0$, while $U^{(1)}_1$, $U^{(2)}_1$, and $\tilde{U}_{1,\theta}$ are bounded linear operators on $L(W)$ with $\lim_{\theta\to 0}\|\tilde{U}_{1,\theta}\|_L = 0$. Using Proposition 4.2.5, we have

$$\begin{aligned} U^n_\theta &= (U^\infty + \theta U^{(1)}_1 + (\theta^2/2)U^{(2)}_1 + \theta^2\tilde{U}_{1,\theta}) \\ &\quad \times \exp[n\{(-\theta^2\sigma^2/2) + c_0\theta^3 + \theta^3\varepsilon(\theta)\}] + T^n_\theta, n \in \mathbb{N}^*, \end{aligned}$$

where c_0 is a constant and $\lim_{\theta\to 0}\varepsilon(\theta) = 0$. Putting $\theta = u/\sigma\sqrt{n}$, the equation $U^\infty 1 = 1$ and Theorem 4.2.4 (iii) enable us to conclude that

$$|U^n_{u/\sigma\sqrt{n}}1 - \exp(-u^2/2)| \leqslant (C_{1,n}(u) + C_{2,n}(u))\exp(-u^2/2) + (c|u|/\sigma\sqrt{n})((1+2\alpha)/3)^{n+1}, \quad n \in \mathbb{N}^*,$$

where

$$C_{1,n}(u) = \left|\exp\left(c_0\frac{u^3}{\sigma^3\sqrt{n}} + \frac{u^3}{\sigma^3\sqrt{n}}\varepsilon\left(\frac{u}{\sigma\sqrt{n}}\right)\right) - 1\right|,$$

$$C_{2,n}(u) = \frac{|u|}{\sigma\sqrt{n}}\left|U^{(1)}_1 1 + \frac{u}{2\sigma\sqrt{n}}U^{(2)}_1 1 + \frac{u}{\sigma\sqrt{n}}\tilde{U}_{1,u/\sigma\sqrt{n}}1\right| \times \exp\left(c_0\frac{u^3}{\sigma^3\sqrt{n}} + \frac{u^3}{\sigma^3\sqrt{n}}\varepsilon\left(\frac{u}{\sigma\sqrt{n}}\right)\right).$$

Suppose a additionally satisfies $2c'_0 a/\sigma^3 < 1/4$ and $|\varepsilon(a)| \leqslant c'_0 = \max(|c_0|, 1)$. Then

$$\left|c_0\frac{u^3}{\sigma^3\sqrt{n}} + \frac{u^3}{\sigma^3\sqrt{n}}\varepsilon\left(\frac{u}{\sigma\sqrt{n}}\right)\right| \leqslant 2c'_0|u|u^2/\sigma^3\sqrt{n} \leqslant \frac{u^2}{4},$$

and the inequality $|e^z - 1| \leqslant |z|e^{|z|}$ yields

$$C_{1,n}(u) \leqslant c_1(|u|^3/\sigma^3\sqrt{n})\exp\left(\frac{u^2}{4}\right), \quad n \in \mathbb{N}^*,$$

with $c_1 = 2c'_0$. Finally, putting

$$c_2 = |U_1^{(1)}1| + \frac{a}{2\sigma}\left(|U_1^{(2)}1| + 2 \sup_{|\theta| < a/\sigma} |\tilde{U}_{1,\theta}1| \right),$$

we have

$$C_{2,n}(u) \leqslant c_2(|u|/\sigma\sqrt{n}) \exp\left(\frac{u^2}{4}\right), \quad n \in \mathbb{N}^*. \qquad \square$$

Theorem 4.2.7. *Assume that $\sigma > 0$. Then there exists a constant $c > 0$ such that for any $w \in W$, $v \in \mathbb{R}$, and $n \in \mathbb{N}^*$ we have*

$$\left| \mathbf{P}_w\left(\frac{S_n - n\mathbf{E}_\infty(f_1)}{\sigma\sqrt{n}} \leqslant v \right) - \Phi(v) \right| \leqslant \frac{c}{\sqrt{n}}.$$

Proof. The result follows from Esséen's inequality (see Feller (1966, p. 516)), which asserts that, for any $\tau > 0$ and $n \in \mathbb{N}^*$,

$$\sup_{v \in \mathbb{R}} \left| \mathbf{P}_w\left(\frac{S_n}{\sigma\sqrt{n}} \leqslant v \right) - \Phi(v) \right|$$

$$\leqslant \frac{24}{\pi\sqrt{2\pi\tau}} + \frac{1}{\pi} \int_{-\tau}^{+\tau} \frac{|\mathbf{E}_w(\exp[(iu/\sigma\sqrt{n})S_n]) - \exp(-u^2/2)|}{|u|} du, \quad w \in W.$$

We have simply to use Propositions 4.2.3 and 4.2.6, and take $\tau = a\sqrt{n}$. $\square$

4.2.4 *An invariance principle for the central limit theorem and the law of the iterated logarithm*

In this subsection, using Theorem A3.12, we give a functional central limit theorem and Strassen's version of the law of the iterated logarithm for the associated MC $(\zeta_n)_{n \in \mathbb{N}}$.

Given a $\mathscr{W}$-measurable real-valued function f on W for which

$$\int_W f^2(w) \mathbf{Q}^\infty(dw) < \infty$$

let us put $f_n = f(\zeta_n)$, $n \in \mathbb{N}^*$. The sequence $(f_n)_{n \in \mathbb{N}^*}$ is strictly stationary under $\mathbf{P}_\infty$. Assume that $\mathbf{E}_\infty(f_1) = 0$ and put $S_0 = 0$, $S_n = \sum_{i=1}^n f_i$, $n \in \mathbb{N}^*$. As in Subsection 4.1.3, define the random processes

$$\boldsymbol{\eta}_n^C : \boldsymbol{\eta}_n^C(t) = \frac{1}{\sigma\sqrt{n}}(S_{[nt]} + (nt - [nt]) f_{[nt]+1}), \quad t \in [0, 1], \quad n \in \mathbb{N}^*,$$

and

$$\eta_n^D : \eta_n^D(t) = \frac{1}{\sigma\sqrt{n}} S_{[nt]}, \quad t \in [0, 1], \quad n \in \mathbb{N}^*,$$

where σ is a positive number which coincides with that occurring in (4.2.8) when $f \in L_r(W)$, and is defined below in the more general case under consideration.

Theorem 4.2.8. *For any f with the properties above, the limit $\lim_{n\to\infty} \mathbf{E}_\infty(S_n^2/n) = \sigma^2 \geqslant 0$ exists. If $\sigma > 0$ and $\sum_{n\in\mathbb{N}^*} \mathbf{E}_\infty^{1/2}[\mathbf{E}_\infty^2(f_{n+1}|\zeta_1)] < \infty$, then $\mathbf{P}_\infty(\eta_n^C)^{-1} \Rightarrow \mathbf{B}$, $\mathbf{P}_\infty(\eta_n^D)^{-1} \Rightarrow \mathbf{B}$, and the sequence $(\eta_n^C/\sqrt{(2\log\log n)})_{n\geqslant 3}$, viewed as a subset of C, is a relatively compact set, whose derived set coincides $\mathbf{P}_\infty$-a.s. with K, the collection of all absolutely continuous functions $\mathbf{x} \in C$ for which $\mathbf{x}(0) = 0$ and $\int_0^1 [\mathbf{x}'(t)]^2 \, dt \leqslant 1$.*

Proof. The results follow from Theorem A3.12 since $\mathbf{E}_{\mathbf{Q}^\infty} = \mathbf{E}_\infty$ (see Subsection 4.2.2). Note that in the case where $f \in L_r(W)$, we have $\mathbf{E}_w(f_n) = O(\alpha^n)$, as proved in Subsection 4.1.3, whence $\mathbf{E}_\infty(f_{n+1}|\zeta_1) = O(\alpha^n)$, $n \in \mathbb{N}^*$. Thus, in this case, the convergence of the series in the statement of the theorem is ensured. □

From Theorem 4.2.8 we can derive the analogues of the corollaries stated in Subsections 4.1.3 and 4.1.5. The details are left to the reader. We draw the reader's attention to the fact that the results in Theorem 4.2.8 are only proved when the underlying probability is $\mathbf{P}_\infty$. The validity under $\mathbf{P}_w$, for an arbitrary $w \in W$, of the functional central limit theorem can also be proved when $f \in L_r(W)$ (see Popescu (1976)). As to Strassen's law of the iterated logarithm, equation (4.2.3) enables us to assert only that $\mathbf{P}_\infty$ can be replaced either by $\mathbf{P}_w$, for all $w \in W$ not belonging to an exceptional set $B \in \mathscr{W}$ for which $\mathbf{Q}^\infty(B) = 0$, or by any probability $\mathbf{Q} \ll \mathbf{P}_\infty$.

Problems

In Problems 1–3 we assume the framework of Subsection 4.1.4, which is a special case of that in Subsection 4.1.1.

1. Prove the following converse to Theorem 4.1.2. If $\sum_{i=1}^n f_i/n$ converges $\mathbf{P}_w$-a.s. to a constant $a \in \mathbb{R}$ for some $w \in W$, then $\mathbf{E}_\infty(f_1) = a$. (Cohn (1965a)).
(*Hint*: Use the following results: (i) let $A_n \in \sigma((\xi_i)_{n\leqslant i\leqslant n+k-1})$, $n \in \mathbb{N}^*$, then $\mathbf{P}_\infty(\limsup_{n\to\infty} A_n) = 0$ iff $\sum_{n\in\mathbb{N}^*} \mathbf{P}_\infty(A_n) < \infty$; (ii) we have

$$\sum_{n\in\mathbb{N}^*} \mathbf{P}_\infty(|f_n| \geqslant n) \leqslant \mathbf{E}_\infty(|f_1|) \leqslant 1 + \sum_{n\in\mathbb{N}^*} \mathbf{P}_\infty(|f_n| \geqslant n).)$$

2. Assume that F is bounded and that $\sum_{n\in\mathbb{N}^*} n^\alpha \varepsilon_n < \infty$, with $\alpha \geqslant 0$. Show that for any $m\in\mathbb{N}^*$ and $w\in W$ we have ($\mathbf{V}$ = variance)
$$\mathbf{V}_w(S_{m,n}) = n(\sigma^2 + \delta_n),$$
where
$$\delta_n = \begin{cases} O(1), & \text{if } \alpha = 0 \\ O(n^{-\alpha/(1+\alpha)}), & \text{if } \alpha > 0 \end{cases}$$
as $n\to\infty$. (Iosifescu & Theodorescu (1969, p. 142))
(*Hint*: Lemma 2.1.1 implies that $|\mathbf{E}_w(f_i^2) - \mathbf{E}_\infty(f_1^2)| \leqslant c^2\varepsilon_i$, $|\mathbf{E}_w(f_i) - \mathbf{E}_\infty(f_1)| \leqslant 2c\varepsilon_i$, $|\mathbf{E}_w(f_i f_j) - \mathbf{E}_\infty(f_1 f_{j-i+1})| \leqslant 2c^2\varepsilon_i, i,j\in\mathbb{N}^*, i<j$ where $c = \sup_{x^{(k)}\in X^k}|F(x^{(k)})|$. Putting $g_i = f_i - \mathbf{E}_w(f_i)$, $h_i = f_i - \mathbf{E}_\infty(f_1)$, Proposition A3.4 yields $|\mathbf{E}_w(g_i g_j)| \leqslant 4c^2\varepsilon_{j-i-k+1}$, $|\mathbf{E}_\infty(h_1 h_{j-i+1})| \leqslant 2c^2\varepsilon_{j-i-k+1}$, $i,j\in\mathbb{N}^*$, $i<j$, with the convention $\varepsilon_u = 1$ if $u\leqslant 0$. Then use the equation
$$\mathbf{V}_w(S_{m,n}) - n\sigma^2 = \sum_{i=m}^{m+n-1} \mathbf{V}_w(f_i) - n\mathbf{V}_\infty(f_1) + 2\left(\sum_{i=m}^{m+n-2} \mathbf{E}_w\left(g_i \sum_{j=i+1}^{m+n-1} g_j\right) - n\sum_{i\in\mathbb{N}^*} \mathbf{E}_\infty(h_1 h_{i+1})\right).\Bigg)$$

3. Prove that under the assumptions of the problem above, if $\alpha > 0$ and $\sigma \neq 0$, then there exist two positive constants $v < 1$ and c such that
$$|\mathbf{P}_w(S_{m,n} - \mathbf{E}_w(S_{m,n}) < a\sqrt{\mathbf{V}_w(S_{m,n})}) - \Phi(a)| \leqslant cn^{-v}$$
for all $w\in W$, $a\in\mathbb{R}$, and $m,n\in\mathbb{N}^*$.
(*Hint*: We have
$$\{S_{m,n} - \mathbf{E}_w(S_{m,n}) < a\sqrt{\mathbf{V}_w(S_{m,n})}\} = \left\{\frac{S_{m,n} - n\mathbf{E}_\infty(f_1)}{\sigma\sqrt{n}} < \frac{a\sqrt{\mathbf{V}_w(S_{m,n})}}{\sigma\sqrt{n}} + \frac{\mathbf{E}_w(S_{m,n}) - n\mathbf{E}_\infty(f_1)}{\sigma\sqrt{n}}\right\}.\Bigg)$$

4. Assume the framework of Subsection 4.1.4, the only difference being that we consider an s-dimensional random vector F on X^k, for a given $k\in\mathbb{N}^*$, and put
$$f_n = F(\xi_n, \ldots, \xi_{n+k-1}) = (f_{n,j})_{1\leqslant j\leqslant s}, \quad n\in\mathbb{N}^*.$$
For $1 \leqslant i,j \leqslant s$ denote
$$\begin{aligned}\sigma_{ij}^2 = {} & \mathbf{E}_\infty[(f_{1,i} - \mathbf{E}_\infty(f_{1,i}))(f_{1,j} - \mathbf{E}_\infty(f_{1,j}))] \\ & + \sum_{n\in\mathbb{N}^*} \mathbf{E}_\infty[(f_{1,i} - \mathbf{E}_\infty(f_{1,i}))(f_{n+1,j} - \mathbf{E}_\infty(f_{1,j}))] \\ & + \sum_{n\in\mathbb{N}^*} \mathbf{E}_\infty[(f_{1,j} - \mathbf{E}_\infty(f_{1,j}))(f_{n+1,i} - \mathbf{E}_\infty(f_{1,i}))].\end{aligned}$$

Give conditions under which the random vector

$$\left(\frac{1}{\sqrt{n}}(f_{n,j} - \mathbf{E}_\infty(f_{1,j}))\right)_{1 \leqslant j \leqslant s}$$

is asymptotically normal with mean vector 0 and covariance matrix $(\sigma_{ij}^2)_{1 \leqslant i,j \leqslant s}$.

(*Hint*: Consider the r.v.s $g_n(u^{(s)}) = \sum_{j=1}^s u_j f_{n,j}$, $n \in \mathbb{N}^*$, where $u^{(s)} = (u_1, \ldots, u_s) \in \mathbb{R}^s$. By the Wold–Cramér theorem the asymptotic normality of the vector considered is equivalent to the asymptotic normality of the r.v.

$$\frac{1}{\sqrt{n}}[g_n(u^{(s)}) - \mathbf{E}_\infty(g_1(u^{(s)}))]$$

for any $u^{(s)} \in \mathbb{R}^s$.)

In Problems 5 and 6 the framework is more general than in Subsections 4.1.4 and 4.1.1, but $S_{m,n}$ and S_n are similarly defined.

5. Let $F_n, n \in \mathbb{N}^*$, be real r.v.s on X^k for a given $k \in \mathbb{N}^*$, and put $f_n = F_n(\xi_n, \ldots, \xi_{n+k-1})$. Show that if $\sup_{n \in \mathbb{N}^*} \mathbf{E}_w(f_n - \mathbf{E}_w(f_n))^2 < \infty$, then the r.v.

$$\frac{1}{n} \sum_{i=1}^n (f_i - \mathbf{E}_w(f_i))$$

converges to 0 as $n \to \infty$ in $\mathbf{P}_w$-square mean. (Iosifescu (1965b))
(*Hint*: Use Proposition A3.4.)

6. Let F_n, $n \in \mathbb{N}^*$, be real r.v.s on $X^{\mathbb{N}^*}$ and put $f_n = F_n(\xi_n, \xi_{n+1}, \ldots)$. Let $(a_n)_{n \in \mathbb{N}^*}$ be a sequence of real numbers and $(b_n)_{n \in \mathbb{N}^*}$ a sequence of positive numbers converging to ∞. Put

$$F_w^{(m,n)}(a) = \mathbf{P}_w\left(\frac{S_{m,n}}{b_n} - a_n < a\right), \quad F_\infty^{(n)}(a) = \mathbf{P}_\infty\left(\frac{S_n}{b_n} - a_n < a\right), \quad a \in \mathbb{R}.$$

Prove that, for arbitrarily fixed $m \in \mathbb{N}^*$ and $w \in W$, if either of the sequences $(F_w^{(m,n)})_{n \in \mathbb{N}^*}$ and $(F_\infty^{(n)})_{n \in \mathbb{N}^*}$ converges weakly to a distribution function F, then the other sequence also converges weakly to F. (Doob (1953, p. 230), Iosifescu (1965b))
(*Hint*: Defining the characteristic functions

$$\varphi_w^{(m,n)}(t) = \int_{\mathbb{R}} e^{ita} \, dF_w^{(m,n)}(a), \quad \varphi_\infty^{(n)}(t) = \int_{\mathbb{R}} e^{ita} \, dF_\infty^{(n)}(a), \quad n \in \mathbb{N}^*,$$

Lemma 2.1.1 leads to the inequality

$$|\varphi_w^{(m,n)}(t) - \varphi_\infty^{(n)}(t)| \leqslant \mathbf{E}_w\left(\left|1 - \exp\left(it\frac{S_{m,r}}{b_n}\right)\right|\right) + \mathbf{E}_\infty\left(\left|1 - \exp\left(it\frac{S_r}{b_n}\right)\right|\right) + 3\varepsilon_{m+r},$$

where $r < n$.)

7. Using Theorem 4.1.5, derive the limit distributions under $\mathbf{P}_w$ of the r.v.s

$$\frac{1}{\sigma n^{3/2}} \sum_{i=1}^{n} |S_i|, \quad \frac{1}{\sigma^2 n^2} \sum_{i=1}^{n} S_i^2 \quad \text{and} \quad \frac{1}{\sigma\sqrt{n}} \min_{\beta n \leqslant i \leqslant n} S_i$$

where $\beta \in (0, 1)$ is given.
(*Hint*: See Billingsley (1968, §16, Problem 1).)

8. Under the assumptions of Subsection 4.2.3, show that the operator $I - T$ on $L(W)$ has an inverse $(I - T)^{-1} = I + T + T^2 + \cdots$. Setting $g = (I - T)^{-1} f$ prove that

$$\sigma^2 = \int_W [g^2 - (Ug)^2] \, d\mathbf{Q}^\infty = \int_W (2fg - f^2) \, d\mathbf{Q}^\infty.$$

(Norman (1972, p. 90))

9. (*Open problem*) Find necessary and sufficient conditions expressed in terms of f, P and u, for $\sigma^2 \neq 0$.

5

Some special systems

5.1 OM chains

5.1.1 Preliminaries

Let m be a fixed natural integer. Consider the simplex

$$\Delta = \left\{ \mathbf{p} = (p_1, \ldots, p_{m+1}) : 0 \leqslant p_i \leqslant 1, 1 \leqslant i \leqslant m+1, \sum_{i=1}^{m+1} p_i = 1 \right\}$$

of all probabilities on the set $\{1, \ldots, m+1\}$. For any $n \in \mathbb{N}$ let $\psi_i^n = (\psi_{i,j}^n)_{1 \leqslant j \leqslant m+1}$, $1 \leqslant i \leqslant m+1$, be $(\mathscr{B}_\Delta, \mathscr{B}_\Delta)$-measurable maps from Δ into itself.

Definition 5.1.1. *An* OM (Onicescu–Mihoc) chain *with state space* $\{1, \ldots, m+1\}$ *and transition maps* $\psi_i^n, 1 \leqslant i \leqslant m+1, n \in \mathbb{N}$, *is the RSCC* $\{(W, \mathscr{W}), (X, \mathscr{X}), (u_n)_{n \in \mathbb{N}}, P\}$ *for which*

$$(W, \mathscr{W}) = (\Delta, \mathscr{B}_\Delta), \quad X = \{1, \ldots, m+1\}, \quad \mathscr{X} = \mathscr{P}(X),$$

$$u_n(\mathbf{p}, i) = \psi_i^n(\mathbf{p}), \quad P(\mathbf{p}, i) = p_i, \quad i \in X, \quad n \in \mathbb{N}, \quad \mathbf{p} \in \Delta.$$

It has been shown in the introduction how an OM chain (there referred to as a chain with complete connections) appears when modelling a real experiment.

It follows from Theorem 1.1.6 that, given an OM chain and a $\mathbf{p} \in \Delta$, there exists a probability space $(\Omega, \mathscr{K}, \mathsf{P}_\mathbf{p})$ and a sequence of X-valued r.v.s $(\xi_n)_{n \in \mathbb{N}^*}$ on Ω such that

$$\mathsf{P}_\mathbf{p}(\xi_1 = i) = p_i, \quad \mathsf{P}_\mathbf{p}(\xi_{n+1} = i \mid \xi_l = i_l, 1 \leqslant l \leqslant n) = \psi^0_{i_1 \ldots i_n, i}(\mathbf{p}),$$

for any $n \in \mathbb{N}^*$ and $i, i_l \in X, 1 \leqslant l \leqslant n$, where $\psi^0_{i_1 \ldots i_n} = \psi^{n-1}_{i_n} \circ \cdots \circ \psi^0_{i_1}$ and $\psi^0_{i_1 \ldots i_n, i}(\mathbf{p})$ is the ith coordinate of $\psi^0_{i_1 \ldots i_n}(\mathbf{p}) \in \Delta$.

It is sometimes more convenient to consider the following alternative method of defining the constituents of an OM chain. The simplex Δ is replaced by the simplex

$$\Delta' = \left\{ \mathbf{p} = (p_1, \ldots, p_m) : 0 \leqslant p_i \leqslant 1, \quad 1 \leqslant i \leqslant m, \sum_{i=1}^{m} p_i \leqslant 1 \right\}$$

while the maps ψ_i^n are replaced by the maps φ_i^n defined on Δ' as

$$\varphi_i^n(p_1,\ldots,p_m) = (\psi_{i,j}^n(p_1,\ldots,p_m,p_{m+1}))_{1 \leq j \leq m},\ p_{m+1} = 1 - \sum_{i=1}^m p_i,$$

for any $i \in X$ and $n \in \mathbb{N}$. In this framework an OM chain is defined as an RSCC $\{(W,\mathscr{W}),(X,\mathscr{X}),(u_n)_{n\in\mathbb{N}},P\}$ for which

$$(W,\mathscr{W}) = (\Delta', \mathscr{B}_{\Delta'}),\quad X = \{1,\ldots,m+1\},\quad \mathscr{X} = \mathscr{P}(X),$$

$$u_n(\mathbf{p},i) = \varphi_i^n(\mathbf{p}),\quad i \in X,$$

$$P(\mathbf{p},i) = \begin{cases} p_i, & \text{if } 1 \leq i \leq m \\ 1 - \sum_{i=1}^m p_i, & \text{if } i = m+1 \end{cases}$$

for any $n \in \mathbb{N}$ and $\mathbf{p} \in \Delta'$. Clearly, $\psi_{i_1\ldots i_n}^0$ should be replaced by the corresponding composition of maps $\varphi_{i_1\ldots i_n}^0 (= (\varphi_{i_1\ldots i_n,j}^0)_{1 \leq j \leq m}) = \varphi_{i_n}^{n-1} \circ \cdots \circ \varphi_{i_1}^0$, while $\varphi_{i_1\ldots i_n,m+1}^0$ is to be understood as $1 - \sum_{j=1}^m \varphi_{i_1\ldots i_n,j}^0$ for any $n \in \mathbb{N}^*$ and $i_l \in X$, $1 \leq l \leq n$. It is also obvious that we can simply start with arbitrarily given $(\mathscr{B}_{\Delta'}, \mathscr{B}_{\Delta'})$-measurable maps $\varphi_i^n, i \in X, n \in \mathbb{N}$, from Δ' into itself. As a special RSCC, an OM chain is said to be homogeneous iff the transition maps ψ_i^n(φ_i^n, respectively) do not depend on $n \in \mathbb{N}$, i.e. $\psi_i^n = \psi_i$($\varphi_i^n = \varphi_i$, respectively), $i \in X, n \in \mathbb{N}$. For a homogeneous OM chain, the transition operator U associated with this special RSCC acts on $B(\Delta, \mathscr{B}_\Delta)$ ($B(\Delta', \mathscr{B}_{\Delta'})$, respectively) and is defined (see equation (1.1.8)) as

$$Uf(\mathbf{p}) = \sum_{i\in X} p_i f(\psi_i(\mathbf{p})),\quad \mathbf{p} \in \Delta.$$

$$\left(Uf(\mathbf{p}) = \sum_{i=1}^m p_i f(\varphi_i(\mathbf{p})) + \left(1 - \sum_{i=1}^m p_i\right) f(\varphi_{m+1}(\mathbf{p})),\quad \mathbf{p} \in \Delta',\quad \text{respectively} \right).$$

We shall now define two important special classes of OM chains.

Definition 5.1.2. *An OM chain is said to be* linear *iff its transition maps are affine functions on Δ (or Δ').*

Definition 5.1.3. *An OM chain is said to be* alternate *iff $m = 1$.*

Let us remark that equation (1.1.12) implies that

$$P_i^{n+1}(\mathbf{p}) = \sum_{j\in X} p_j P_i^n(\psi_j(\mathbf{p})) \tag{5.1.1}$$

for any $i \in X, n \in \mathbb{N}^*$, and $\mathbf{p} \in \Delta$, or, equivalently,

$$P_i^{n+1}(\mathbf{p}) = \sum_{j=1}^m p_j P_i^n(\varphi_j(\mathbf{p})) + \left(1 - \sum_{k=1}^m p_k\right) P_i^n(\varphi_{m+1}(\mathbf{p})), \tag{5.1.1'}$$

for any $i \in X, n \in \mathbb{N}^*$, and $\mathbf{p} \in \Delta'$, where

$$P_i^n(\mathbf{p}) = P_1^n(\mathbf{p}, \{i\}) = \mathsf{P}_{\mathbf{p}}(\xi_n = i).$$

More generally,

$$P_{i_1 \ldots i_r}^{n+1}(\mathbf{p}) = \sum_{j \in X} p_j P_{i_1 \ldots i_r}^n(\psi_j(\mathbf{p}))$$

for any $r, n \in \mathbb{N}^*, i_l \in X, 1 \leqslant l \leqslant r$, and $\mathbf{p} \in \Delta$, where

$$P_{i_1 \ldots i_r}^n(\mathbf{p}) = P_r^n(\mathbf{p}, \{(i_1, \ldots, i_r)\}) = \mathsf{P}_{\mathbf{p}}(\xi_n = i_1, \ldots, \xi_{n+r-1} = i_r),$$

with the obvious corresponding variant for transition maps from Δ' into itself. Clearly, $P_{i_1 \ldots i_r}^{n+1}(\mathbf{p}) = U^n P_{i_1 \ldots i_r}^1(\mathbf{p})$, $r, n \in \mathbb{N}^*, i_l \in X, 1 \leqslant l \leqslant r, \mathbf{p} \in \Delta$ (or $\mathbf{p} \in \Delta'$).

In the case of an alternate OM chain, equation (5.1.1′) can be written (with $\mathbf{p} = p_1 = p$) as

$$P^{n+1}(p) = pP^n(\varphi_1(p)) + (1-p)P^n(\varphi_2(p)), \quad p \in [0, 1],$$

where $P^n(p) = P_1^n(p)$; if the alternate OM chain is also linear, then

$$P^{n+1}(p) = pP^n(ap+b) + (1-p)P^n(cp+d), \quad p \in [0, 1], \qquad (5.1.2)$$

with $0 \leqslant b, d \leqslant 1, 0 \leqslant a+b, c+d \leqslant 1$. The asymptotic behaviour of the probability $P^n(p)$ as $n \to \infty$ will be examined in Subsection 5.1.3.

To close these preliminary considerations let us note that the associated Δ-valued MC corresponding to an OM chain is defined by the equations

$$\zeta_0 = \mathbf{p}, \quad \zeta_n = \psi_{\xi^{(n)}}^0(\mathbf{p}), \quad n \in \mathbb{N}^*.$$

Therefore ζ_n may take on one of the $(m+1)^n$ values $\psi_{i_1 \ldots i_n}^0(\mathbf{p}), i_r \in X$, $1 \leqslant r \leqslant n$, with probability

$$^0P_{i_1 \ldots i_n}(\mathbf{p}) = \begin{cases} p_{i_1}, & \text{if } n = 1 \\ p_{i_1} \psi_{i_1, i_2}^0(\mathbf{p}), & \text{if } n = 2 \\ p_{i_1} \psi_{i_1, i_2}^0(\mathbf{p}) \cdots \psi_{i_{n-1}, i_n}^{n-2}(\psi_{i_1 \ldots i_{n-2}}^0(\mathbf{p})), & \text{if } n > 2 \end{cases}$$

for any fixed $\mathbf{p} \in \Delta$. The associated Δ'-valued MC can be written similarly.

5.1.2 *Ergodicity*

All the ergodicity concepts considered in Chapter 2 obviously apply to an OM chain, which is a special RSCC. For instance, in the homogeneous case, uniform ergodicity amounts to the existence of the limits

$$\lim_{n \to \infty} P_{i_1 \ldots i_r}^n(\mathbf{p}) = P_{i_1 \ldots i_r}^\infty (= \mathsf{P}_r^\infty(\{(i_1, \ldots, i_r)\})), \quad i_1, \ldots, i_r \in X,$$

independent of $\mathbf{p}\in\Delta$ (or $\mathbf{p}\in\Delta'$), such that (see Section A1.4)

$$\sum_{i_1,\ldots,i_r\in X} |P^n_{i_1\ldots i_r}(\mathbf{p}) - P^\infty_{i_1\ldots i_r}| \to 0$$

as $n\to\infty$ uniformly w.r.t. $r\in\mathbb{N}^*$ and $\mathbf{p}\in\Delta$ (or $\mathbf{p}\in\Delta'$).

A few definitions are needed for stating results concerning ergodicity for OM chains.

If $\mathbf{f}=(f_1,\ldots,f_m)$ is a map from Δ' into itself and d is a metric on Δ', then the norm $\mu_d(\mathbf{f})$ will be defined as

$$\mu_d(\mathbf{f}) = \sup_{\substack{\mathbf{p},\mathbf{q}\in\Delta'\\ \mathbf{p}\neq\mathbf{q}}} \frac{d(\mathbf{f}(\mathbf{p}),\mathbf{f}(\mathbf{q}))}{d(\mathbf{p},\mathbf{q})}.$$

Clearly, if $\mathbf{g}$ is another map from Δ' into itself, then we have $\mu_d(\mathbf{f}\circ\mathbf{g}) \leqslant \mu_d(\mathbf{f})\mu_d(\mathbf{g})$.

The metric d_0 on Δ' defined as

$$d_0(\mathbf{p},\mathbf{q}) = \sum_{i=1}^m |p_i - q_i|, \quad \mathbf{p},\mathbf{q}\in\Delta',$$

is of special importance since it is possible to give an upper bound for $\mu_0 = \mu_{d_0}$.

Proposition 5.1.4. *If the functions $f_i:\Delta'\to[0,1]$, $1\leqslant i\leqslant m$, have bounded first partial derivatives, then*

$$\mu_0(\mathbf{f}) \leqslant \max_{1\leqslant j\leqslant m} \sup_{\mathbf{p}\in\Delta'} \sum_{i=1}^m \left|\frac{\partial f_i(\mathbf{p})}{\partial p_j}\right|.$$

Proof. Fix $\mathbf{p},\mathbf{q}\in\Delta'$. By the intermediate value formula applied to the function

$$h(t) = \sum_{i=1}^m f_i(t\mathbf{p} + (1-t)\mathbf{q})\,\mathrm{sgn}\,(f_i(\mathbf{p}) - f_i(\mathbf{q})), \quad t\in[0,1],$$

we can write for some $\theta\in(0,1)$

$$\begin{aligned} d_0(\mathbf{f}(\mathbf{p}),\mathbf{f}(\mathbf{q})) &= \sum_{i=1}^m |f_i(\mathbf{p}) - f_i(\mathbf{q})| = h(1) - h(0) \\ &= \sum_{i=1}^m \sum_{j=1}^m (p_j - q_j)\left(\frac{\partial f_i(\mathbf{r})}{\partial r_j}\right)\Bigg|_{\mathbf{r}=\theta\mathbf{p}+(1-\theta)\mathbf{q}} \mathrm{sgn}\,(f_i(\mathbf{p}) - f_i(\mathbf{q})) \\ &\leqslant d_0(\mathbf{p},\mathbf{q}) \max_{1\leqslant j\leqslant m} \sup_{\mathbf{p}\in\Delta'} \sum_{i=1}^m \left|\frac{\partial f_i(\mathbf{p})}{\partial p_j}\right|, \end{aligned}$$

whence the stated result follows. □

First, let us consider the non-homogeneous case.

Theorem 5.1.5. *Assume that*

(i) *there exist* $k_0 \in X$ *and* $\alpha > 0$ *such that*

$$\varphi^n_{i,k_0}(\mathbf{p}) \geqslant \alpha, \quad \text{if } k_0 \neq m+1$$

or

$$\sum_{j=1}^{m} \varphi^n_{i,j}(\mathbf{p}) \leqslant 1 - \alpha, \quad \text{if } k_0 = m+1$$

for any $i \in X, n \in \mathbb{N}$, *and* $\mathbf{p} \in \Delta'$;

(ii) *the functions* $\varphi^n_{i,j}, i \in X, 1 \leqslant j \leqslant m, n \in \mathbb{N}$, *have first partial derivatives and*

$$\sum_{j=1}^{m} \left| \frac{\partial \varphi^n_{i,j}(\mathbf{p})}{\partial p_k} \right| \leqslant \begin{cases} 1, & \text{if } i \neq k_0 \\ b < 1, & \text{if } i = k_0 \end{cases}$$

for any $1 \leqslant k \leqslant m, n \in \mathbb{N}$, *and* $\mathbf{p} \in \Delta'$.
Then uniform weak ergodicity holds and the rate of convergence to 0 *is* $O(\exp(-cn^a))$ *with* $c > 0, 0 < a < \frac{1}{2}$.

Proof. Proposition 5.1.4 and assumption (ii) imply $a'_n \leqslant 2b^n, n \in \mathbb{N}^*$, so that Conditions FLS'$(\{k_0\}, 2)$ and $\mathrm{K}'_1(\{k_0\}, 1)$ hold (see Subsection 2.3.1). The stated result follows by Proposition 2.3.5, using Theorem 2.3.4. □

Remark. The conclusion of Theorem 5.1.5 is not altered if (i) and (ii) only hold for $n \geqslant n_0$, where $n_0 \in \mathbb{N}$ is arbitrary but fixed.

Now, let us consider the homogeneous case.

Theorem 5.1.6. *Assume that the metric d on* Δ' *makes* Δ' *into a compact metric space. Assume also that*

(i) *there exists* $k_0 \in X$ *such that*

$$\varphi_{i,k_0}(\mathbf{p}) > 0, \quad \text{if } k_0 \neq m+1$$

or

$$\sum_{j=1}^{m} \varphi_{i,j}(\mathbf{p}) < 1, \quad \text{if } k_0 = m+1$$

for any $i \in X$ *and* $\mathbf{p} \in \Delta'$;

(ii) $\mu_d(\varphi_i) \leqslant 1$ *if* $i \neq k_0$, *while* $\mu_d(\varphi_{k_0}) < 1$.

Then the OM chain is an RSCC with contraction, whose transition operator U is regular w.r.t. $L(\Delta')$ *(see Definitions 3.1.15 and 3.1.4). In particular (see Theorem 3.4.5), uniform ergodicity holds and the rate of convergence to the limit probability is exponential.*

Proof. By the definition of an OM chain we have that $r_1 < \infty$ and

$R_1 < \infty$. Assumption (ii) implies that the integrand in the definition of r_2 is $\leqslant 1$. Since it is $\leqslant \mu_d(\boldsymbol{\varphi}_{k_0}) < 1$ on the set $X \times \{k_0\}$, for which, by (i), we have $\inf_{\mathbf{p}\in\Delta'} P_2(\mathbf{p}, X \times \{k_0\}) > 0$, it follows easily that $r_2 < 1$, i.e. our OM chain is an RSCC with contraction.

To prove regularity, by Theorem 3.2.13, it is sufficient to show that there exists $\mathbf{p}_0 \in \Delta'$ such that $\lim_{n\to\infty} d(\Sigma_n(\mathbf{p}), \mathbf{p}_0) = 0$ for any $\mathbf{p}\in\Delta'$. This can be done as follows. Since $\mu_d(\boldsymbol{\varphi}_{k_0}) < 1$, the map $\boldsymbol{\varphi}_{k_0}$ has a fixed point, say $\mathbf{p}_0$, and, clearly,

$$d(\boldsymbol{\varphi}_{\underbrace{k_0 \ldots k_0}_{n \text{ times}}}(\mathbf{p}), \mathbf{p}_0) = d(\boldsymbol{\varphi}_{\underbrace{k_0 \ldots k_0}_{n \text{ times}}}(\mathbf{p}), \boldsymbol{\varphi}_{\underbrace{k_0 \ldots k_0}_{n \text{ times}}}(\mathbf{p}_0)) \leqslant (\mu_d(\boldsymbol{\varphi}_{k_0}))^n \delta$$

for any $n \in \mathbb{N}^*$ and $\mathbf{p}\in\Delta'$, where δ is the diameter of Δ'. But

$$\boldsymbol{\varphi}_{\underbrace{k_0 \ldots k_0}_{n \text{ times}}}(\mathbf{p}) \in \Sigma_n(\mathbf{p}),$$

so that

$$d(\Sigma_n(\mathbf{p}), \mathbf{p}_0) \leqslant d(\boldsymbol{\varphi}_{\underbrace{k_0 \ldots k_0}_{n \text{ times}}}(\mathbf{p}), \mathbf{p}_0) \to 0$$

as $n \to \infty$. □

Corollary 5.1.7. *Uniform ergodicity holds for an alternate homogeneous OM chain when either of the following conditions is fulfilled:*

(i) *φ_1 and φ_2 are differentiable, and $\varphi_1(p) > 0$, $\varphi_2(p) > 0$, $|\varphi_1'(p)| \leqslant \alpha < 1$, $|\varphi_2'(p)| \leqslant 1$ for any $p \in [0, 1]$;*

(ii) *φ_1 and φ_2 are differentiable, and $\varphi_1(p) < 1$, $\varphi_2(p) < 1$, $|\varphi_1'(p)| \leqslant 1$, $|\varphi_2'(p)| \leqslant \alpha < 1$ for any $p \in [0, 1]$.*

Proof. This is an immediate consequence of Theorem 5.1.6, taking the metric d_0 as metric d on $[0, 1]$. □

An important special case of Corollary 5.1.7 is Corollary 5.1.8.

Corollary 5.1.8. *Uniform ergodicity holds for an alternate linear homogeneous OM chain with transition maps*

$$\varphi_1(p) = ap + b, \quad \varphi_2(p) = cp + d, \quad p \in [0, 1],$$

where $0 \leqslant b, d \leqslant 1, 0 \leqslant a + b, c + d \leqslant 1$, when either of the following conditions is fulfilled:

(i) $|a| < 1, \quad 0 < b, \quad 0 < a + b, \quad |c| < 1, \quad 0 < d, \quad 0 < c + d$;

(ii) $|a| < 1, \quad b < 1, a + b < 1, \quad |c| < 1, d < 1, \quad c + d < 1$.

We think it instructive to present the original proof of Fortet's (1938) ergodic theorem, which is weaker than Theorem 5.1.6. This proof is essentially based on the ideas of Onicescu & Mihoc's (1935b) proof of the first ergodic theorem for OM chains. This gives an idea of the methodology used at the beginning of dependence-with-complete-connections-theory, and enables us to assess the progress achieved over fifty years.

Theorem 5.1.6′. *Assume conditions* (i) *and* (ii) *of Theorem* 5.1.6 *hold with* $d = d_0$ *in the last one*†. Let $f \in B(\Delta', \mathscr{B}_{\Delta'})$ *be such that*

(iii) $|f(\mathbf{p}) - f(\mathbf{q})| \leqslant c d_0(\mathbf{p}, \mathbf{q})$

for any $\mathbf{p}, \mathbf{q} \in \Delta'$, *where* $c > 0$ *is a constant.*

Then $f_{n+1} = U^n f$, $n \in \mathbb{N}$, *converges uniformly on* Δ' *to a constant* $U^\infty f$.

Proof. We shall assume that $k_0 \neq m + 1$. The alterations needed to suit the case $k_0 = m + 1$ are left to the reader. Remark that condition (ii) implies that

$$d_0(\underbrace{\varphi_{k_0 \ldots k_0}}_{r \text{ times}}(\mathbf{p}), \underbrace{\varphi_{k_0 \ldots k_0}}_{r \text{ times}}(\mathbf{q})) \leqslant (\mu_0(\varphi_{k_0}))^r d_0(\mathbf{p}, \mathbf{q}) \tag{5.1.3}$$

for any $\mathbf{p}, \mathbf{q} \in \Delta'$ and $r \in \mathbb{N}^*$.

To begin with, we shall show that the functions $U^n f, n \in \mathbb{N}$, are equicontinuous on

$$\Delta'_a = \{\mathbf{p} : \mathbf{p} = (p_1, \ldots, p_m) \in \Delta', p_{k_0} \geqslant a\},$$

where

$$a = \min_{i \in X} \min_{\mathbf{p} \in \Delta'} \varphi_{i, k_0}(\mathbf{p}) > 0$$

(that a is positive follows from condition (i)). By condition (ii), the maps φ_i, $i \in X$, take Δ' into Δ'_a and Δ'_a into Δ'_a.

We have

$$\begin{aligned} f_{n+1}(\mathbf{p}) - f_{n+1}(\mathbf{q}) &= U f_n(\mathbf{p}) - U f_n(\mathbf{q}) \\ &= \sum_{i \in X} (p_i - q_i) f_n(\varphi_i(\mathbf{p})) + \sum_{i \in X} q_i (f_n(\varphi_i(\mathbf{p})) - f_n(\varphi_i(\mathbf{q}))) \end{aligned}$$

for any $\mathbf{p}, \mathbf{q} \in \Delta'$ and $n \in \mathbb{N}^*$, where $p_{m+1} = 1 - \sum_{j=1}^m p_j$, $q_{m+1} = 1 - \sum_{j=1}^m q_j$.

Since $|f_n| \leqslant |f|$, $n \in \mathbb{N}^*$,‡ we can write

$$\left| \sum_{i \in X} (p_i - q_i) f_n(\varphi_i(\mathbf{p})) \right| \leqslant 2|f| d_0(\mathbf{p}, \mathbf{q}) \tag{5.1.4}$$

for any $\mathbf{p}, \mathbf{q} \in \Delta'$ and $n \in \mathbb{N}^*$.

† Any other metric equivalent to d_0 can also be considered.

‡ We remind the reader that $|f| = \sup_{\mathbf{p} \in \Delta'} |f(\mathbf{p})|$.

Assume that for some $n \in \mathbb{N}^*$ we have

$$|f_n(\mathbf{p}) - f_n(\mathbf{q})| \leqslant c_1 d_0(\mathbf{p}, \mathbf{q}), \quad \mathbf{p}, \mathbf{q} \in \Delta'_a, \tag{5.1.5}$$

where

$$c_1 = \max\left(c, \frac{2|f|}{a(1 - \mu_0(\varphi_{k_0}))}\right). \tag{5.1.6}$$

Let us show that (5.1.5) also holds when n is replaced by $n+1$. It follows from (5.1.5) that for any $i \in X$

$$|f_n(\varphi_i(\mathbf{p})) - f_n(\varphi_i(\mathbf{q}))| \leqslant c_1 d_0(\varphi_i(\mathbf{p}), \varphi_i(\mathbf{q})) \leqslant c_1 \mu_0(\varphi_i) d_0(\mathbf{p}, \mathbf{q}),$$

whence

$$\left|\sum_{i \in X} q_i(f_n(\varphi_i(\mathbf{p})) - f_n(\varphi_i(\mathbf{q})))\right| \leqslant c_1 d_0(\mathbf{p}, \mathbf{q}) \sum_{i \in X} q_i \mu_0(\varphi_i). \tag{5.1.7}$$

Condition (ii) implies that

$$\sum_{i \in X} q_i \mu_0(\varphi_i) \leqslant 1 - q_{k_0} + q_{k_0} \mu_0(\varphi_{k_0}). \tag{5.1.8}$$

Using equations (5.1.4) to (5.1.8), we obtain

$$\begin{aligned}
|f_{n+1}(\mathbf{p}) - f_{n+1}(\mathbf{q})| &\leqslant d_0(\mathbf{p}, \mathbf{q})(2|f| + c_1(1 - q_{k_0} + q_{k_0}\mu_0(\varphi_{k_0}))) \\
&\leqslant c_1 d_0(\mathbf{p}, \mathbf{q})(1 - q_{k_0} + a - (a - q_{k_0})\mu_0(\varphi_{k_0})) \\
&= c_1 d_0(\mathbf{p}, \mathbf{q})(1 - (1 - \mu_0(\varphi_{k_0}))(q_{k_0} - a)) \\
&\leqslant c_1 d_0(\mathbf{p}, \mathbf{q}), \quad \mathbf{p}, \mathbf{q} \in \Delta'_a,
\end{aligned}$$

which proves our assertion.

Since, by condition (ii), (5.1.5) holds for $n = 1$, it follows by induction that (5.1.5) holds for any $n \in \mathbb{N}^*$, which implies the equicontinuity of the $f_n, n \in \mathbb{N}^*$ on Δ'_a.

Now, since the $f_n, n \in \mathbb{N}^*$, are equally bounded and equicontinuous by the Arzelà–Ascoli Theorem A1.16 there exists a subsequence $(f_{n_k})_{k \in \mathbb{N}^*}$ which converges uniformly on Δ'_a to a continuous function, say f_∞. Setting

$$\bar{f}_n = \sup_{\mathbf{p} \in \Delta'_a} f_n(\mathbf{p}), \quad \underline{f}_n = \inf_{\mathbf{p} \in \Delta'_a} f_n(\mathbf{p}), \quad n \in \mathbb{N}^*,$$

it is obvious that the sequences $(\bar{f}_n)_{n \in \mathbb{N}^*}$ and $(\underline{f}_n)_{n \in \mathbb{N}^*}$ are non-increasing and non-decreasing, respectively. Let

$$\bar{f} = \lim_{n \to \infty} \bar{f}_n, \quad \underline{f} = \lim_{n \to \infty} \underline{f}_n.$$

Remark that $\bar{f}$ and $\underline{f}$ are upper and lower bounds on Δ'_a of the limit function f_∞ and of any other limit function of a subsequence of $(f_n)_{n \in \mathbb{N}^*}$. In fact, starting from the subsequence $(f_{n_k})_{k \in \mathbb{N}^*}$ we can obtain infinitely many other convergent subsequences. Indeed, for any $r \in \mathbb{N}^*$ we have

$$f_{n+r}(\mathbf{p}) = \sum_{i_1, \ldots, i_r \in X} P_{i_1 \ldots i_r}(\mathbf{p}) f_n(\varphi_{i_1 \ldots i_r}(\mathbf{p})),$$

where

$$P_{i_1 \ldots i_r}(\mathbf{p}) = P^1_{i_1 \ldots i_r}(\mathbf{p}) = \mathbf{P}_{\mathbf{p}}(\xi_1 = i_1, \ldots, \xi_r = i_r).$$

Since

$$\sum_{i_1, \ldots, i_r \in X} P_{i_1 \ldots i_r}(\mathbf{p}) = 1, \quad \mathbf{p} \in \Delta',$$

the subsequence $(f_{n_k+r})_{k \in \mathbb{N}^*}$ converges uniformly on Δ'_a to the continuous function $f_{r,\infty}$ defined by the equation

$$f_{r,\infty}(\mathbf{p}) = \sum_{i_1, \ldots, i_r \in X} P_{i_1 \ldots i_r}(\mathbf{p}) f_\infty(\boldsymbol{\varphi}_{i_1 \ldots i_r}(\mathbf{p})), \quad \mathbf{p} \in \Delta'_a. \tag{5.1.9}$$

We can now prove that $\bar{f} = \underline{f}$, i.e.

$$\lim_{n \to \infty} f_n(\mathbf{p}) = \bar{f} = \underline{f}$$

uniformly w.r.t. $\mathbf{p} \in \Delta'_a$. Let $\mathbf{u}$ and $\mathbf{v}$ be points in Δ'_a at which the function $f_{r,\infty}$ reaches its maximum and minimum, respectively. Since

$$P_{\underbrace{k_0 \ldots k_0}_{r \text{ times}}}(\mathbf{p}) > 0, \quad \mathbf{p} \in \Delta'_a,$$

it follows from (5.1.9) that

$$\bar{f} = f_{r,\infty}(\mathbf{u}) = f_\infty(\boldsymbol{\varphi}_{\underbrace{k_0 \ldots k_0}_{r \text{ times}}}(\mathbf{u})),$$

$$\underline{f} = f_{r,\infty}(\mathbf{v}) = f_\infty(\boldsymbol{\varphi}_{\underbrace{k_0 \ldots k_0}_{r \text{ times}}}(\mathbf{v})),$$

for any $r \in \mathbb{N}^*$. Using (5.1.3), the equation $\bar{f} = \underline{f}$ follows immediately.

To complete the proof we must show that

$$\lim_{n \to \infty} f_n(\mathbf{p}) = \bar{f}$$

for $\mathbf{p} \in \Delta'$. To do this it is sufficient to note that, since $\boldsymbol{\varphi}_i(\mathbf{p}) \in \Delta'_a$ for any $i \in X$ and $\mathbf{p} \in \Delta'$, we have

$$\lim_{n \to \infty} f_n(\boldsymbol{\varphi}_i(\mathbf{p})) = \bar{f}, \quad i \in X, \quad \mathbf{p} \in \Delta'.$$

Consequently

$$\lim_{n \to \infty} f_n(\mathbf{p}) = \sum_{i \in X} p_i \lim_{n \to \infty} f_{n-1}(\boldsymbol{\varphi}_i(\mathbf{p})) = \bar{f}$$

for any $\mathbf{p} \in \Delta'$. □

Remark. In the original statement of Theorem 5.1.6′ and in subsequent work by other authors, the convergence of the sequence $(f_n)_{n \in \mathbb{N}^*}$ to a constant has been asserted 'excepting, possibly, the points $\mathbf{p}$ for which $p_{k_0} = 0$'. We have just seen that such a possibility cannot occur.

5.1.3 Alternate linear OM chains

An interesting and important problem is that of determining effectively the limit probabilities $P^{\infty}_{i_1\ldots i_r}$, $r\in\mathbb{N}^*$, $i_1,\ldots,i_r\in X$, for uniformly ergodic OM chains. This appears to be possible only in special cases. Even for alternate homogeneous linear OM chains the solution is far from being complete, but the effective determination of the limit probabilities is possible when (see Corollary 5.1.8)

$$\varphi_1(p)=ap+b,\quad \varphi_2(p)=ap+d.$$

Indeed, for $r=1$ and $n\geqslant 2$, we can write (using the notation of Subsection 5.1.1)

$$\begin{aligned}P^{n+1}(p)&=\sum_{i_1,\ldots,i_n=1}^{2}P_{i_1\ldots i_n}(p)\varphi_{i_1\ldots i_n}(p)\\&=\sum_{i_1,\ldots,i_{n-1}=1}^{2}(P_{i_1\ldots i_{n-1}1}(p)\varphi_{i_1\ldots i_{n-1}1}(p)+P_{i_1\ldots i_{n-1}2}(p)\varphi_{i_1\ldots i_{n-1}2}(p))\\&=\sum_{i_1,\ldots,i_{n-1}=1}^{2}(P_{i_1\ldots i_{n-1}}(p)\varphi_{i_1\ldots i_{n-1}}(p)(a\varphi_{i_1\ldots i_{n-1}}(p)+b)\\&\quad+P_{i_1\ldots i_{n-1}}(p)(1-\varphi_{i_1\ldots i_{n-1}}(p))(a\varphi_{i_1\ldots i_{n-1}}(p)+d))\\&=d+(a+b-d)P^n(p).\end{aligned}$$

A direct computation shows that the above equation also holds for $n=1$.

It follows that if $|a+b-d|<1$,[†] then

$$\left|P^n(p)-\frac{d}{1-a-b+d}\right|\leqslant|a+b-d|^{n-1}$$

for any $p\in[0,1]$ and $n\in\mathbb{N}^*$, i.e.,

$$P^{\infty}_1=\frac{d}{1-a-b+d}.$$

Remark that $P^{n+1}(p)$ is the mean value $\mathsf{E}_p(\zeta_n)$ of the $(n+1)$th variable, $n\in\mathbb{N}$, of the associated $[0,1]$-valued MC.

The case $r\geqslant 2$ requires knowledge of the asymptotic values as $n\to\infty$, of the moments

$$M_r^{n+1}(p)=\mathsf{E}_p(\zeta_n^r),\quad n\in N,\quad r\in\mathbb{N}^*.$$

Similarly to the computation above, for any $r\in\mathbb{N}^*$ and $n\geqslant 2$ we can write

$$M_r^{n+1}(p)=\sum_{i_1,\ldots,i_n=1}^{2}P_{i_1\ldots i_n}(p)\varphi^r_{i_1\ldots i_n}(p)$$

[†] If this condition is fulfilled, then by Corollary 5.1.8, uniform ergodicity holds. If $|a+b-d|=1$, uniform ergodicity does not hold. Indeed, if $a+b-d=1$, then $a+b=1$ and $d=0$, whence $P^{n+1}(p)=P^n(p), n\in\mathbb{N}^*$, i.e. $P^n(p)=P^1(p)=p$ for any $p\in[0,1], n\in\mathbb{N}^*$. If $a+b-d=-1$, then $d=1$ and $a+b=0$, whence $P^{n+1}(p)=1-P^n(p), n\in\mathbb{N}^*$, i.e. $P^{2n-1}(p)=p$, $P^{2n}(p)=1-p$, for any $p\in[0,1], n\in\mathbb{N}^*$.

$$= \sum_{i_1,\dots,i_{n-1}=1}^{2} (P_{i_1\dots i_{n-1}}(p)\varphi_{i_1\dots i_{n-1}}(p)(a\varphi_{i_1\dots i_{n-1}}(p)+b)^r$$
$$+ P_{i_1\dots i_{n-1}}(p)(1-\varphi_{i_1\dots i_{n-1}}(p))(a\varphi_{i_1\dots i_{n-1}}(p)+d)^r)$$
$$= d^r + \sum_{l=1}^{r} c_{rl} M_l^n(p),$$

where

$$c_{rl} = \binom{r}{l-1} a^{l-1}(b^{r-l+1} - d^{r-l+1}) + \binom{r}{l} d^{r-l} a^l, \quad 1 \leqslant l \leqslant r.$$

A direct computation shows that the above equation also holds for $n = 1$.

In particular

$$M_2^{n+1}(p) = a(a+2b-2d)M_2^n(p) + (2ad+b^2-d^2)M_1^n(p) + d^2.$$

It follows that if $|a+b-d|<1$ (an inequality implying that $|a(a+2b-2d)|<1$) then the limit

$$\lim_{n\to\infty} M_2^n(p) = \frac{(2ad+b^2-d^2)P_1^\infty + d^2}{1-a(a+2b-2d)}$$

exists.

For example, to calculate P_{11}^∞ we note that $P_{11}^{n+1}(p) = U^n f(p)$, where $f(p) = p\varphi_1(p) = ap^2 + bp$. Therefore

$$P_{11}^{n+1}(p) = \sum_{i_1,\dots,i_n=1}^{2} P_{i_1\dots i_n}(p)(a\varphi_{i_1\dots i_n}^2(p) + b\varphi_{i_1\dots i_n}(p))$$
$$= aM_2^n(p) + bM_1^n(p),$$

whence

$$P_{11}^\infty = \lim_{n\to\infty} P_{11}^n(p) = a \lim_{n\to\infty} M_2^n(p) + bP_1^\infty.$$

Similarly, $P_{21}^{n+1}(p) = U^n f(p)$, where $f(p) = (1-p)\varphi_1(p) = (1-p)(ap+b) = -ap^2 + (a-b)p + b$, and we have

$$P_{21}^{n+1}(p) = -aM_2^n(p) + (a-b)M_1^n(p) + b,$$

whence

$$P_{21}^\infty = \lim_{n\to\infty} P_{21}^n(p) = -a \lim_{n\to\infty} M_2^n(p) + (a-b)P_1^\infty + b.$$

Clearly, $P_{12}^\infty = P_1^\infty - P_{11}^\infty$ and $P_{22}^\infty = P_2^\infty - P_{21}^\infty = 1 - P_1^\infty - P_{21}^\infty$.

In this manner we can determine step by step the limit probabilities $P_{i_1\dots i_r}^\infty$, $r = 1, 2, \dots$. Nevertheless, we are far from a formula which will give them explicitly.

The asymptotic behaviour as $n \to \infty$ of the probability $P^n(p)(= P^n(p, a, b, c, d))$ for an arbitrary alternate linear homogeneous OM chain has been exhaustively studied by Fortet (1938). The cases which do not fulfil the conditions of Corollary 5.1.8 will be reviewed below. We should add to that list the cases which are obtained by replacing a by c, b by $1-c-d$, c by a, and d by $1-a-b$ (see Problem 6). (This transformation arises when $p=1-q$ is substituted in equation (5.1.2) and we put $\bar{P}^n(q)=P^n(1-q)$. This actually amounts to interchanging labelling of the elements of X.)

In any of the supplementary cases the asymptotic behaviour of the probability $P^n(p, c, 1-c-d, a, 1-a-b)$ is described by $1-h(1-p, c, 1-c-d, a, 1-a-b)$, where $h(p, a, b, c, d)$ denotes the limit value associated with the asymptotic behaviour of $P^n(p, a, b, c, d)$ as $n \to \infty$.

In the cases

1° $a=b=d=0, \quad 0 \leqslant c \leqslant 1$

2° $a=-1, \quad b=1, \quad c=d=0$

3° $0<a<1, \quad b=d=0, \quad 0 \leqslant c \leqslant 1$

4° $|a|<1, \quad b>0, \quad 0 \leqslant a+b<1, \quad 0 \leqslant c<1, \quad d=0$

we have

$$P_1^\infty = \lim_{n\to\infty} P^n(p)=0, \quad 0 \leqslant p \leqslant 1.$$

(In fact, in case 1° we have $P^n(p)=0, n \geqslant 2$, for $a=b=c=d=0$, and $P^n(p)=p(1-p)^{n-1}, n \in \mathbb{N}^*$, for $a=b=d=0, c=1$. In case 2° we have $P^{2n-1}(p)=p^n(1-p)^{n-1}, P^{2n}(p)=p^n(1-p)^n, n \in \mathbb{N}^*$.)

In the cases

5° $-1<a<0, \quad b=-a, \quad -1<c<0, \quad d=-c$

6° $0 \leqslant a<1, \quad b=0, \quad |c|<1, \quad d>0, \quad c+d \geqslant 0$

7° $-1<a<0, \quad b=-a, \quad |c|<1, \quad d>0, \quad c+d>0$

8° $|a|<1, \quad b>0, \quad a+b>0, \quad -1<c<0, \quad d=-c$

9° $a=-1, \quad b=1, \quad |c|<1, \quad 0<d<1, \quad 0<c+d<1$

the limit $P_1^\infty = \lim_{n\to\infty} P^n(p)$ exists, and is independent of $p \in [0,1]$; however, we are unable to give its precise value.

In the cases

10° $a=-1, \quad b=c=1, \quad d=0$

11° $a=c=1, \quad b=d=0$

we have $P^n(p)=2p(1-p), n \geqslant 2$, and $P^n(p)=p, n \in \mathbb{N}^*$, respectively.

In the case

12° $a=1, \quad b=d=0, \quad 0 \leqslant c<1$

we have

$$\lim_{n\to\infty} P^n(p)=\begin{cases} 0, & \text{if } 0 \leqslant p<1 \\ 1, & \text{if } p=1. \end{cases}$$

(For $a=1, b=c=d=0$, we have $P^n(p)=p^n$, $n\in\mathbb{N}^*$.)

In the case

$$13°\quad -1<a<0,\quad b=-a,\quad c=1,\quad d=0$$

we have

$$\lim_{n\to\infty} P^n(p)=\begin{cases}\dfrac{a}{a-1}, & \text{if } 0<p<1\\ 0, & \text{if } p=0 \text{ or } 1.\end{cases}$$

In the case

$$14°\quad a=1,\quad b=0,\quad |c|<1,\quad 0\leqslant c+d<1,\quad 0<d<1$$

we have

$$\lim_{n\to\infty} P^n(p)=\begin{cases}\dfrac{d}{1-c}, & \text{if } 0\leqslant p<1\\ 1, & \text{if } p=1.\end{cases}$$

In the case

$$15°\quad 0\leqslant a<1,\quad b=1-a,\quad 0\leqslant c<1,\quad d=0$$

the limit $\lim_{n\to\infty} P^n(p)$ exists, and is dependent on $p\in[0,1]$; however, we are unable to give its precise value (except for $a=c=d=0, b=1$, when $P^n(p)=p, n\in\mathbb{N}^*$).

In the case

$$16°\quad a=b=0,\quad c=-1,\quad d=1$$

we have

$$P^{2n-1}(p)=\frac{p(1-(p-p^2)^n)}{1-p+p^2},$$

$$P^{2n}(p)=\frac{(1-p)^2(1-(p-p^2)^n)}{1-p+p^2},\quad n\in\mathbb{N}^*,$$

whence

$$\lim_{n\to\infty} P^{2n-1}(p)=\frac{p}{1-p+p^2},\quad \lim_{n\to\infty} P^{2n}(p)=\frac{(1-p)^2}{1-p+p^2},\quad p\in[0,1].$$

In the case

$$17°\quad a=c=-1,\quad b=d=1$$

we have $P^{2n-1}(p)=p, P^{2n}(p)=1-p, n\in\mathbb{N}^*$.

In the case

$$18°\quad a=-1,\quad b=1,\quad 0<c<1,\quad d=0$$

the sequence $(P^n(p))_{n\in\mathbb{N}^*}$ is Césaro convergent to 0, i.e.

$$\lim_{n\to\infty}\frac{1}{n}\sum_{m=1}^{n} P^m(p)=0,\quad p\in[0,1].$$

In the cases

$$19^\circ \quad 0<a<1, \quad b=0, \quad c=-1, \quad d=1$$

$$20^\circ \quad a=-1, \quad b=1, \quad -1<c<0, \quad d=-c$$

the Césaro limit of the sequence $(P^n(p))_{n\in\mathbb{N}^*}$ exists, and is independent of $p\in[0,1]$; however, we are unable to give its precise value.

In the case

$$21^\circ \quad -1<a<0, \quad b=-a, \quad c=-1, \quad d=1$$

the sequences $(P^{2n-1}(p))_{n\in\mathbb{N}^*}$ and $(P^{2n}(p))_{n\in\mathbb{N}^*}$ are Césaro convergent to distinct limits, which are dependent on $p\in(0,1)$. (It is easy to see that $P^{2n-1}(0)=1-P^{2n}(0)=P^{2n}(1)=1-P^{2n-1}(1)=0, n\in\mathbb{N}^*$.)

Remark. It is still an open problem to single out a case in which $\lim_{n\to\infty} P^n(p)$ exists and is independent of $p\in[0,1]$, but uniform ergodicity does not hold.

5.1.4 The case of an arbitrary state space

Definition 5.1.1 can be generalized to the case of an arbitrary state space. Let $(X,\mathscr{X})$ be a measurable space and let $\mathfrak{M}=\mathrm{pr}(X,\mathscr{X})$ denote the collection of all probabilities on $\mathscr{X}$. Let $\mathscr{M}$ be a σ-algebra of subsets of $\mathfrak{M}$ such that for any $A\in\mathscr{X}$ the map $p\to p(A)$ is an r.v. on $\mathfrak{M}$ (i.e. it is $\mathscr{M}$-measurable). For any $n\in\mathbb{N}$ let $(T_x^n)_{x\in X}$ be a family of maps from $\mathfrak{M}$ into itself such that the map $(p,x)\to T_x^n p$ from $\mathfrak{M}\times X$ into $\mathfrak{M}$ is $(\mathscr{M}\otimes\mathscr{X},\mathscr{M})$-measurable.

An OM chain with state space $(X,\mathscr{X})$ and transition maps $T_x^n, x\in X, n\in\mathbb{N}$, is the RSCC $\{(W,\mathscr{W}),(X,\mathscr{X}),(u_n)_{n\in\mathbb{N}},P\}$ for which $(W,\mathscr{W})=(\mathfrak{M},\mathscr{M})$, $u_n(p,x)=T_x^n p$, $p\in\mathfrak{M}$, $x\in X$, $n\in\mathbb{N}$, $P(p,A)=p(A)$, $p\in\mathfrak{M}$, $A\in\mathscr{X}$. In the homogeneous case $T_x^n=T_x$ for any $n\in\mathbb{N}$ and $x\in X$.

Obviously, all the results proved in the previous chapters for general RSCCs can be stated, in particular, for OM chains with an arbitrary state space.

Attention has recently been given to homogeneous OM chains with arbitrary state space having transition maps of the form

$$T_x p(A)=\alpha(x)p(A)+(1-\alpha(x))\Lambda(x,A), \quad x\in X, \quad A\in\mathscr{X},$$

where α is an $\mathscr{X}$-measurable $[0,1]$-valued function on X and Λ a t.p.f. on $(X,\mathscr{X})$. Clearly, these are a special case of what should be called a linear OM chain with arbitrary state space (see Problem 8).

5.2 The continued fraction expansion

5.2.1 Preliminaries

In this section we shall look at the material presented in Example 6 in Section 1.2. For the fundamentals of the theory of continued fractions the reader may consult Khinchin (1964).

From now on λ will denote Lebesgue measure on the σ-algebra $\mathscr{B}_I$ of Borel subsets of the unit interval $I = [0, 1]$. Following traditional usage, for any probability $\mu \equiv \lambda$, instead of μ-a.s. we shall write a.e. (almost everywhere).

As we have noted, any irrational number $y \in I$ can be uniquely expressed as an infinite (simple) continued fraction of the form

$$y = \cfrac{1}{a_1(y) + \cfrac{1}{a_2(y) + \cdots}} = [a_1(y), a_2(y), \ldots].$$

The natural integers $a_n(y)$, $n \in \mathbb{N}^*$, which are called partial quotients or digits, are determined as follows. Consider the continued fraction transformation τ of I defined as $\tau(t) = 1/t (\bmod 1) = \{1/t\}$ (the fractional part of $1/t$) if $t \neq 0$ and $\tau(0) = 0$. Then $a_1(y) = 1/y - \tau(y) = [1/y]$ (the integer part of $1/y$) and $a_{n+1}(y) = a_n(\tau(y)) = a_1(\tau^n(y))$, $n \in \mathbb{N}^*$. Here τ^n denotes the nth iterate of τ. For any $n \in \mathbb{N}^*$, writing

$$[x_1, \ldots, x_n] = \cfrac{1}{x_1 + \cfrac{}{\ddots + \cfrac{1}{x_n}}}$$

for arbitrary indeterminates x_i, $1 \leqslant i \leqslant n$, we clearly have

$$y = \frac{1}{a_1(y) + \tau(y)} = [a_1(y), \ldots, a_{n-1}(y), a_n(y) + \tau^n(y)] \qquad (5.2.1)$$

for all $n \geqslant 2$. The equation $y = [a_1(y), a_2(y), \ldots]$ means that

$$y = \lim_{n \to \infty} [a_1(y), \ldots, a_n(y)].$$

Finite continued fractions correspond to the rational numbers from I. As we shall only consider infinite continued fractions, in what follows 'continued fraction' will mean 'infinite continued fraction'.

Here are a few standard facts about continued fractions that we shall need. Given that $y = [a_1(y), a_2(y), \ldots]$, let us define $\mathbb{N}^*$-valued functions $p_n(y)$ and $q_n(y)$ by the recursions

$$p_n(y) = a_n(y) p_{n-1}(y) + p_{n-2}(y), \quad q_n(y) = a_n(y) q_{n-1}(y) + q_{n-2}(y), \quad n \in \mathbb{N}^*,$$

with $p_{-1}(y) = q_0(y) = 1, q_{-1}(y) = p_0(y) = 0$ for any irrational number $y \in I$. In matrix notation, dropping the variable y, we have

$$\begin{pmatrix} p_n & p_{n-1} \\ q_n & q_{n-1} \end{pmatrix} = \begin{pmatrix} p_{n-1} & p_{n-2} \\ q_{n-1} & q_{n-2} \end{pmatrix} \begin{pmatrix} a_n & 1 \\ 1 & 0 \end{pmatrix}, \quad n \in \mathbb{N}^*,$$

whence

$$\begin{pmatrix} p_n & p_{n-1} \\ q_n & q_{n-1} \end{pmatrix} = \begin{pmatrix} 0 & 1 \\ 1 & 0 \end{pmatrix} \begin{pmatrix} a_1 & 1 \\ 1 & 0 \end{pmatrix} \cdots \begin{pmatrix} a_n & 1 \\ 1 & 0 \end{pmatrix}, \quad n \in \mathbb{N}^*.$$

Hence, taking determinants, we get

$$p_n q_{n-1} - p_{n-1} q_n = (-1)^{n+1}, \quad n \in \mathbb{N}^*. \tag{5.2.2}$$

Now, remark that for arbitrary indeterminates $x_1, \ldots, x_n$, the components α and β of the vector

$$\begin{pmatrix} \alpha \\ \beta \end{pmatrix} = \begin{pmatrix} 0 & 1 \\ 1 & 0 \end{pmatrix} \begin{pmatrix} x_1 & 1 \\ 1 & 0 \end{pmatrix} \cdots \begin{pmatrix} x_{n-1} & 1 \\ 1 & 0 \end{pmatrix} \begin{pmatrix} x_n \\ 1 \end{pmatrix}$$

are the numerator and denominator of $[x_1, \ldots, x_n]$, respectively. Thus, multiplying the matrix equation above on the right by the vector $(1, u)'$, $u \geqslant 0$, we get

$$\frac{p_n + u p_{n-1}}{q_n + u q_{n-1}} = [a_1, \ldots, a_{n-1}, a_n + u], \quad n \in \mathbb{N}^*. \tag{5.2.1'}$$

(Clearly, for $n = 1$, $[a_1, \ldots, a_{n-1}, a_n + u]$ is $1/(a_1 + u)$.) Hence, for $u = 0$, we have

$$[a_1, \ldots, a_n] = \frac{p_n}{q_n}, \quad n \in \mathbb{N}^*,$$

while equations (5.2.1) and (5.2.1') imply that

$$y = \frac{p_n(y) + \tau^n(y) p_{n-1}(y)}{q_n(y) + \tau^n(y) q_{n-1}(y)}, \quad n \in \mathbb{N}$$

for any irrational number $y \in I$. Putting

$$r_n(y) = 1/\tau^{n-1}(y) = a_n(y) + [a_{n+1}(y), a_{n+2}(y), \ldots], \quad n \in \mathbb{N}^*,$$

the above equation can be also written as

$$y = \frac{p_n(y) r_{n+1}(y) + p_{n-1}(y)}{q_n(y) r_{n+1}(y) + q_{n-1}(y)}, \quad n \in \mathbb{N}. \tag{5.2.3}$$

It follows that the set $E_{i_1 \ldots i_n}$ of irrational numbers $y = [a_1(y), a_2(y), \ldots]$, for which $a_1(y) = i_1, \ldots, a_n(y) = i_n$, is the set of irrational numbers in the interval with end points p_n/q_n and $(p_n + p_{n-1})/(q_n + q_{n-1})$. (Clearly, the values assumed here by p_{n-1}, q_{n-1}, p_n, and q_n are those corresponding to the values $i_1, \ldots, i_n$ of the digits $a_1, \ldots, a_n$.) By (5.2.2) the first fraction above is smaller or greater than the second one depending on whether n is even or odd.

As we have already remarked, the a_n, $n\in\mathbb{N}^*$, can be viewed as random variables on $(I, \mathscr{B}_I)$. They are defined almost surely w.r.t. any probability on $\mathscr{B}_I$ which assigns probability 0 to the set of rational numbers in I (thus, in particular, w.r.t. Lebesgue measure λ). We obviously have

$$\lambda(r_1 > t) = \lambda([0, 1/t]) = 1/t, \quad t \geqslant 1,$$

and, using (5.2.2) and (5.2.3),

$$\lambda(r_{n+1} > t \mid a_1 = i_1, \ldots, a_n = i_n) = \frac{\lambda(\{r_{n+1} > t\} \cap E_{i_1 \ldots i_n})}{\lambda(E_{i_1 \ldots i_n})}$$

$$= \frac{\left|\dfrac{p_n t + p_{n-1}}{q_n t + q_{n-1}} - \dfrac{p_n}{q_n}\right|}{\left|\dfrac{p_n + p_{n-1}}{q_n + q_{n-1}} - \dfrac{p_n}{q_n}\right|} = \frac{q_n + q_{n-1}}{q_n t + q_{n-1}} = \frac{s_n + 1}{s_n + t},$$

where $s_n = q_{n-1}/q_n$, $n\in\mathbb{N}^*$. (The above equations have been called the Brodén–Borel–Lévy formula.) Hence, since the random event $\{a_n = i\}$ is equivalent to the random event $\{i < r_n < i+1\}$, we have

$$\lambda(a_1 = i) = \lambda(r_1 > i) - \lambda(r_1 > i+1) = \frac{1}{i} - \frac{1}{i+1} = \frac{1}{i(i+1)},$$

$$\begin{aligned} &\lambda(a_{n+1} = i \mid a_1 = i_1, \ldots, a_n = i_n) \\ &= \lambda(r_{n+1} > i \mid a_1 = i_1, \ldots, a_n = i_n) - \lambda(r_{n+1} > i+1 \mid a_1 = i_1, \ldots, a_n = i_n) \\ &= \frac{s_n + 1}{s_n + i} - \frac{s_n + 1}{s_n + i + 1} = \frac{s_n + 1}{(s_n + i)(s_n + i + 1)}, \quad n\in\mathbb{N}^*. \end{aligned} \tag{5.2.4}$$

Note that the equation $q_n = a_n q_{n-1} + q_{n-2}$ implies that $1/s_n = a_n + s_{n-1}$, whence

$$s_n = \frac{1}{a_n + s_{n-1}}, \quad n\in\mathbb{N}^*, \tag{5.2.5}$$

with $s_0 = 0$, i.e.

$$s_n = [a_n, \ldots, a_1], \quad n\in\mathbb{N}^*.$$

Equations (5.2.4) and (5.2.5) lead us to the RSCC defined in Example 6 in Section 1.2. It is clear that (see Theorem 1.1.2):

(i) for $w = 0$, the sequences $(\xi_n)_{n\in\mathbb{N}^*}$ and $(\zeta_n)_{n\in\mathbb{N}}$ associated with this RSCC under $\mathbf{P}_0$ are equivalent to the sequences $(a_n)_{n\in\mathbb{N}^*}$ and $(s_n)_{n\in\mathbb{N}}$, respectively, under λ;

(ii) for a rational $0 \neq w = [i_r, \ldots, i_1] \in I$, the sequences $(\xi_n)_{n\in\mathbb{N}^*}$ and $(\zeta_n)_{n\in\mathbb{N}}$ under $\mathbf{P}_w$ are equivalent to the sequences $(a_{r+n})_{n\in\mathbb{N}^*}$ and $(s_{r+n})_{n\in\mathbb{N}}$, respectively, under $\lambda(\cdot \mid E_{i_1 \ldots i_r})$.

Remark. An interpretation of $\mathbf{P}_w$ for the irrational values of w will be given in Subsection 5.5.3.

Proposition 5.2.1. *The sequence $(a_n)_{n\in \mathbb{N}^*}$ is strictly stationary under Gauss' probability γ on $\mathscr{B}_I$ defined as*

$$\gamma(A) = \frac{1}{\log 2}\int_A \frac{dt}{1+t}, \quad A \in \mathscr{B}_I.$$

Proof. We should prove that the value of the probability $\gamma(a_m = i_1, \dots, a_{m+r-1} = i_r)$ does not depend on $m \in \mathbb{N}^*$ for any $r \in \mathbb{N}^*$ and $i_1, \dots, i_r \in \mathbb{N}^*$. It is easy to see that by the definition of the a_n we have

$$\begin{aligned}&\{y : a_m(y) = i_1, \dots, a_{m+r-1}(y) = i_r\}\\ &= \tau^{-m+1}(\{y : a_1(y) = i_1, \dots, a_r(y) = i_r\}) = \tau^{-m+1}(E_{i_1 \dots i_r}).\end{aligned}$$

So, it is enough to prove that τ is γ-preserving (see Section A2.11) and for this it is enough to show that γ preserves the measure of any interval $(0, u], 0 < u \leqslant 1$. As

$$\tau^{-1}((0,u]) = \bigcup_{i \in \mathbb{N}^*}\left[\frac{1}{i+u}, \frac{1}{i}\right),$$

we only need to verify that

$$\int_0^u \frac{dt}{1+t} = \sum_{i \in \mathbb{N}^*}\int_{1/(i+u)}^{1/i} \frac{dt}{1+t},$$

which follows immediately. □

Remark. The expectation of a_1 under γ is infinite. Indeed

$$\frac{1}{\log 2}\int_0^1 \frac{a_1(t)}{1+t}dt = \frac{1}{\log 2}\sum_{i \in \mathbb{N}^*} i \int_{1/(i+1)}^{1/i} \frac{dt}{1+t} = \infty.$$

Proposition 5.2.2. *The RSCC associated with the continued fraction expansion is an RSCC with contraction for which the transition operator U is regular w.r.t. $L(I)$.*

Proof. We have

$$\begin{aligned}\frac{d}{dw}P(w,x) &= \frac{x^2 - x - (w+1)^2}{(w+x)^2(w+x+1)^2},\\ \frac{d}{dw}u(w,x) &= -\frac{1}{(w+x)^2}, \quad w \in I, \quad x \in \mathbb{N}^*,\end{aligned}$$

so that

$$\sup_{w \in I} \left| \frac{\mathrm{d}}{\mathrm{d}w} P(w,x) \right| < \frac{1}{x^2},$$

$$\sup_{w \in I} \left| \frac{\mathrm{d}}{\mathrm{d}w} u(w,x) \right| \leqslant \frac{1}{x^2}, \quad x \in \mathbb{N}^*.$$

Hence the requirements of Definition 3.1.15 are met with $k = 1$.

To prove the regularity of U w.r.t. $L(I)$ let us define recursively $w_{n+1} = (w_n + 2)^{-1}, n \in \mathbb{N}$, with $w_0 = w$. Clearly, $w_{n+1} \in \Sigma_1(w_n)$ and therefore, Lemma 3.2.14 and an induction argument lead to the conclusion that $w_n \in \Sigma_n(w), n \in \mathbb{N}^*$. But $\lim_{n\to\infty} w_n = \sqrt{2} - 1$ for any $w \in I$. Hence

$$d(\Sigma_n(w), \sqrt{2} - 1) \leqslant |w_n - \sqrt{2} + 1| \to 0$$

as $n \to \infty$. Now, the regularity of U w.r.t. $L(I)$ follows from Theorem 3.2.13.

□

Remarks. 1. An alternative proof of the regularity of U w.r.t. $L(I)$ can be obtained by using Theorem 3.1.20.

2. For a more general result see Problem 11.

For the RSCC under consideration it is possible to identify the limit probability $\mathbf{Q}^\infty$ which appears in Theorem 3.4.5.

Proposition 5.2.3. *The probability $\mathbf{Q}^\infty$ coincides with Gauss' probability γ.*

Proof. On account of the uniqueness of $\mathbf{Q}^\infty$ we have to show that, with the usual notation

$$Q(w,B) = \sum_{\{x:(w+x)^{-1} \in B\}} P(w,x), \quad w \in I, \quad B \in \mathscr{B}_I,$$

we have

$$\int_0^1 Q(w,B)\gamma(\mathrm{d}w) = \gamma(B)$$

for any $B \in \mathscr{B}_I$. Since the intervals $[0,u) \subset I$ generate $\mathscr{B}_I$, it is sufficient to check the above equation just for $B = [0,u), 0 < u \leqslant 1$. We have

$$Q(w,[0,u)) = \sum_{x \geqslant [u^{-1} - w] + 1} P(w,x) = \frac{w+1}{w + [u^{-1} - w] + 1},$$

so that

$$\int_0^1 Q(w,[0,u))\gamma(\mathrm{d}w) = \frac{1}{\log 2} \int_0^1 \frac{\mathrm{d}w}{w + [u^{-1} - w] + 1}$$

$$= \frac{1}{\log 2}\left(\int_0^{u^{-1}-[u^{-1}]} \frac{\mathrm{d}w}{w+[u^{-1}]+1} + \int_{u^{-1}-[u^{-1}]}^{1} \frac{\mathrm{d}w}{w+[u^{-1}]}\right)$$

$$= \frac{\log(1+u)}{\log 2} = \gamma([0,u)). \qquad \square$$

5.2.2 Gauss' problem: Lévy's approach

We are now able to solve Gauss' problem, i.e. to estimate the error

$$\lambda(r_n > t) - \frac{1}{\log 2}\log\frac{t+1}{t}, \quad t \geqslant 1.$$

For any real number $t \geqslant 1$ consider the function f_t defined on I by

$$f_t(w) = \frac{w+1}{w+t}.$$

It is easy to prove that

$$\lambda(r_n > t) = U^{n-1} f_t(0), \tag{5.2.6}$$

$$\lambda(r_{n+m} > t \mid a_1, \ldots, a_m) = U^{n-1} f_t(s_m),$$

where $s_m = [a_m, \ldots, a_1]$, for any $t \geqslant 1$ and $n, m \in \mathbb{N}^*$. Indeed, for $n = 1$ the above equations reduce to the Brodén–Borel–Lévy formula. The general case is obtained by induction w.r.t. n and using equation (5.2.5).

In fact, we shall solve a more general problem than the Gauss problem. For this we must compute $|f_s - f_t|$ and $\mathrm{s}(f_s - f_t)$ for any $1 \leqslant s < t$. Here is a summary of an elementary but tedious computation. Put

$$\begin{aligned}
g(s,t) &= (s-1)(t-1),\\
h(s,t) &= g^{\frac{1}{3}}(s,t)((s-1)^{\frac{1}{3}} + (t-1)^{\frac{1}{3}}),\\
D_1 &= \{(s,t) : 1 \leqslant s < t, g(s,t) \leqslant 1\},\\
D_{11} &= D_1 \cap \{(s,t) : 1 \leqslant s < t, h(s,t) \leqslant 1\},\\
D_{12} &= D_1 \cap \{(s,t) : 1 \leqslant s < t, 1 < h(s,t) < 2\},\\
D_{13} &= D_1 \cap \{(s,t) : 1 \leqslant s < t, h(s,t) \geqslant 2\},\\
D_2 &= \{(s,t) : 1 \leqslant s < t, 1 < g(s,t) < 4\},\\
D_3 &= \{(s,t) : 1 \leqslant s < t, g(s,t) \geqslant 4\}.
\end{aligned}$$

Then

$$|f_s - f_t| = \sup_{w \in I} |f_s(w) - f_t(w)|$$

$$= (t-s) \times \begin{cases} 1/st, & \text{if } (s,t) \in D_1\\ 1/((s-1)^{\frac{1}{2}} + (t-1)^{\frac{1}{2}}), & \text{if } (s,t) \in D_2\\ 2/(s+1)(t+1), & \text{if } (s,t) \in D_3. \end{cases}$$

$$s(f_s - f_t) = \sup_{w \in I} |f_s'(w) - f_t'(w)|$$

$$= (t-s) \times \begin{cases} \dfrac{1-(s-1)(t-1)}{s^2t^2}, & \text{if } (s,t) \in D_{11} \\ \dfrac{(s-1)^{\frac{1}{2}}(t-1)^{\frac{1}{2}}}{(s+h(s,t)-1)^{\frac{3}{2}}(t+h(s,t)-1)^{\frac{3}{2}}}, & \text{if } (s,t) \in D_{12} \\ \dfrac{4-(s-1)(t-1)}{(s+1)^2(t+1)^2}, & \text{if } (s,t) \in D_{13} \\ \max\left(\dfrac{(s-1)(t-1)-1}{s^2t^2}, \dfrac{4-(s-1)(t-1)}{(s+1)^2(t+1)^2}\right), & \text{if } (s,t) \in D_2 \\ \dfrac{(s-1)(t-1)-1}{s^2t^2}, & \text{if } (s,t) \in D_3, \end{cases}$$

and

$$s(f_s - f_t)/|f_s - f_t| < 1. \tag{5.2.7}$$

Proposition 5.2.4. *There exist positive constants $q < 1$ and c such that*

$$U^n(f_s - f_t)(w) = (1 + \theta q^n)U^\infty(f_s - f_t)$$

for all $1 \leqslant s < t$, $n \in \mathbb{N}^*$ and $w \in I$, where $\theta = \theta(s,t,n,w)$ with $|\theta| \leqslant c$.

Proof. We have

$$U^\infty(f_s - f_t) = \int_0^1 (f_s(w) - f_t(w))\gamma(dw)$$
$$= \frac{1}{\log 2}\left(\log\frac{s+1}{s} - \log\frac{t+1}{t}\right),$$

and it is easy to see that

$$|f_s - f_t| \leqslant (3\log 2)U^\infty(f_s - f_t), \quad 1 \leqslant s < t. \tag{5.2.8}$$

By Lemma 3.1.19 there exist two positive constants $q < 1$ and c_0 such that $\| U^n - U^\infty \|_L \leqslant c_0 q^n$ for all $n \in \mathbb{N}^*$. Therefore, using (5.2.7) and (5.2.8), we can write

$$|U^n(f_s - f_t) - U^\infty(f_s - f_t)| \leqslant \| U^n(f_s - f_t) - U^\infty(f_s - f_t) \|_L$$
$$\leqslant c_0 q^n(|f_s - f_t| + s(f_s - f_t))$$
$$\leqslant (6\log 2)c_0 q^n U^\infty(f_s - f_t), \quad 1 \leqslant s < t,$$

which implies the stated result. □

Corollary 5.2.5. (Solution of Gauss' problem) *There exist positive* *constants $q<1$ and c such that*

$$\lambda(s<r_n\leqslant t)=\frac{1}{\log 2}(1+\theta_0 q^n)\left(\log\frac{s+1}{s}-\log\frac{t+1}{t}\right),$$

$$\lambda(s<r_{n+m}\leqslant t\,|\,a_1,\ldots,a_m) \tag{5.2.9}$$

$$=\frac{1}{\log 2}(1+\theta_m q^n)\left(\log\frac{s+1}{s}-\log\frac{t+1}{t}\right),$$

for all $1\leqslant s<t$ and $m,n\in\mathbb{N}^*$, where $\theta_0=\theta_0(n,s,t)$, $\theta_m=\theta_m(n,s,t,[a_m,\ldots,a_1])$ with $|\theta_0|,|\theta_m|\leqslant c$, $m\in\mathbb{N}^*$.

The proof follows from equations (5.2.6) and Proposition 5.2.4.

Remark. The equation

$$\lambda(r_n>s)=\frac{1}{\log 2}\log\frac{s+1}{s}+\theta_0 q^{\sqrt{n}},\quad n\in\mathbb{N}^*,\quad s\geqslant 1,$$

with $|\theta_0|\leqslant c$ for some positive constants $q<1$ and c was obtained by Kuzmin (1928), i.e., 116 years after Gauss had stated his problem. Lévy (1929), using a different approach, obtained the stronger result

$$\lambda(r_n>s)=\frac{1}{\log 2}\log\frac{s+1}{s}+\theta_0 q^n,\quad n\in\mathbb{N}^*,\quad s\geqslant 1,$$

with $|\theta_0|<1$ and $q<0.7$. Szüsz (1961), using Kuzmin's approach, claimed to have lowered the Lévy estimate for q to 0.4. Actually, Szüsz's argument, the reconstruction of which is not at all easy, yields just 0.485 rather than 0.4. The optimal value of q in (5.2.9) was determined by Wirsing (1974), who found that it equalled 0.303 663 002 898 732 658 60.... Babenko (1978) obtained what might be called the 'complete' solution of Gauss' problem, namely to express $\lambda(r_n>s)$ in terms of the eigenvalues and eigenfunctions of a linear operator related to the operator U (see Remarks 3 and 4 following Proposition 5.2.8).

Corollary 5.2.6. *For any $m,n,r,i_1,\ldots,i_r\in\mathbb{N}^*$ we have*

$$\lambda(a_n=i_1,\ldots,a_{n+r-1}=i_r)$$

$$=\frac{1}{\log 2}(1+\theta_0 q^n)\log\frac{1+v(i_1,\ldots,i_r)}{1+u(i_1,\ldots,i_r)},$$

$$\lambda(a_{m+n}=i_1,\ldots,a_{m+n+r-1}=i_r\,|\,a_1,\ldots,a_m) \tag{5.2.10}$$

$$=\frac{1}{\log 2}(1+\theta_m q^n)\log\frac{1+v(i_1,\ldots,i_r)}{1+u(i_1,\ldots,i_r)},$$

where

$$u(i_1,\ldots,i_r)=\begin{cases}\dfrac{p_r+p_{r-1}}{q_r+q_{r-1}}, & \text{if } r \text{ is odd}\\[2mm] \dfrac{p_r}{q_r}, & \text{if } r \text{ is even}\end{cases}$$

$$v(i_1,\ldots,i_r)=\begin{cases}\dfrac{p_r}{q_r}, & \text{if } r \text{ is odd}\\[2mm] \dfrac{p_r+p_{r-1}}{q_r+q_{r-1}}, & \text{if } r \text{ is even,}\end{cases}$$

while p_n *and* q_n *assume the values corresponding to the values* $i_1,\ldots,i_n$ *of the digits* $a_1,\ldots,a_n$, $n\in\mathbb{N}^*$.

Proof. The random event $\{a_{m+n}=i_1,\ldots,a_{m+n+r-1}=i_r\}$ amounts to the inclusion of $1/r_{m+n}$ in the interval $[u(i_1,\ldots,i_r),\ v(i_1,\ldots,i_r)]$. Equations (5.2.10) then follow from equations (5.2.9). □

Remarks. 1. The strict stationarity of the sequence $(a_n)_{n\in\mathbb{N}^*}$ under γ (see Proposition 5.2.1) leads us to the equations

$$\begin{aligned}\gamma(a_n=i_1,\ldots,a_{n+r-1}=i_r)&=\gamma([u(i_1,\ldots,i_r),v(i_1,\ldots,i_r)])\\&=\frac{1}{\log 2}\log\frac{1+v(i_1,\ldots,i_r)}{1+u(i_1,\ldots,i_r)}\end{aligned}\qquad(5.2.11)$$

for all $n,r,i_1,\ldots,i_r\in\mathbb{N}^*$.

2. Corollary 5.2.6 shows that for the RSCC associated with the continued fraction expansion the limit probabilities $\mathbf{P}_r^\infty(\{(i_1,\ldots,i_r)\})$ are given by

$$\mathbf{P}_r^\infty(\{(i_1,\ldots,i_r)\})=\frac{1}{\log 2}\log\frac{1+v(i_1,\ldots,i_r)}{1+u(i_1,\ldots,i_r)}$$

for all $r,i_1,\ldots,i_r\in\mathbb{N}^*$. Clearly, the quantities ε_n defined in Subsection 4.1.1 are of the form $O(a^n)$ with $0<a<1$, $n\in\mathbb{N}^*$.

Corollary 5.2.7. *The sequence* $(a_n)_{n\in\mathbb{N}^*}$ *is exponentially* ψ*-mixing under both* λ *and* γ.

Proof. The ψ-mixing under λ follows at once from equations (5.2.10) which imply

$$\frac{\lambda(a_{m+n}=i_1,\ldots,a_{m+n+r-1}=i_r\mid a_1,\ldots,a_m)}{\lambda(a_{m+n}=i_1,\ldots,a_{m+n+r-1}=i_r)}-1=\frac{(\theta_m-\theta_0q^n)q^n}{1+\theta_0q^{m+n}}.$$

The ψ-mixing under γ can be obtained by the argument used in the proof of Theorem 2.1.5 (iii) (to prove φ-mixing under $\mathbf{P}_\infty$) using equations (5.2.10) and (2.1.11) (for $w=0$). The details can be found in Iosifescu (1989), where effective upper bounds for the mixing coefficients are also given. □

Remark. It can be proved, more generally, that the sequence $(a_n)_{n\in\mathbb{N}^*}$ is exponentially ψ-mixing under a whole class of probabilities $\mu \ll \lambda$. See Lemma 1 in Philipp (1970). See also Theorem 5.4.2.

5.2.3 Gauss' problem: Kuzmin's approach

In this subsection we shall present Kuzmin's approach to Gauss' problem, which uses an operator related to the transition operator U of the RSCC associated with the continued fraction expansion.

Let μ be an arbitrary non-atomic probability on $\mathscr{B}_I$ and define

$$F_n(w) = F_n(w,\mu) = \mu(r_{n+1}^{-1} < w), \quad n\in\mathbb{N}, \quad w\in I.$$

Clearly, $F_0(w) = \mu([0,w])$. Since $0 < r_{n+2}^{-1} < w$ iff $(w+a_{n+1})^{-1} < r_{n+1}^{-1} < a_{n+1}^{-1}$, we can write Gauss' equation

$$F_{n+1}(w) = \sum_{i\in\mathbb{N}^*}\left(F_n\left(\frac{1}{i}\right) - F_n\left(\frac{1}{w+i}\right)\right).$$

Assuming that the derivative F_0' exists everywhere in I and is bounded, it is easy to see by induction that F_n' exists and is bounded for all $n\in\mathbb{N}^*$ and we have

$$F_{n+1}'(w) = \sum_{i\in\mathbb{N}^*}\frac{1}{(w+i)^2}F_n'\left(\frac{1}{w+i}\right), \quad n\in\mathbb{N}, \quad w\in I. \tag{5.2.12}$$

Putting $f_n(w) = (w+1)F_n'(w)$, $n\in\mathbb{N}$, we get

$$f_{n+1}(w) = \sum_{i\in\mathbb{N}^*}\frac{w+1}{(w+i)(w+i+1)}f_n\left(\frac{1}{w+i}\right),$$

i.e. $f_{n+1} = Uf_n$, $n\in\mathbb{N}$. Hence

$$F_n(w) = \int_0^w \frac{U^n f_0(u)}{u+1}\,du, \quad n\in\mathbb{N}, \quad w\in I. \tag{5.2.13}$$

By Propositions 5.2.2 and 5.2.3 and Lemma 3.1.19 we have $U^n = U^\infty + T^n$, $n\in\mathbb{N}^*$, where $U^\infty f = \int_0^1 f(w)\gamma(dw)$, $f\in L(I)$, and

$$\| T^n f\|_L = O(q^n)\| f\|_L \tag{5.2.14}$$

as $n\to\infty$, for some positive number $q<1$, the constant implied in O being independent of $f \in L(I)$. Hence, since $L(I)$ is dense in $C(I)$, we get

$$\lim_{n\to\infty} |T^n f| = 0 \tag{5.2.15}$$

for all $f \in C(I)$. Finally, by Proposition A1.21, equation (5.2.15) implies that $\lim_{n\to\infty} T^n f(w) = 0$ for all $w \in I$ and any $f \in B(I)$ which is continuous a.e., i.e. for any Riemann integrable function f on I. Using (5.2.13) we can therefore state Proposition 5.2.8.

Proposition 5.2.8. *If the density $F_0' = d\mu/d\lambda$ is Riemann integrable, then*

$$\lim_{n\to\infty} \mu(r_n > t) = \frac{1}{\log 2} \log \frac{t+1}{t}, \quad t \geqslant 1.$$

If the density $F_0' \in L(I)$, then there exist positive constants $q < 1$ and c such that

$$\mu(s < r_n \leqslant t) = \frac{1}{\log 2}(1 + \theta q^n)\left(\log\frac{s+1}{s} - \log\frac{t+1}{t}\right)$$

for all $1 \leqslant s < t$ and $n \in \mathbb{N}^*$, where $\theta = \theta(\mu, n, s, t)$ with $|\theta| \leqslant c$.

Remarks. 1. It is possible to derive from Theorem 2.2.7 an estimate of the rate of convergence in (5.2.15) for $f \in C(I)$.

2. For a generalization of Proposition 5.2.8 see Proposition 5.4.1.

3. Wirsing (1971, 1974) studied the operator T in greater depth. He considered the operator T_1 on $C(I)$ defined by the equation $T_1 g = -(TG)'$, where $G' = g$. More precisely

$$(T_1 g)(w) = \sum_{i \in \mathbb{N}^*} \left[\frac{i}{(w+i+1)^2} \int_{1/(w+i+1)}^{1/(w+i)} g(y)\,dy + \frac{w+1}{(w+i)^3 (w+i+1)} g\left(\frac{1}{w+i}\right) \right], \quad g \in C(I).$$

Wirsing showed that T_1 has an eigenvalue

$$\alpha = 0.303\,663\,002\,898\,732\,658\,60\ldots$$

with a corresponding positive eigenfunction $\psi \in C(I)$, and for any $g \in C(I)$, as $n \to \infty$, we have

$$T_1^n g = \alpha^n \psi F(g) + O(\beta^n |g|),$$

where F is a positive bounded linear functional on $C(I)$, and $0 < \beta \leqslant \alpha - 0.0031$. Hence, for any $f \in L(I)$ with continuous derivative f', we have

$$T^n f(w) = (-\alpha)^n (w+1)\Psi'(w) F(f') + O(\beta^n |f'|), \tag{5.2.16}$$

where the function Ψ is defined by $((w+1)\Psi'(w))' = \psi(w)$, $w \in I$,

$\Psi(0)=\Psi(1)=0$. (Clearly, like F, both ψ and Ψ are determined up to a multiplicative constant.) Actually, the functions ψ and Ψ are analytic and their analytic continuations are holomorphic in the complex plane with a cut along the negative real axis from $-\infty$ to -1, which constitutes their natural boundary. The function Ψ satisfies the functional equation $\Psi(z)-\Psi(z+1)=\alpha^{-1}\Psi(1/(1+z))$. From equation (5.2.16) we conclude that the spectral radius $r_L(T)=\alpha$. Then the special case $f(w)=w+1$, with a suitable norming of F, yields ($t=1/w$, $n\to\infty$) the estimate

$$\lambda(r_n>t)-\frac{1}{\log 2}\log\frac{t+1}{t}=(-\alpha)^n\Psi\left(\frac{1}{t}\right)+O\left(\frac{1}{t}\left(1-\frac{1}{t}\right)\beta^n\right),\quad t\geqslant 1. \tag{5.2.17}$$

Thus, the optimal value of q in (5.2.14), and also in (5.2.9), equals α. It is easy to see (by taking $f(w)=(w+1)F_0'(w)$ in (5.2.16)) that (5.2.17) still holds when λ is replaced by any probability $\mu\ll\lambda$, with density F_0' which has a continuous derivative. An estimate for

$$\mu(r_n>t)-\frac{1}{\log 2}\log\frac{t+1}{t}$$

in the case where $F_0'\in C(I)$ was referred to in Remark 1 above. An estimate of the same difference in the case where μ is assumed only to be absolutely continuous w.r.t. λ can be obtained from Lemma 1 in Philipp (1970) (see the Remark following Corollary 5.2.7).

4. Kuzmin (1928) did not use the operator U but equation (5.2.12), which amounts to considering the Frobenius–Perron operator P_λ associated with the continued fraction transformation τ defined (see Section A2.11 and Subsection 5.4.2) as

$$P_\lambda f(w)=\sum_{i\in\mathbb{N}^*}\frac{1}{(w+i)^2}f\left(\frac{1}{w+i}\right),\quad f\in C(I). \tag{5.2.18}$$

Kuzmin produced the elementary proof that $P_\lambda^n f_0$ converges uniformly to

$$\frac{(\log 2)^{-1}\int_0^1 f_0(w)\,dw}{1+w}$$

as $n\to\infty$ for any function $f_0\in C(I)$ with bounded derivative, the rate of convergence being $O(q^{\sqrt{n}})$ for some positive number $q<1$.

Babenko (1978) considered the collection of complex-valued functions f which are regular in the half-plane $\operatorname{Re} z>-\frac{1}{2}$. This is a Banach space, to be denoted H_2, under the norm

$$|||f|||=\left(\sup_{-1/2<x<\infty}\frac{1}{2\pi}\int_{-\infty}^{\infty}|f(x+iy)|^2\,dy\right)^{\frac{1}{2}}.$$

Then the linear operator B_λ defined, similarly to (5.2.18), as

$$B_\lambda f(z) = \sum_{i\in\mathbb{N}^*} \frac{1}{(z+i)^2} f\left(\frac{1}{z+i}\right), \quad f\in H_2, \quad \operatorname{Re} z > -\tfrac{1}{2},$$

is a compact operator on H_2 (see Section A2.8). Babenko was able to prove that the eigenvalues of B_λ are all real. Multiple eigenvalues counted multiply, let σ_i, $i\in\mathbb{N}^*$, be the eigenvalues of B_λ, arranged in order of decreasing absolute values. Then $1 = \sigma_1 > \alpha = -\sigma_2 > |\sigma_3| \geqslant \cdots$ and, for any $\varepsilon > 0$, we have $\sum_{i\in\mathbb{N}^*} |\sigma_i|^\varepsilon < \infty$. If $\psi_i \in H_2$ is the eigenfunction, suitably normed, corresponding to σ_i, $i\in\mathbb{N}^*$, then $\psi_1(z) = (\log 2)^{-1/2}(1+z)^{-1}$, and ψ_i, $i \geqslant 2$, is regular in the complex plane with a cut along the negative real-axis from $-\infty$ to -1, which constitutes its natural boundary, and we have

$$\sum_{i\in\mathbb{N}^*} |\psi_i(z)|^2 = \sum_{i\in\mathbb{N}^*} \frac{1}{(2\operatorname{Re} z + i)^2}$$

for any $z\in\mathbb{C}$ with $\operatorname{Re} z > -\frac{1}{2}$. When $\mu = \lambda$, it follows that, for any $w\in I$ and $n\in\mathbb{N}^*$, $F_n(w)$ can be written as

$$F_n(w) = \frac{\log(w+1)}{\log 2} + \sum_{i\geqslant 2} \sigma_i^{n-1} \psi_i(0) \int_0^w \psi_i(u)\,du. \tag{5.2.18}$$

Numerical investigations (Babenko & Jur'ev (1978) have suggested that:

(a) any eigenvalue σ_i, $i\in\mathbb{N}^*$, is simple;

(b) we have $|\psi_i(0)| = \max_{w\in I} |\psi_i(w)|$, $i\in\mathbb{N}^*$.

It is thus clear that, although (5.2.18) was called the 'complete' solution of Gauss' problem, new problems were raised whose solutions do not seem easy.

5.2.4 *Limit theorems*

The results obtained previously in this section allow us to state certain limit theorems proved in Chapter 4 for the sequences $(a_n)_{n\in\mathbb{N}^*}$, $(s_n)_{n\in\mathbb{N}}$, and related sequences. This allows us to obtain in a unifying way results originally proved by *ad hoc* methods by A. Ya. Khinchin, P. Lévy and A. Denjoy (see Khinchin (1964) and Lévy (1954, Ch.IX)) as well as improvements of them.

In Theorems 4.1.2 and 4.1.5, in the present context, we have to consider a measurable real-valued function h on I such that $\int_0^1 |h(w)|\,dw < \infty$ (for Theorem 4.1.2) or $\int_0^1 h^2(w)\,dw < \infty$ (for Theorem 4.1.5). Then, taking $F(i_1, i_2, \ldots) = h([i_1, i_2, \ldots]), (i_1, i_2, \ldots) \in (\mathbb{N}^*)^{\mathbb{N}^*}$, we have $f_n = h\circ\tau^{n-1}$, $n\in\mathbb{N}^*$, where τ is the continued fraction transformation. Next

$$\mathsf{E}_\infty f_1 = \int_0^1 h(w)\gamma(dw) \ (=a),$$

$$\mathsf{E}_\infty(f_1|(\xi_i)_{1\leqslant i\leqslant n})(y)=\frac{\displaystyle\int_{E_{i_1\ldots i_n}}h(u)\gamma(du)}{\gamma(E_{i_1\ldots i_n})}$$

for $y\in E_{i_1\ldots i_n}$, and

$$\sigma^2=\int_0^1 h^2(w)\gamma(dw)-a^2+2\sum_{n\in\mathbb{N}^*}\left(\int_0^1 h(w)h(\tau^n(w))\gamma(dw)-a^2\right).$$

Let us consider two special cases. First, if $h=\chi_{\{a_1=i\}}$, $i\in\mathbb{N}^*$, then

$$a=\mathsf{E}_\infty(f_1)=\frac{1}{\log 2}\int_{1/(i+1)}^{1/i}\frac{dw}{1+w}=\frac{1}{\log 2}\log\frac{(i+1)^2}{i(i+2)}$$

while $f_n=h\circ\tau^{n-1}=\chi_{\{a_n=i\}}$, $n\in\mathbb{N}^*$. It follows from Theorem 4.1.2 that the asymptotic relative frequency of i among the digits $a_1(y), a_2(y),\ldots$ of the continued fraction expansion of y equals $(\log 2)^{-1}\log(i+1)^2/i(i+2)$ a.e.

Second, if $h=\log a_1$, then

$$a=\mathsf{E}_\infty(f_1)=\frac{1}{\log 2}\sum_{i\in\mathbb{N}^*}\int_{1/(i+1)}^{1/i}\frac{\log i}{1+u}du=\log\prod_{i\in\mathbb{N}^*}\left(1+\frac{1}{i^2+2i}\right)^{\log i/\log 2}$$

while $f_n=h\circ\tau^{n-1}=\log a_n$, $n\in\mathbb{N}^*$. It follows that

$$\lim_{n\to\infty}\sqrt[n]{a_1(y)\cdots a_n(y)}=\prod_{i\in\mathbb{N}^*}\left(1+\frac{1}{i^2+2i}\right)^{\log i/\log 2}$$

a.e. The above infinite product equals 2.685 452 001 065 306 445 30... (see Wrench, Jr (1960), where the first 155 decimals are given; see also Wrench, Jr & Shanks (1966)).

Theorem 4.1.5 leads to functional central limit theorems for sequences constructed from the sequence $(a_n)_{n\in\mathbb{N}^*}$ under λ (thus, by Theorem 4.1.10, under any $\mu\ll\lambda$). Clearly, all the corollaries of Theorem 4.1.5 also hold, as well as Theorem 4.1.11.

In Theorems 4.1.12 and 4.1.13, in the present context, we have to consider a real-valued function g on $(\mathbb{N}^*)^k$ for some fixed $k\in\mathbb{N}^*$, for which

$$\sum_{i_1,\ldots,i_k\in\mathbb{N}^*}|g(i_1,\ldots,i_k)|^{2+\delta}\log\frac{1+v(i_1,\ldots,i_k)}{1+u(i_1,\ldots,i_k)}<\infty.$$

Then

$$\mathsf{E}_\infty(f_1^r)=\frac{1}{\log 2}\sum_{i_1,\ldots,i_k\in\mathbb{N}^*}g^r(i_1,\ldots,i_k)\log\frac{1+v(i_1,\ldots,i_k)}{1+u(i_1,\ldots,i_k)}$$

with $r=1$ or 2, and

$$\sigma^2=\mathsf{E}_\infty(f_1^2)-\mathsf{E}_\infty^2(f_1)+2\sum_{n\in\mathbb{N}^*}\sum_{i_1,\ldots,i_{n+k}\in\mathbb{N}^*}\Bigg[(\log 2)^{-1}g(i_1,\ldots,i_k)$$
$$\times\, g(i_{n+1},\ldots,i_{n+k})\log\frac{1+v(i_1,\ldots,i_{n+k})}{1+u(i_1,\ldots,i_{k+n})}-\mathsf{E}_\infty^2(f_1)\Bigg].$$

Theorem 4.1.12 yields an estimate of the rate of convergence in the central limit theorem for the sequence $(g(a_{n+r},\ldots,a_{n+r+k-1}))_{n\in\mathbb{N}^*}$ under λ for $r=0$, or under $\lambda(\cdot|E_{i_1\ldots i_r})$ for $r, i_1,\ldots,i_r\in\mathbb{N}^*$. (Note that Theorem A3.5 does the same thing under γ.)

Theorem 4.1.13 leads to Strassen's law of the iterated logarithm for the sequence $(g(a_n,\ldots,a_{n+k-1}))_{n\in\mathbb{N}^*}$ under λ (and therefore under any $\mu\ll\lambda$). It can be shown (see Gordin (1971a)) that in the case $k=1$ we have $\sigma=0$ iff $g=$ constant. We note that the classical law of the iterated logarithm was obtained by Stackelberg (1966) in the special case $k=1$, $g(i)=\log i$, $i\in\mathbb{N}^*$, and by Szüsz (1971) in the special case $k=1$, $g(i)=O(i^{1/2-\varepsilon})$, $\varepsilon>0$, $i\in\mathbb{N}^*$.

Regarding the sequence $(s_n)_{n\in\mathbb{N}}$, for any function $f\in C(I)$ Theorem 4.2.1 leads to a strong law of large numbers for the sequence $(f(s_n))_{n\in\mathbb{N}}$ under λ (and therefore under any $\mu\ll\lambda$). Let us consider two special cases. First, taking $f(w)=w$, $w\in I$, we deduce that $n^{-1}\sum_{k=1}^n s_k$ converges a.e. as $n\to\infty$ to

$$\frac{1}{\log 2}\int_0^1 \frac{w}{w+1}\,\mathrm{d}w = \frac{1}{\log 2} - 1.$$

Second, taking $f(w)=w^\varepsilon\log w$, $w\in I$, with $\varepsilon>0$, we deduce that $n^{-1}\sum_{k=1}^n s_k^\varepsilon \log s_k$ converges a.e. as $n\to\infty$ to

$$\frac{1}{\log 2}\int_0^1 \frac{w^\varepsilon \log w}{w+1}\,\mathrm{d}w.$$

Letting $\varepsilon\to 0$, we conclude that $n^{-1}\sum_{k=1}^n \log s_k$ converges a.e. as $n\to\infty$ to

$$\begin{aligned}\frac{1}{\log 2}\int_0^1 \frac{\log w}{w+1}\,\mathrm{d}w &= \frac{1}{\log 2}\left(\log(w+1)\log w\Big|_0^1 - \int_0^1 \frac{\log(w+1)}{w}\,\mathrm{d}w\right)\\ &= -\frac{1}{\log 2}\sum_{k\in\mathbb{N}}\frac{(-1)^k}{k+1}\int_0^1 w^k\,\mathrm{d}w\\ &= -\frac{1}{\log 2}\sum_{k\in\mathbb{N}}\frac{(-1)^k}{(k+1)^2} = -\frac{\pi^2}{12\log 2}.\end{aligned}$$

(The details are left to the reader.)

Noting that $\sum_{k=1}^n \log s_k = -\log q_n$, it follows that $\sqrt[n]{q_n}$ converges a.e. as $n\to\infty$ to $\exp(\pi^2/12\log 2)$. (For another proof of this result see Billingsley (1965, pp. 45–46).)

For any real-valued function $f\in L(I)$, Theorem 4.2.7 yields the optimal rate of convergence $O(n^{-\frac{1}{2}})$ in the central limit theorem for the sequence $(f(s_{n+r}))_{n\in\mathbb{N}}$ under λ, for $r=0$, or under $\lambda(\cdot|E_{i_1\ldots i_r})$ for $r, i_1,\ldots,i_r\in\mathbb{N}^*$. (For a different and in some respects, more general result, see Theorem 5.3.24.) Note that in the special case $f(w)=\log w$, $w\in I$, leading to $\sum_{k=1}^n f(s_k)=-\log q_n$, the assumption $f\in L(I)$ is not fulfilled. Nevertheless, Misevičius

(1981) was able to prove an $O(n^{-\frac{1}{2}}\log n)$ rate of convergence in the central limit theorem for the sequence $(\log q_n)_{n\in\mathbb{N}}$ under λ.

For any continuous real-valued function f on I, Theorem 4.2.8 leads to Strassen's law of the iterated logarithm for the sequence $(f(s_n))_{n\in\mathbb{N}}$ under any $\mu \ll \lambda$. Again in the special case $f(w)=\log w$, $w\in I$, the assumption on f is not fulfilled. Nevertheless, Gordin & Reznik (1970) and Philipp & Stackelberg (1969) were able to prove the classical law of the iterated logarithm for $(\log q_n)_{n\in\mathbb{N}}$, namely, for some $b>0$,

$$\lambda\left(\limsup_{n\to\infty}\frac{\log q_n - n\pi^2/12\log 2}{b\sqrt{2n\log\log n}}=1\right)=1.$$

5.2.5 Generalizations

The continued fraction transformation τ is the prototype of a general class of transformations to be studied in Section 5.4. (An even more general class will be studied in Section 5.3.)

A straightforward generalization of τ which has recently received attention is the so called α-continued fraction transformation τ_α of $[\alpha-1,\alpha]$, $\frac{1}{2}\leqslant\alpha\leqslant 1$, defined as

$$\tau_\alpha(x)=\begin{cases}|1/x|-[|1/x|+1-\alpha], & \text{if } x\neq 0\\ 0, & \text{if } x=0.\end{cases}$$

Clearly, $\tau_1=\tau$. It is a remarkable fact that, for any $\frac{1}{2}\leqslant\alpha\leqslant 1$, there exists a unique probability $\gamma_\alpha \ll \lambda$ such that τ_α is γ_α-preserving, and the density h_α of γ_α can be written in closed form (see Nakada (1981) and the references therein) as follows:

$$h_\alpha(x)=\frac{1}{\log G}\times\begin{cases}1/(x+G+1), & \text{if } x\in[\alpha-1,(1-2\alpha)/\alpha]\\ 1/(x+2), & \text{if } x\in((1-2\alpha)/\alpha,(2\alpha-1)/(1-\alpha))\\ 1/(x+G), & \text{if } x\in[(2\alpha-1)/(1-\alpha),\alpha],\end{cases}$$

for $\frac{1}{2}\leqslant\alpha\leqslant 1/G=(\sqrt{5}-1)/2$, and

$$h_\alpha(x)=\frac{1}{\log(1+\alpha)}\times\begin{cases}1/(x+2), & \text{if } x\in[\alpha-1,(1-\alpha)/\alpha]\\ 1/(x+1), & \text{if } x\in((1-\alpha)/\alpha,\alpha],\end{cases}$$

for $1/G<\alpha\leqslant 1$.

Special attention has been paid to the case $\alpha=\frac{1}{2}$, which leads to the so-called nearest integer continued fraction expansion of irrational numbers in $[-\frac{1}{2},\frac{1}{2}]$. Any irrational number $y\in[-\frac{1}{2},\frac{1}{2}]$ can be written in a unique way as

$$y=\cfrac{\varepsilon_1(y)}{b_1(y)+\cfrac{\varepsilon_2(y)}{b_2(y)+\ddots}}$$

where $\varepsilon_n(y) \in \{-1, 1\}$ and $2 \leqslant b_n(y) \in \mathbb{N}^*$, $b_n(y) + \varepsilon_{n+1}(y) \geqslant 2$, $n \in \mathbb{N}^*$. The integers $\varepsilon_n(y)$ and $b_n(y)$, $n \in \mathbb{N}^*$, are determined as follows: $\varepsilon_1(y) = \operatorname{sgn} y$, $b_1(y) = |1/y| - \tau_{\frac{1}{2}}(y)$, and $\varepsilon_{n+1}(y) = \varepsilon_1(\tau_{\frac{1}{2}}^n(y))$, $b_{n+1}(y) = b_1(\tau_{\frac{1}{2}}^n(y))$, $n \in \mathbb{N}^*$. The name '*nearest integer*' continued fraction comes from the equation

$$\tau_{\frac{1}{2}}^n(y) = -b_n(y) + \frac{\varepsilon_n(y)}{\tau_{\frac{1}{2}}^{n-1}(y)}$$

showing that $b_n(y)$ is the nearest integer to $\varepsilon_n(y)/\tau_{\frac{1}{2}}^{n-1}(y)$, $n \in \mathbb{N}^*$. For various aspects of the metric theory of the nearest integer continued fraction expansion, the reader is referred to Adams (1979), Jager (1985), Kraaikamp (1987), Nakada (1981), Rieger (1979), and Rockett (1980).

A dependence-with-complete-connection approach in the spirit of the previous subsections of this section, has been developed by Kalpazidou (1985a, 1986b, c).

5.3 Piecewise monotonic transformations

5.3.1 Preliminaries

In this section we study the iterates of certain transformations of the unit interval. The motivation comes from the material presented in Example 7 in Section 1.2.

A many-to-one map τ from the unit interval $I = [0, 1]$ into itself is called a C^s *piecewise monotonic transformation* (pw.m.t.), $s \in \mathbb{N}$, iff there exists a denumerable collection $(I(a))_{a \in \mathfrak{A}}$ of pairwise disjoint open subintervals of I such that

(i) $S = I - \bigcup_{a \in \mathfrak{A}} I(a)$ has Lebesgue measure 0 (this implies that the closure of $\bigcup_{a \in \mathfrak{A}} I(a)$ is I);

(ii) for any $a \in \mathfrak{A}$ the restriction of τ to $I(a)$ is strictly monotonic and extends to a C^s-function τ_a on the closure $\overline{I(a)}$ of $I(a)$, with

$$\tau_a(\overline{I(a)}) = I. \tag{5.3.1}$$

The more general case where (5.3.1) does not hold will be discussed later (see Subsection 5.3.6).

For any $a \in \mathfrak{A}$ let f_a denote the inverse function of τ_a, and thus a one-to-one map of I onto $\overline{I(a)}$. For any $a^{(n)} \in \mathfrak{A}^n$, $n \in \mathbb{N}^*$, put $f_{a^{(n)}} = f_{a_1} \circ \cdots \circ f_{a_n}$. Clearly, $f_{a^{(n)}}$ is the inverse function of $\tau_{a_n} \circ \cdots \circ \tau_{a_1}$, and maps I one-to-one onto the closure $\overline{I(a^{(n)})}$ of the open interval $I(a^{(n)})$, with endpoints $f_{a^{(n)}}(0)$ and $f_{a^{(n)}}(1)$. For any $n \in \mathbb{N}^*$, the $\overline{I(a^{(n)})}$, $a^{(n)} \in \mathfrak{A}^n$, which are called intervals of rank n, exhaust I up to a set of Lebesgue measure

0, and any two of them can have only an endpoint in common. The collection of the interiors $I(a^{(n)})$, $a^{(n)} \in \mathfrak{A}^n$, of the intervals of rank n consists of all sets of the form $\bigcap_{i=0}^{n-1} \tau^{-i}(I(a_{i+1}))$, $a_j \in \mathfrak{A}$, $1 \leqslant j \leqslant n$, with $\tau^0 =$ the identity map. Hence, for any $n \in \mathbb{N}^*$, the intervals of rank $n+1$ are obtained by partitioning the intervals of rank n. The collection of the intervals of all ranks $n \in \mathbb{N}^*$ is known as the class of fundamental intervals. The numbers $f_{a^{(n)}}(0), f_{a^{(n)}}(1), a^{(n)} \in \mathfrak{A}^n, n \in \mathbb{N}^*$, will be referred to as fundamental endpoints. For any $n \in \mathbb{N}^*$ put

$$\Delta_n = \max |f_{a^{(n)}}(1) - f_{a^{(n)}}(0)|,$$

where the maximum is taken over all $a^{(n)} \in \mathfrak{A}^n$. Thus Δ_n is the length of the largest interval of rank n.

With τ we associate the label sequence $(a_n(\cdot))_{n \in \mathbb{N}^*}$ defined by $a_n(t) = a$ iff $\tau^{n-1}(t) \in I(a), t \in I, a \in \mathfrak{A}$, whenever this is possible. Clearly, with the possible exception of the points of a set of Lebesgue measure 0, any point $t \in I$ has a label sequence $(a_n(t))_{n \in \mathbb{N}^*}$. Note that the $a_n, n \in \mathbb{N}^*$, are $\mathfrak{A}$-valued random variables on $(I, \mathscr{B}_I)$, where $\mathscr{B}_I$ is the σ-algebra of Borel sets in I. It is easy to see that, for any $n \in \mathbb{N}^*$, the set of points t having a label sequence $(a_n(t))_{n \in \mathbb{N}^*}$ such that $a_i(t) = a_i$, $1 \leqslant i \leqslant n$, is included in $I(a^{(n)})$, $a^{(n)} = (a_1, \ldots, a_n)$, and differs from it by a countable set. Hence distinct points will have different label sequences iff

$$\lim_{n \to \infty} \Delta_n = 0. \tag{5.3.2}$$

For, if (5.3.2) holds and $t', t'' \in I, t' \neq t''$, then choosing $n \in \mathbb{N}^*$ such that $\Delta_n < |t' - t''|$ shows that the sequences $(a_n(t'))_{n \in \mathbb{N}^*}$ and $(a_n(t''))_{n \in \mathbb{N}^*}$ cannot agree for the first n terms. Conversely, if (5.3.2) does not hold, then it is easy to see that there exists a whole interval whose points have the same label sequence.

Notice that equation (5.3.2) is equivalent to the fact that the fundamental intervals generate the σ-algebra $\mathscr{B}_I$.

Consider the following conditions.

Condition (C)

$$\operatorname*{ess\,sup}_{t \in I} |f'_{a^{(n)}}(t)| \Big/ \operatorname*{ess\,inf}_{t \in I} |f'_{a^{(n)}}(t)| \leqslant C, \quad a^{(n)} \in \mathfrak{A}^n, \quad n \in \mathbb{N}^*,$$

where C is a constant $\geqslant 1$.

Condition (E_m)

$$\operatorname*{ess\,sup}_{t \in I} |f'_{a^{(m)}}(t)| \leqslant \alpha, \quad a^{(m)} \in \mathfrak{A}^m,$$

where α is a constant < 1.

Condition (A)

$$\sup_{a\in\mathfrak{A}} \operatorname*{ess\,sup}_{t\in I} \left|\frac{f''_a(t)}{f'_a(t)}\right| < \infty. \tag{5.3.3}$$

Condition (C) is known as Rényi's condition and was first used in the special case of f-expansions (see Section 5.4). Condition (E_m) is an expansiveness condition, and is clearly equivalent to

$$\operatorname*{ess\,inf}_{t\in I(a^{(m)})} |(\tau_{a_m}\circ\cdots\circ\tau_{a_1})'(t)| \geqslant 1/\alpha > 1, \quad a^{(m)}\in\mathfrak{A}^m.$$

Finally, Condition (A) limits the departure of the f_a, $a\in\mathfrak{A}$, from linearity.

Note that Conditions (C) and (E_m) make sense for any C^0 pw.m.t. for which the f_a, $a\in\mathfrak{A}$, are absolutely continuous in I, while Condition (A) makes sense for any C^1 pw.m.t. for which the second derivatives f''_a, $a\in\mathfrak{A}$, exist a.e. in I.

If Conditions (C) or (E_m) are assumed for a C^s pw.m.t. with $s\geqslant 1$, then, clearly, $\operatorname{ess\,sup}_{t\in I}$ and $\operatorname{ess\,inf}_{t\in I}$ coincide with $\sup_{t\in I}$ and $\inf_{t\in I}$, respectively. Also if Condition (A) is assumed for a C^s pw.m.t. with $s\geqslant 2$, then $\operatorname{ess\,sup}_{t\in I}$ coincides with $\sup_{t\in I}$.

Proposition 5.3.1. *Let τ be a C^0 pw.m.t. for which the f_a, $a\in\mathfrak{A}$, are absolutely continuous in I. If Condition (E_m) holds for some $m\in\mathbb{N}^*$, then*

$$\Delta_n \leqslant \alpha^{[n/m]}\Delta_u, \quad n\in\mathbb{N}^*,$$

where $u\equiv n (\operatorname{mod} m)$, $0\leqslant u<m$, with $\Delta_0=1$.

Proof. Note first that, by the assumptions made on τ, the derivatives we write below do exist a.e. in I. We use the formula

$$f'_{ij}(t) = f'_i(f_j(t))f'_j(t), \tag{5.3.4}$$

which is valid for any u-tuple $i\in\mathfrak{A}^u$ and any v-tuple $j\in\mathfrak{A}^v$, for any $u, v\in\mathbb{N}^*$. Here ij denotes the $(u+v)$-tuple, whose components are those of i followed by those of j. Since

$$f_{a^{(n)}}(1) - f_{a^{(n)}}(0) = \int_0^1 f'_{a^{(n)}}(t)\,dt, \quad a^{(n)}\in\mathfrak{A}^n, \quad n\in\mathbb{N}^*,$$

by (5.3.4), using (E_m), for any $n\in\mathbb{N}^*$ we have

$$\begin{aligned}\Delta_n &= \max_{a^{(n)}\in\mathfrak{A}^n} \int_0^1 |f'_{a^{(n)}}(t)|\,dt \\ &\leqslant \alpha^{[n/m]} \times \begin{cases} 1, & \text{if } n\equiv 0 (\operatorname{mod} m) \\ \max_{a^{(u)}\in\mathfrak{A}^u} \int_0^1 |f'_{a^{(u)}}(t)|\,dt, & \text{if } n\equiv u (\operatorname{mod} m), \quad 1\leqslant u<m \end{cases} \\ &= \alpha^{[n/m]}\Delta_u. \end{aligned}$$

□

Proposition 5.3.2. *Let τ be a C^0 pw.m.t. for which the $f_a, a \in \mathfrak{A}$, are absolutely continuous in I. Then Condition* (C) *implies Condition* (E_m) *for some $m \in \mathbb{N}^*$.*

Proof. (*cf*. Halfant (1977)) We show first that Condition (C) implies equation (5.3.2).

Assume that this is not the case. Then $\lim_{n\to\infty} \Delta_n = d > 0$ since the sequence $(\Delta_n)_{n\in\mathbb{N}^*}$ is non-increasing. It is easy to see that there exists an interval $[u, v] \subset I$, with $v - u = d$, such that the open interval (u, v) contains no fundamental endpoints. It is possible that (a) neither u nor v is a fundamental endpoint, or (b) one of them is such an endpoint. We shall only consider alternative (a), since for alternative (b) the same kind of argument applies. By the definition of $[u, v]$, for any $\varepsilon > 0$, there exist $n \in \mathbb{N}^*$ and $a^{(n)} \in \mathfrak{A}^n$ such that $[u, v] \subset I(a^{(n)})$, the corresponding endpoints of these intervals being separated by less than ε. Hence $f_{a^{(n)}}(u) \in I(a^{(n)}) \subset [u - \varepsilon, v + \varepsilon]$. We cannot have $f_{a^{(n)}}(u) \in (u, v)$, for then also $f_{a^{(n)}}(t) \in (u, v)$, where t is a fundamental endpoint sufficiently close to u. But $f_{a^{(n)}}(t)$ is itself a fundamental endpoint, contradicting the definition of $[u, v]$. Consequently, we should have either (i) $f_{a^{(n)}}(u) \in [u - \varepsilon, u]$ or (ii) $f_{a^{(n)}}(u) \in [v, v + \varepsilon]$. To make a choice, assume that $f_{a^{(n)}}$ is increasing. For alternative (i) we then have

$$0 < f_{a^{(n)}}(u) - f_{a^{(n)}}(0) = \int_0^u f'_{a^{(n)}}(t)\,\mathrm{d}t \leqslant \varepsilon$$

and

$$d \leqslant f_{a^{(n)}}(1) - f_{a^{(n)}}(u) = \int_u^1 f'_{a^{(n)}}(t)\,\mathrm{d}t,$$

whence

$$\frac{\operatorname*{ess\,sup}_{t\in I} f'_{a^{(n)}}(t)}{\operatorname*{ess\,inf}_{t\in I} f'_{a^{(n)}}(t)} \geqslant \frac{du}{\varepsilon(1-u)},$$

which, since ε is arbitrary, violates Condition (C). Similarly, for alternative (ii), we have

$$d \leqslant f_{a^{(n)}}(u) - f_{a^{(n)}}(0) = \int_0^u f'_{a^{(n)}}(t)\,\mathrm{d}t,$$

$$0 < f_{a^{(n)}}(1) - f_{a^{(n)}}(u) = \int_u^1 f'_{a^{(n)}}(t)\,\mathrm{d}t \leqslant \varepsilon,$$

whence

$$\frac{\operatorname*{ess\,sup}_{t\in I} f'_{a^{(n)}}(t)}{\operatorname*{ess\,inf}_{t\in I} f'_{a^{(n)}}(t)} \geqslant \frac{d(1-u)}{\varepsilon u},$$

again violating Condition (C). The proof of (5.3.2) is thus complete.

Now, for any $n\in\mathbb{N}^*$ and $a^{(n)}\in\mathfrak{A}^n$, we have $\Delta_n \geqslant |f_{a^{(n)}}(1)-f_{a^{(n)}}(0)| \geqslant \operatorname{ess\,inf}_{t\in I}|f'_{a^{(n)}}(t)|$. Hence Condition (C) implies that $\operatorname{ess\,sup}_{t\in I}|f'_{a^{(n)}}(t)| \leqslant C\Delta_n$ for all $n\in\mathbb{N}^*$ and $a^{(n)}\in\mathfrak{A}^n$. By (5.3.2) we can choose $m\in\mathbb{N}^*$ such that $\Delta_m < \alpha/C$ with $\alpha<1$, implying that Condition (E_m) holds. □

Proposition 5.3.3. *Let τ be a C^1 pw.m.t. If Condition* (E_m) *holds for some $m\in\mathbb{N}^*$, then*

$$\sup_{t\in I, a\in\mathfrak{A}} |f'_a(t)| < \infty. \tag{5.3.5}$$

Proof. The assumption made on τ implies that $i(a)=\inf_{t\in I}|f'_a(t)|>0$ for any $a\in\mathfrak{A}$. Now, if $m=1$, then (5.3.5) obviously holds. Assume therefore that Condition (E_m) holds with $m>1$. Fix $a^{(m-1)}=(a_1,\ldots,a_{m-1})$ and assume that (5.3.5) does not hold. Then there exist $t_0\in I$ and $a\in\mathfrak{A}$ such that $|f'_a(t_0)|>1/i(a_1)\cdots i(a_{m-1})$. Setting $a_m=a$, $a^{(m)}=(a_1,\ldots,a_{m-1},a_m)$ we can write

$$f'_{a^{(m)}}(t) = f'_{a_1}(f_{(a_2,\ldots,a_m)}(t))\cdots f'_{a_{m-1}}(f_{a_m}(t))f'_{a_m}(t),$$

whence $|f'_{a^{(m)}}(t_0)| \geqslant i(a_1)\cdots i(a_{m-1})|f'_a(t_0)|>1$, thus contradicting Condition (E_m). □

Proposition 5.3.4. *Let τ be a C^1 pw.m.t. for which the derivatives f'_a, $a\in\mathfrak{A}$, are absolutely continuous in I. Then Condition* (E_m) *for some $m\in\mathbb{N}^*$ and Condition* (A) *imply Condition* (C).

Proof. Note that, by the assumptions made on τ, the second derivatives we write below do exist a.e. in I.

We use the formula

$$\frac{f''_{k_1\cdots k_p}(t)}{f'_{k_1\cdots k_p}(t)} = \sum_{i=1}^{p-1}\frac{f''_{k_i}(f_{k_{i+1}\cdots k_p}(t))}{f'_{k_i}(f_{k_{i+1}\cdots k_p}(t))} + \frac{f''_{k_p}(t)}{f'_{k_p}(t)},$$

which is valid for any u_i-tuple $k_i\in\mathfrak{A}^{u_i}$, for any $u_i\in\mathbb{N}^*$, $1\leqslant i\leqslant p$, $p\geqslant 2$. This is easily obtained starting with the relation of logarithmic differentiation $f''_{k_1\cdots k_p}(t)/f'_{k_1\cdots k_p}(t) = \mathrm{d}\log|f'_{k_1\cdots k_p}(t)|/dt$, and using equation (5.3.4).

First, using Proposition 5.3.3 and Condition (A), we deduce that

$$\max_{1\leqslant p\leqslant m}\ \sup_{a^{(p)}\in\mathfrak{A}^p}\ \operatorname*{ess\,sup}_{t\in I}\left|\frac{f''_{a^{(p)}}(t)}{f'_{a^{(p)}}(t)}\right| = M_1 < \infty. \tag{5.3.6}$$

Second, writing $u\equiv n \pmod m$, $0\leqslant u<m$, and using (5.3.6), Proposition 5.3.3, and Condition (E_m) we get

$$\operatorname*{ess\,sup}_{t\in I}\left|\frac{f''_{a^{(n)}}(t)}{f'_{a^{(n)}}(t)}\right| \leqslant \begin{cases} M_1 \sum_{j=0}^{[n/m]-1}\alpha^j, & \text{if } u=0 \\ M_1 + M_0 M_1 \sum_{j=0}^{[n/m]-1}\alpha^j, & \text{if } u>0 \end{cases}$$

$$\leqslant \begin{cases} M_1/(1-\alpha), & \text{if } u=0 \\ M_1 + M_0M_1/(1-\alpha), & \text{if } u>0 \end{cases}$$

for all $n>m$ and $a^{(n)}\in\mathfrak{A}^n$, where

$$M_0 = \max\left(1, \left(\sup_{t\in I, a\in\mathfrak{A}} |f'_a(t)|\right)^{m-1}\right) < \infty.$$

Now, for any $n\in\mathbb{N}^*$, $a^{(n)}\in\mathfrak{A}^n$ and $t_1, t_2\in I$, we have

$$\left|\log\frac{f'_{a^{(n)}}(t_2)}{f'_{a^{(n)}}(t_1)}\right| = \left|\int_{t_1}^{t_2}\frac{f''_{a^{(n)}}(t)}{f'_{a^{(n)}}(t)}\,\mathrm{d}t\right| \leqslant \log C,$$

where $C=\exp(M_1+M_0M_1/(1-\alpha))$. This clearly implies Condition (C).

□

5.3.2 The invariant probability: existence and uniqueness

When studying a transformation τ it is of prime importance that there exists an invariant probability ρ, i.e. one such that $\rho\tau^{-1}=\rho$, which actually means that τ is ρ-preserving (see Section A2.11). In this subsection we prove the existence and uniqueness of an invariant probability for certain C^0 pw.m.t.s. The invariant probability will be studied in more depth in Subsection 5.3.4.

Theorem 5.3.5. *Let τ be a C^0 pw.m.t. for which the f_a, $a\in\mathfrak{A}$, are absolutely continuous in I. Assume that Condition* (C) *holds. Then there exists a unique probability ρ such that τ is ρ-preserving. Moreover, ρ is equivalent to Lebesgue measure λ, the density $r=\mathrm{d}\rho/\mathrm{d}\lambda$ satisfying the inequalities $1/C\leqslant r\leqslant C$ a.e. in I and the endomorphism (τ,ρ) is exact (see Section* A2.11).

Proof. First, note that the absolute continuity of the f_a, $a\in\mathfrak{A}$, implies that τ is λ-non-singular.

For any subinterval $[u, v]$ of I, for any $n \in \mathbb{N}^*$ and $a^{(n)} \in \mathfrak{A}^n$, we can write

$$\frac{\lambda(I(a^{(n)}) \cap \tau^{-n}([u, v]))}{\lambda(I(a^{(n)}))} = \frac{f_{a^{(n)}}(v) - f_{a^{(n)}}(u)}{f_{a^{(n)}}(1) - f_{a^{(n)}}(0)}.$$

But

$$|f_{a^{(n)}}(v) - f_{a^{(n)}}(u)| = \int_u^v |f'_{a^{(n)}}(t)| \mathrm{d}t$$

and

$$(v-u) \operatorname*{ess\,inf}_{t \in I} |f'_{a^{(n)}}(t)| \leqslant \int_u^v |f'_{a^{(n)}}(t)| \mathrm{d}t \leqslant (v-u) \operatorname*{ess\,sup}_{t \in I} |f'_{a^{(n)}}(t)|$$

for all $0 \leqslant u < v \leqslant 1$. Hence, by Condition (C)

$$\frac{v-u}{C} \leqslant \frac{\lambda(I(a^{(n)}) \cap \tau^{-n}([u, v]))}{\lambda(I(a^{(n)}))} \leqslant C(v-u).$$

It is clear that we may replace the interval $[u, v]$ by an arbitrary Borel set $A \subset I$ to get

$$\frac{\lambda(I(a^{(n)}))\lambda(A)}{C} \leqslant \lambda(I(a^{(n)}) \cap \tau^{-n}(A)) \leqslant C\lambda(I(a^{(n)}))\lambda(A) \qquad (5.3.7)$$

for all $n \in \mathbb{N}^*$ and $a^{(n)} \in \mathfrak{A}^n$. Finally, summing (5.3.7) over $a^{(n)} \in \mathfrak{A}^n$, we get

$$\frac{\lambda(A)}{C} \leqslant \lambda(\tau^{-n}(A)) \leqslant C\lambda(A), A \in \mathscr{B}_I, n \in \mathbb{N}^*. \qquad (5.3.8)$$

Note that, in particular, (5.3.8) shows once more that τ is λ-non-singular.

We use (5.3.7) and (5.3.8) to prove that the tail σ-algebra $\mathscr{T} = \bigcap_{n \in \mathbb{N}} \tau^{-n}(\mathscr{B}_I)$ is trivial under λ, i.e. for any $A \in \mathscr{T}$ we have either $\lambda(A) = 0$ or $\lambda(A^c) = 0$. Indeed, let $A \in \mathscr{T}$. Then for any $p \in \mathbb{N}$ we have $A = \tau^{-p}(A_p)$ for some $A_p \in \mathscr{B}_I$, and (5.3.7) and (5.3.8) yield

$$\frac{\lambda(I(a^{(n)}) \cap A)}{\lambda(I(a^{(n)}))\lambda(A)} = \frac{\lambda(I(a^{(n)}) \cap \tau^{-n}(A_n))}{\lambda(I(a^{(n)}))\lambda(\tau^{-n}(A_n))} \geqslant \frac{\lambda(I(a^{(n)}) \cap \tau^{-n}(A_n))}{C\lambda(I(a^{(n)}))\lambda(A_n)} \geqslant \frac{1}{C^2} > 0,$$

whence

$$C^2 \lambda(I(a^{(n)}) \cap A) \geqslant \lambda(I(a^{(n)}))\lambda(A)$$

for all $n \in \mathbb{N}^*$ and $a^{(n)} \in \mathfrak{A}^n$. As the fundamental intervals generate $\mathscr{B}_I$ it is possible to approximate (to any ε) any set in $\mathscr{B}_I$, in particular A^c, by unions of fundamental intervals. Therefore

$$0 = C^2 \lambda(A^c \cap A) \geqslant \lambda(A^c)\lambda(A) \pm \varepsilon,$$

whence either $\lambda(A) = 0$ or $\lambda(A^c) = 0$.

We are now able to prove the existence of ρ with the stated properties. For any $A\in\mathscr{B}_I$ and $n\in\mathbb{N}^*$ put

$$\rho_n(A)=\frac{1}{n}\sum_{k=0}^{n-1}\lambda(\tau^{-k}(A))=\int_0^1\left(\frac{1}{n}\sum_{k=0}^{n-1}\chi_A(\tau^k(t))\right)dt.$$

Using (5.3.8) the integrand approaches a limit a.e. as $n\to\infty$ (see Dunford & Schwartz (1958, p. 683)). (Since $\mathscr{T}$ is λ-trivial, this limit should be a constant a.e.) Thus by dominated convergence the limit $\lim_{n\to\infty}\rho_n(A)=\rho(A)$ exists for any $A\in\mathscr{B}_I$, and the Vitali–Hahn–Saks Theorem A1.1 shows that ρ is a probability on $\mathscr{B}_I$. It follows from the definition of ρ_n that, for any $A\in\mathscr{B}_I$,

$$\rho_n(\tau^{-1}(A))=\frac{n+1}{n}\rho_n(A)-\frac{\lambda(A)}{n},$$

$$\frac{\lambda(A)}{C}\leqslant\rho_n(A)\leqslant C\lambda(A), n\in\mathbb{N}^*.$$

By letting $n\to\infty$ the equation above yields $\rho(\tau^{-1}(A))=\rho(A), A\in\mathscr{B}_I$, and Proposition A2.5 shows that ρ is the unique probability with this property. (Note that the λ-triviality of $\mathscr{T}$ implies that τ is ergodic w.r.t. λ.) By letting $n\to\infty$ the inequalities above imply that $\rho\equiv\lambda$, and that the density $r=d\rho/d\lambda$ enjoys the stated property.

Finally, the exactness of the endomorphism (τ,ρ) follows from the fact that $\rho\equiv\lambda$ and the λ-triviality of $\mathscr{T}$. □

5.3.3 The Frobenius–Perron operator

In this subsection we consider a C^1 λ-non-singular pw.m.t. (Note that in general the λ-non-singularity of a C^0 pw.m.t. means that, for any $a\in\mathfrak{A}$, the set of points $t\in I$ for which $\tau_a'(t)=0$ has Lebesgue measure 0, or, equivalently, that the set of points $u\in I$ for which $|f_a'(u)|=\infty$ has Lebesgue measure 0 (i.e. f_a is absolutely continuous). In particular, by Proposition 5.3.3, Condition (E_m) for some $m\in\mathbb{N}^*$ is sufficient to ensure the λ-non-singularity of τ.)

It is easy to check that the Frobenius–Perron operator P_λ associated with τ under λ (see Section A2.11) can be expressed as[†]

$$P_\lambda h=\sum_{a\in\mathfrak{A}}|f_a'|(h\circ f_a), h\in L^1(I,\mathscr{B}_I,\lambda).$$

Remark that, by Beppo Levi's Theorem A1.6, the series on the right (which

[†]Actually, this holds for any pw.m.t. for which the $f_a, a\in\mathfrak{A}$, are absolutely continuous.

is a finite sum when $\mathfrak{A}$ is finite) converges absolutely a.e. in I since

$$\sum_{a\in\mathfrak{A}}\int_I |f'_a||h\circ f_a|\mathrm{d}\lambda = \sum_{a\in\mathfrak{A}}\int_{I(a)}|h|\mathrm{d}\lambda = \int_I |h|\mathrm{d}\lambda < \infty.$$

It can be seen either directly or by considering the interpretation of the nth iterate P^n_λ of P_λ that

$$P^n_\lambda h = \sum_{a^{(n)}\in\mathfrak{A}^n} |f'_{a^{(n)}}|(h\circ f_{a^{(n)}}), n\in\mathbb{N}^*, h\in L^1(I,\mathscr{B}_I,\lambda).$$

Our goal is to study the operator P_λ. The main result, Theorem 5.3.11 below, shows that, under suitable assumptions, P_λ belongs to a class of operators described in Section A2.9. First of all, some definitions are needed.

Put $S_n = \bigcup_{k=0}^{n-1}\tau^{-k}(S), n\in\mathbb{N}^*$. Clearly, $S_1 = S(=I - \bigcup_{a\in\mathfrak{A}} I(a))$, and since, by definition, $\lambda(S)=0$, we also have $\lambda(S_n)=0$ for all $n\in\mathbb{N}^*$.

Let g be a non-negative function of bounded variation defined on I such that $g(t)=0, t\in S$. Define on I the functions $g_n, n\in\mathbb{N}^*$, by

$$g_n(t) = \begin{cases} \prod_{k=0}^{n-1} g(\tau^k(t)), & \text{if } t\notin S_n \\ 0, & \text{if } t\in S_n. \end{cases} \tag{5.3.9}$$

Clearly, $g_1 = g$.

In what follows we use the notation introduced in Section A2.10(3) writing L^1_λ and BV_λ for $L^1(I,\mathscr{B}_I,\lambda)$ and $BV(I,\mathscr{B}_I,\lambda)$, respectively.

Proposition 5.3.6. *For any complex-valued functions of bounded variation h and h_n on I such that $h_n(t)=0, t\in S_n$, we have*

$$\operatorname{var}(hh_n) = \sum_{a^{(n)}\in\mathfrak{A}^n} \operatorname{var}_{\overline{I(a^{(n)})}}(hh_n), n\in\mathbb{N}^*.$$

Proof. The equation above is a simple consequence of the fact that the restriction of h_n to S_n is 0 and the $I(a^{(n)}), a^{(n)}\in\mathfrak{A}^n$, exhaust I up to a set of Lebesgue measure 0. □

Proposition 5.3.7. *For any $n\in\mathbb{N}^*$ we have* $\operatorname{var} g_n \leq 2^{n-1}(\operatorname{var} g)^n$.

Proof. For $n=1$ the proposition is obvious. Assume it holds for some $n\geq 1$. Then by Proposition 5.3.6 we have

$$\operatorname{var} g_{n+1} = \operatorname{var}((g_n\circ\tau)g_1) = \sum_{a\in\mathfrak{A}} \operatorname{var}_{\overline{I(a)}}((g_n\circ\tau)g_1).$$

Now, for any $a\in\mathfrak{A}$, we can write

$$\begin{aligned}\operatorname{var}_{\overline{I(a)}}((g_n\circ\tau)g_1) &\leqslant |g_n| \operatorname{var}_{\overline{I(a)}} g_1 + |g_1\chi_{\overline{I(a)}}| \operatorname{var}_{\overline{I(a)}}(g_n\circ\tau)\\ &\leqslant \operatorname{var} g_n \operatorname{var}_{\overline{I(a)}} g + \operatorname{var}_{\overline{I(a)}} g \operatorname{var} g_n\\ &= 2\operatorname{var} g_n \operatorname{var}_{\overline{I(a)}} g.\end{aligned}$$

Using Proposition 5.3.6 again, with $h=1, n=1$, and $h_1=g$, we conclude that $\operatorname{var} g_{n+1}\leqslant 2\operatorname{var} g_n \operatorname{var} g \leqslant 2^n(\operatorname{var} g)^{n+1}$. □

Lemma 5.3.8. *For any $h\in B(I)$ and $A\in\mathscr{B}_I$ for which $\lambda(A)>0$ we have*

$$|h\chi_A|\leqslant \frac{1}{\lambda(A)}\left|\int_A h\,\mathrm{d}\lambda\right| + \operatorname{var}_A h.$$

Proof. For any $t\in A$ we can write

$$\begin{aligned}|h(t)| - \frac{1}{\lambda(A)}\left|\int_A h\,\mathrm{d}\lambda\right| &\leqslant \left|h(t) - \frac{1}{\lambda(A)}\int_A h\,\mathrm{d}\lambda\right|\\ &= \left|\frac{1}{\lambda(A)}\int_A (h(t)-h(u))\lambda(\mathrm{d}u)\right|\\ &\leqslant \operatorname{var}_A h,\end{aligned}$$

from which the stated inequality follows at once. □

Proposition 5.3.9. *Let $0=t_0<t_1\cdots<t_p=1$ be a division of I and put $A(i)=[t_i,t_{i+1}], 0\leqslant i\leqslant p-1$. For any complex-valued function h of bounded variation on I and any $n\in\mathbb{N}^*$ we have*

$$\operatorname{var}(hg_n)\leqslant q_n \operatorname{var} h + Q_n\sum_{i=0}^{p-1}\left|\int_{t_i}^{t_{i+1}} h\,\mathrm{d}\lambda\right|,$$

where $Q_n=\max_{0\leqslant i\leqslant p-1}\operatorname{var}_{A(i)} g_n/\lambda(A(i))$ and $q_n=|g_n|+\max_{0\leqslant i\leqslant p-1}\operatorname{var}_{A(i)} g_n$.

Proof. We have

$$\operatorname{var}_{A(i)}(hg_n)\leqslant |h\chi_{A(i)}|\operatorname{var}_{A(i)} g_n + |g_n|\operatorname{var}_{A(i)} h, 0\leqslant i\leqslant p-1.$$

The stated result follows from the equations

$$\sum_{i=0}^{p-1}\operatorname{var}_{A(i)}(hg_n)=\operatorname{var}(hg_n),\quad \sum_{i=0}^{p-1}\operatorname{var}_{A(i)} h=\operatorname{var} h$$

by using Lemma 5.3.8. □

Now, let us define our function g as

$$g(t)=\begin{cases}|f'_a(\tau_a(t))|, & \text{if } t\in I(a), a\in\mathfrak{A},\\ 0, & \text{if } t\in S.\end{cases}\tag{5.3.10}$$

Then it is easy to see that the functions $g_n, n\in\mathbb{N}^*$, associated with this special g by (5.3.9) are given by

$$g_n(t) = \begin{cases} |f'_{a^{(n)}}(\tau_{a_n} \circ \cdots \circ \tau_{a_1}(t))|, & \text{if } t \in I(a^{(n)}), a^{(n)} \in \mathfrak{A}^n, \\ 0, & \text{if } t \in S_n \end{cases}$$

for all $n \in \mathbb{N}^*$. Next, for any $h \in L^1_\lambda, n \in \mathbb{N}^*$, and $a^{(n)} \in \mathfrak{A}^n$, we have

$$P^n_\lambda(h\chi_{I(a^{(n)})}) \circ (\tau_{a_n} \circ \cdots \circ \tau_{a_1}) = hg_n$$

on $\overline{I(a^{(n)})}$. Assuming that h and g defined by (5.3.10) are functions of bounded variation on I, it follows that $\operatorname{var} P^n_\lambda(h\chi_{I(a^{(n)})}) = \operatorname{var}_{\overline{I(a^{(n)})}} hg_n$. Hence,by Proposition 5.3.6,

$$\sum_{a^{(n)} \in \mathfrak{A}^n} \operatorname{var} P^n_\lambda(h\chi_{I(a^{(n)})}) = \operatorname{var}(hg_n). \tag{5.3.11}$$

Let us introduce Condition (BV).

Condition (BV)

$$\sum_{a \in \mathfrak{A}} \operatorname{var} f'_a < \infty.$$

Note that, since a derivative can be a function of bounded variation only if it is continuous, Condition (BV) is consistent with our assumptions on τ.

We can now state Proposition 5.3.10.

Proposition 5.3.10. *Assume that Conditions* (E_m) *for some* $m \in \mathbb{N}^*$ *and* (BV) *hold. Then there exists* $k \in \mathbb{N}^*$ *such that*

$$\mathrm{v}(P^{km}_\lambda h) \leqslant q\,\mathrm{v}(h) + Q\|h\|_1, h \in BV_\lambda,$$

with suitable positive constants $q < 1$ *and* Q.

Proof. Let us first check that $\operatorname{var} g < \infty$, where g is given by (5.3.10). By Proposition 5.3.6, with $h = 1$, we have

$$\begin{aligned} \operatorname{var} g &= \sum_{a \in \mathfrak{A}} \operatorname{var}_{\overline{I(a)}}((f'_a \circ \tau_a)\chi_{I(a)}) \\ &= \sum_{a \in \mathfrak{A}} \operatorname{var} f'_a + \sum_{a \in \mathfrak{A}} (|f'_a(0)| + |f'_a(1)|). \end{aligned}$$

It follows from Lemma 5.3.8 that, for any $h \in B(I)$ and $u, v \in I, u < v$, we have

$$|h(u)| + |h(v)| \leqslant \frac{2}{v-u}\int_u^v |h|\,d\lambda + \operatorname{var}_{[u,v]} h.$$

(Apply Lemma 5.3.8 twice, first with $A = [u, (u+v)/2]$, then with $A = [(u+v)/2, v]$.) Therefore

$$\sum_{a\in\mathfrak{A}} (|f'_a(0)|+|f'_a(1)|) \leqslant \sum_{a\in\mathfrak{A}} \left(2\int_0^1 |f'_a| \mathrm{d}\lambda + \operatorname{var} f'_a\right)$$

$$= 2\sum_{a\in\mathfrak{A}} \lambda(I(a)) + \sum_{a\in\mathfrak{A}} \operatorname{var} f'_a = 2 + \sum_{a\in\mathfrak{A}} \operatorname{var} f'_a,$$

so that

$$\operatorname{var} g \leqslant 2\left(1 + \sum_{a\in\mathfrak{A}} \operatorname{var} f'_a\right) < \infty.$$

Next, remark that the jumps of g_m do not exceed

$$\sup_{t\in I, a^{(m)}\in\mathfrak{A}^m} |f'_{a^{(m)}}(t)| \leqslant \alpha < 1.$$

It follows that the jumps of g_{km} do not exceed $\alpha^k, k\in\mathbb{N}^*$. Choose k such that $3\alpha^k < 1$. Clearly, we can choose the division $0 = t_0 < t_1 < \cdots < t_p = 1$ in Proposition 5.3.9 such that $\max_{0\leqslant 1 \leqslant p-1} \operatorname{var}_{A(i)} g_{km} < 2\alpha^k$.

Finally, choose a version $\tilde{h}$ of h for which $\mathrm{v}(h) = \operatorname{var} \tilde{h}$ (see Section A2.10(3)). By Proposition 5.3.9, we have

$$\operatorname{var}(g_{km}\tilde{h}) \leqslant q \operatorname{var} \tilde{h} + Q\|h\|_1,$$

or

$$\operatorname{var}(g_{km}\tilde{h}) \leqslant q\mathrm{v}(h) + Q\|h\|_1,$$

with $q = 3\alpha^k$ and $Q = 2\alpha^k/\min_{0\leqslant i\leqslant p-1}\lambda(A(i))$.

The result stated follows by using equation (5.3.11) and the fact that $\sum_{a^{(km)}\in\mathfrak{A}^{km}} P_\lambda^{km}(\tilde{h}\chi_{I(a^{(km)})})$ is a version of $P_\lambda^{km}\tilde{h}$, hence of $P_\lambda^{km}h$, since

$$\operatorname{var} \sum_{a^{(km)}\in\mathfrak{A}^{km}} P_\lambda^{km}(\tilde{h}\chi_{I(a^{(km)})}) \leqslant \sum_{a^{(km)}\in\mathfrak{A}^{km}} \operatorname{var} P_\lambda^{km}(\tilde{h}\chi_{I(a^{(km)})})$$

and

$$\mathrm{v}(P_\lambda^{km}h) = \mathrm{v}(P_\lambda^{km}\tilde{h}) \leqslant \operatorname{var} \sum_{a^{(km)}\in\mathfrak{A}^{km}} P_\lambda^{km}(\tilde{h}\chi_{I(a^{(km)})}). \qquad \square$$

Theorem 5.3.11. *Let τ be a C^1 pw.m.t. Assume that Conditions* (E_m) *for some $m\in\mathbb{N}^*$ and* (BV) *hold. Then the Frobenius–Perron operator P_λ associated with τ under λ belongs to the class $\mathscr{L}_{km}(BV_\lambda, L_\lambda^1)$ (see Section A2.9). Here k is the natural integer whose existence has been proved in Proposition 5.3.10.*

Proof. We must show that the Ionescu Tulcea–Marinescu Theorem A2.4 applies to P_λ for the special case of the spaces $(X, |\cdot|) = (L_\lambda^1, \|\cdot\|_1)$ and $(Y, \|\cdot\|) = (BV_\lambda, \|\cdot\|_v)$ (see Sections A2.9 and A2.10(3)).

That P_λ takes BV_λ into itself is a consequence of equation (5.3.11) and Proposition 5.3.9. Next, Condition (ITM_2) is immediate, with $H = 1$, since

$\|P_\lambda\|_1 \leqslant 1$ (see Section A2.11). Also Condition (ITM_3) is true, for the inequality $\|P_\lambda^{km}h\|_1 \leqslant \|h\|_1, h \in BV_\lambda$, and Proposition 5.3.10 yield

$$\|P_\lambda^{km}h\|_v = v(P_\lambda^{km}h) + \|P_\lambda^{km}h\|_1 \leqslant q\|h\|_v + (Q+1-q)\|h\|_1.$$

Finally, to check Condition (ITM_4), let A be a bounded subset of BV_λ i.e. $\sup_{h\in A}\|h\|_v < \infty$. We have to prove that $P_\lambda^{km}A$ is relatively compact in L_λ^1. Since P_λ^{km} is a bounded linear operator on BV_λ, we have in fact to prove that any sequence $(h_n)_{n\in\mathbb{N}^*} \subset L_\lambda^1$ for which $\sup_{n\in\mathbb{N}^*}\|h_n\|_v = M < \infty$ contains a subsequence that converges in L_λ^1 to some h for which $\|h\|_v \leqslant M$. To proceed, let $\tilde{h}_n$ be a version of h_n such that $v(h_n) = \operatorname{var}\tilde{h}_n, n\in\mathbb{N}^*$ (see Section A2.10(3)). Then we have both $\operatorname{var}\tilde{h}_n \leqslant M$ and $\|\tilde{h}_n\|_1 \leqslant M, n\in\mathbb{N}^*$, which by Lemma 5.3.8, imply that $|\tilde{h}_n| \leqslant 2M, n\in\mathbb{N}^*$. Now, the well-known Helly theorem (see, e.g. Natanson (1961, Ch. VIII, §4)) ensures the existence of a function h of bounded variation on I and a subsequence $(\tilde{h}_{n_k})_{k\in\mathbb{N}^*}$ of $(\tilde{h}_n)_{n\in\mathbb{N}^*}$ such that $\lim_{k\to\infty}\tilde{h}_{n_k}(t) = h(t)$ for all $t\in I$. Clearly, $\lim_{k\to\infty}\|\tilde{h}_{n_k} - h\|_1 = 0$ and $\lim_{k\to\infty}\|\tilde{h}_{n_k}\|_1 = \|h\|_1$. Next, for any $\varepsilon > 0$, for suitable $p\in\mathbb{N}^*$ and $0 = t_0 < t_1 < \cdots < t_p = 1$ we have

$$\begin{aligned} v(h) &\leqslant \operatorname{var} h \leqslant \varepsilon + \sum_{i=1}^{p} |h(t_i) - h(t_{i-1})| \\ &= \varepsilon + \lim_{k\to\infty}\sum_{i=1}^{p} |\tilde{h}_{n_k}(t_i) - \tilde{h}_{n_k}(t_{i-1})| \leqslant \varepsilon + \limsup_{k\to\infty} \operatorname{var}\tilde{h}_{n_k}. \\ &= \varepsilon + \limsup_{k\to\infty} v(h_{n_k}). \end{aligned}$$

Hence $\|h\|_v \leqslant M + \varepsilon$, so that, since ε is arbitrary, $\|h\|_v \leqslant M$. □

Theorem 5.3.11, the Ionescu Tulcea–Marinescu Theorem A2.4, and Theorem A2.4′ lead to the description of the spectral properties of P_λ as given below.

Theorem 5.3.12. *Let τ be a C^1 pw.m.t. Assume that Conditions* (E_m) *for some $m\in\mathbb{N}^*$ and* (BV) *hold. Then the Frobenius–Perron operator P_λ associated with τ under λ has the following properties.*

(i) *$P: L_\lambda^1 \to L_\lambda^1$ has a finite number of eigenvalues $\sigma_1, \ldots, \sigma_p$ of modulus* 1.

(ii) *$E_i = \{h\in L_\lambda^1 : P_\lambda h = \sigma_i h\} \subset BV_\lambda$ and E_i is finite dimensional, $1 \leqslant i \leqslant p$.*

(iii) *$P_\lambda^n = \sum_{i=1}^{p}\sigma_i^n P_{\lambda,i} + T_\lambda^n$ for any $n\in\mathbb{N}^*$, where, for any $1\leqslant i\leqslant p$, $P_{\lambda,i}$ is a projection on E_i (that is, $P_{\lambda,i}L_\lambda^1 \subset E_i$ and $P_{\lambda,i}^2 = P_{\lambda,i}$), which is bounded as an operator from L_λ^1 into BV_λ and has $\|P_{\lambda,i}\|_1 \leqslant 1$, T_λ is a linear operator on L_λ^1 with $T_\lambda(BV_\lambda) \subset BV_\lambda$, $\sup_{n\in\mathbb{N}^*}\|T_\lambda^n\|_1 < \infty$, $\|T_\lambda^n\|_v = O(q^n)$ as $n\to\infty$, for some $0 < q < 1$, and $\lim_{n\to\infty} T_\lambda^n h = 0$ for any $h\in L_\lambda^1$. Moreover, $P_{\lambda,i}P_{\lambda,j} = 0$ for any $1\leqslant i\neq j\leqslant p$, and $P_{\lambda,i}T_\lambda = T_\lambda P_{\lambda,i} = 0$ for any $1\leqslant i\leqslant p$.*

Corollary 5.3.13. *Under the assumptions of Theorem 5.3.12, $\sigma_1 = 1$ is an eigenvalue of P_λ. Setting $h = P_{\lambda,1}1$ and $d\mu = h\,d\lambda$ we have $0 \leqslant h \in BV_\lambda$, $\int_I h\,d\lambda = 1$, and $P_\lambda h = h$, i.e. τ is μ-preserving. For any probability $\nu \ll \lambda$ such that τ is ν-preserving we have $\nu \ll \mu$.*

Proof. For any $n \in \mathbb{N}^*$ set $h_n = \sum_{j=0}^{n-1} P_\lambda^j 1/n$ (see Section A2.9). From the properties of P_λ (see Section A2.11) we have $h_n \geqslant 0$ and $\int_I h_n\,d\lambda = 1$. It follows by Theorem 5.3.12(iii) that h_n converges in L_λ^1 as $n \to \infty$ to a non-negative limit h for which, clearly, $\int_I h\,d\lambda = 1$. Since $P_\lambda h_n = (n+1)h_{n+1}/n - 1/n, n \in \mathbb{N}^*$, we also have $P_\lambda h = h$, which shows that $\sigma_1 = 1$ is an eigenvalue of P_λ, and that, in fact, $h = P_{\lambda,1}1 \in BV_\lambda$.

Finally, if τ is ν-preserving with $d\nu = \hat{h}\,d\lambda$, then $P_\lambda \hat{h} = \hat{h}$, and by Theorem 5.3.12 (ii), we have $\hat{h} \in BV_\lambda$, and hence $c = \operatorname{ess\,sup}|\hat{h}| < \infty$. It is easy to see that $P_\lambda^j \hat{h} \leqslant c P_\lambda^j 1$ for all $j \in \mathbb{N}$. Therefore $\hat{h} = \sum_{j=0}^{n-1} P_\lambda^j \hat{h}/n \leqslant c h_n$, $n \in \mathbb{N}^*$, hence $\hat{h} \leqslant ch$ and $\nu \ll \mu$. □

Remarks. 1. In what follows we shall be interested in the special case where the only eigenvalue of modulus 1 of P_λ is 1 and this eigenvalue is simple (a P_λ with this property is said to be *weak-mixing*). Sufficient conditions for P_λ to be weak-mixing are given below in Proposition 5.3.14. This limitation of our interest to weak-mixing operators P_λ is justified by the situation arising in the general case.

In the general case, Theorem 5.3.12 leads to a decomposition of τ into ergodic components (see Section A2.11), as described below.

First of all, as a consequence of the theory of positive operators, the set E of eigenvalues of modulus 1 of P_λ is a union of full cyclic groups. Next, let m be any natural integer for which $\sigma^m = 1$ for all $\sigma \in E$. There exist non-negative functions $h_1, \ldots, h_s \in BV_\lambda$ and pairwise disjoint sets $C_1, \ldots, C_s \in \mathscr{B}_I$ with the following properties.

(i) $\lambda(I - \bigcup_{i=1}^s C_i) = 0, P^m h_i = h_i, \lambda(\tau^m(C_i)\Delta C_i) = 0, 1 \leqslant i \leqslant s$, and $\int_{C_i} h_j\,d\lambda = \delta_{ij}, 1 \leqslant i,j \leqslant s$

(ii) $\lim_{n\to\infty} \| P_\lambda^{nm} h - \sum_{i=1}^s h_i \int_{C_i} h\,d\lambda \|_1 = 0$ for any $h \in L_\lambda^1$, and $\| P_\lambda^{nm} h - \sum_{i=1}^s h_i \int_{C_i} h\,d\lambda \|_v = O(q^n) \| h \|_v$ for some $0 \leqslant q < 1$, as $n \to \infty$, for any $h \in BV_\lambda$

(iii) There exists a permutation π of $\{1, \ldots, s\}$ such that $P_\lambda h_i = h_{\pi(i)}$, $\lambda(\tau(C_{\pi(i)})\Delta C_i) = 0, 1 \leqslant i \leqslant s$

(iv) Let $\pi_1, \ldots, \pi_\theta$ be the independent cycles of π and let l_j be the length of $\pi_j, 1 \leqslant j \leqslant \theta$. Put $\tilde{C}_j = \bigcup_{i\in\pi_j} C_i, 1 \leqslant j \leqslant \theta$. Then E is a union of cyclic groups of ranks $l_j, 1 \leqslant j \leqslant \theta$. For any arbitrarily fixed j, $1 \leqslant j \leqslant \theta$, the following hold. First, $\lambda(\tau(\tilde{C}_j)\Delta\tilde{C}_j) = 0$ and τ permutes cyclically the $C_i, i \in \pi_j$, modulo λ-null sets. Next, the restriction of τ to $\tilde{C}_j$ is ergodic w.r.t. λ. (Note that the restriction of τ to $\tilde{C}_j$ is ν_j-preserving with $d\nu_j = (l_j^{-1}\sum_{i\in\pi_j} h_i)\,d\lambda$.) Finally, under $h_i d\lambda$, the restriction of τ^{l_j} to C_i is an exact endomorphism (of C_i), $i\in\pi_j$. (Clearly, $\lambda(\tau^{l_j}(C_i)\Delta C_i) = 0, i\in\pi_j$.)

For the proof of the above assertions see Rychlik (1983) and also Kowalski (1979b). It should be noted that, under more stringent assumptions, the ergodic decomposition of τ has been essentially described in Kosjakin & Sandler (1972), Li & Yorke (1978), and Wagner (1979).

2. For a penetrating study of the eigenvalues of P_λ contained in the interior of the unit disk we refer the reader to Keller (1984).

Proposition 5.3.14. *Let τ be a C^1 pw.m.t. Assume that Conditions* (BV) *and* (C) *hold. Then the only eigenvalue of modulus* 1 *of P_λ is $\sigma = 1$ and this eigenvalue is simple (i.e. P_λ is weak-mixing).*

Proof. Note first that, by Proposition 5.3.2, Condition (C) implies Condition (E_m) for some $m\in\mathbb{N}^*$. Thus our assumptions imply those of Theorem 5.3.12, and more particularly those of Theorem 5.3.5.

Next, by Theorem 5.3.5, there exists a unique probability ρ on $\mathscr{B}_I$ such that τ is ρ-preserving; ρ is equivalent to λ and the density $r = d\rho/d\lambda$ satisfies $1/C \leqslant r \leqslant C$ a.e. in I. Thus (see Section A2.11) we have $P_\lambda r = r$, and by Theorem 5.3.12, under the present assumptions, $r\in BV_\lambda$. The uniqueness of ρ implies that $\sigma = 1$ is a simple eigenvalue (i.e. the corresponding eigenspace is one-dimensional).

Finally, to prove the uniqueness of $\sigma = 1$ as an eigenvalue of modulus 1 of P_λ, we again use Theorem 5.3.5, which asserts that (τ, ρ) is an exact endomorphism. By Proposition A2.7 this is equivalent to

$$\lim_{n\to\infty} \int_I \left| P_\rho^n h - \int_I h\,d\rho \right| d\rho = 0 \tag{5.3.12}$$

for all $h\in L_\rho^1 = L^1(I, \mathscr{B}_I, \rho)$, where P_ρ is the Frobenius–Perron operator associated with τ under ρ, i.e. P_ρ satisfies

$$\int_A P_\rho h\,d\rho = \int_{\tau^{-1}(A)} h\,d\rho, A\in\mathscr{B}_I,$$

for all $h\in L_\rho^1$. It is easy to see that $P_\rho h = P_\lambda(hr)/r$, whence $P_\rho^n h = P_\lambda^n(hr)/r$, $h\in L_\rho^1, n\in\mathbb{N}^*$. (Note that the equivalence of ρ and λ implies that the elements of L_ρ^1 coincide with those of L_λ^1.) Hence equation (5.3.12) is equivalent to

$$\lim_{n\to\infty} \int_I \left| P_\lambda^n(hr) - r\int_I hr\,d\lambda \right| d\lambda = 0, h\in L_\lambda^1. \tag{5.3.13}$$

Now, if $P_\lambda \hat{h} = \sigma \hat{h}$ with $|\sigma| = 1, \sigma \neq 1$, for some $\hat{h} \in L_\lambda^1, \hat{h} \neq 0$, then we have $\sigma \int_I \hat{h}\, d\lambda = \int_I P_\lambda \hat{h}\, d\lambda = \int_I \hat{h}\, d\lambda$, whence $\int_I \hat{h}\, d\lambda = 0$. Taking $h = \hat{h}/r$, (5.3.13) implies that $\hat{h} = 0$ a.e. in I. This contradiction completes the proof. □

Corollary 5.3.15. *Under the assumptions of Proposition* 5.3.14 *we have* $P_\lambda^n h = r \int_I h\, d\lambda + T_\lambda^n h$, *for any* $n \in \mathbb{N}^*$ *and* $h \in L_\lambda^1$, *where* T_λ *is a linear operator on* L_λ^1 *with* $T_\lambda(BV_\lambda) \subset BV_\lambda$, $\sup_{n \in \mathbb{N}^*} \| T_\lambda^n \|_1 < \infty$, $\| T_\lambda^n \|_v = O(q^n)$, *as* $n \to \infty$, *for some* $0 < q < 1$, *and* $\lim_{n \to \infty} T_\lambda^n h = 0$ *for any* $h \in L_\lambda^1$.

The proof follows immediately from Proposition 5.3.14 and Theorem 5.3.12.

Remark. The conclusions of Corollary 5.3.15 obviously hold for any C^1 pw.m.t. τ satisfying Conditions (E_m) for some $m \in \mathbb{N}^*$ and (BV), whenever the Frobenius–Perron operator P_λ associated with τ under λ is weak-mixing. In this case, r should be replaced by the density of the unique probability μ such that τ is μ-preserving (see Corollary 5.3.13).

5.3.4 The invariant probability: differential properties of its density

We saw in Subsection 5.3.2 that for a C^0 pw.m.t. τ for which the f_a, $a \in \mathfrak{A}$, are absolutely continuous, Condition (C) is sufficient to ensure the existence of a unique probability $\rho \equiv \lambda$ such that τ is ρ-preserving. Moreover, the density $r = d\rho/d\lambda$ is bounded away from 0 and ∞ (see Theorem 5.3.5). In Subsection 5.3.3 (see the proof of Proposition 5.3.14) it was shown that for a C^1 pw.m.t. for which Conditions (C) and (BV) hold, the density $r \in BV_\lambda$, i.e. it possesses a version of bounded variation.

In this subsection we deepen the study of the extent to which the differential properties of τ are transferred to r.

We begin by giving sufficient conditions for r to be a Lipschitz function. Before proceeding, we need some preparations.

First, note that since for any $n \in \mathbb{N}^*$ the intervals $I(a^{(n)})$, $a^{(n)} \in \mathfrak{A}^n$, exhaust I up to a set of Lebesgue measure 0, i.e.

$$\sum_{a^{(n)} \in \mathfrak{A}^n} | f_{a^{(n)}}(1) - f_{a^{(n)}}(0) | = 1,$$

Condition (C) implies that

$$\sup_{t \in I} \sum_{a^{(n)} \in \mathfrak{A}^n} | f'_{a^{(n)}}(t) | \leq C, \quad n \in \mathbb{N}^*, \tag{5.3.14}$$

for any pw.m.t. for which the derivatives f'_a, $a \in \mathfrak{A}$, exist everywhere in I. Indeed, for any $t \in I$, $n \in \mathbb{N}^*$ and $a^{(n)} \in \mathfrak{A}^n$ we can write

$$\frac{|f'_{a^{(n)}}(t)|}{C} \leqslant \frac{\sup_{t\in I} |f'_{a^{(n)}}(t)|}{C} \leqslant \inf_{t \in I} |f'_{a^{(n)}}(t)| \leqslant |f_{a^{(n)}}(1) - f_{a^{(n)}}(0)|,$$

which leads to (5.3.14).

Then, assuming that the derivatives f''_a, $a \in \mathfrak{A}$, exist everywhere in I, consider Condition ($\mathrm{BD}^{(2)}$).

Condition ($\mathrm{BD}^{(2)}$)

$$\sup_{t \in I} \sum_{a \in \mathfrak{A}} |f''_a(t)| < \infty.$$

Condition ($\mathrm{BD}^{(2)}$) implies Condition (BV). To check this use the equation $\operatorname{var} f'_a = \int_I |f''_a(t)| \mathrm{d}t$, $a \in \mathfrak{A}$.

Lemma 5.3.16. *If Conditions* (C) *and* ($\mathrm{BD}^{(2)}$) *hold then*

$$\sup_{n \in \mathbb{N}^*} \sup_{t \in I} \sum_{a^{(n)} \in \mathfrak{A}^n} |f''_{a^{(n)}}(t)| < \infty.$$

Proof. We use the formula

$$f''_{ij}(t) = f''_i(f_j(t))(f'_j(t))^2 + f'_i(f_j(t)) f''_j(t), \tag{5.3.15}$$

which is valid for any u-tuple $i \in \mathfrak{A}^u$ and any v-tuple $j \in \mathfrak{A}^v$, for any $u, v \in \mathbb{N}^*$. Using (5.3.14) it is easy to show that

$$D_n = \sup_{t \in I} \sum_{a^{(n)} \in \mathfrak{A}^n} |f''_{a^{(n)}}(t)| < \infty$$

for any $n \in \mathbb{N}^*$. It remains to prove that the sequence $(D_n)_{n \in \mathbb{N}^*}$ is bounded.

By Proposition 5.3.2, Condition (C) implies Condition (E_m) for some $m \in \mathbb{N}^*$. Choose a natural integer k such that $C\alpha^k = q < 1$. Using (5.3.14) it follows from (5.3.15), with $u = (s-k)m + l$, $s \geqslant k$, $l < m$, and $v = km$, that

$$D_{sm+l} \leqslant C\alpha^k D_{(s-k)m+l} + CD_{km}.$$

Hence

$$D_n \leqslant D = \max_{1 \leqslant p < km} D_p + \frac{CD_{km}}{1-q}, \quad n \in \mathbb{N}^*, \tag{5.3.16}$$

thus completing the proof. □

Now we are ready to prove Theorem 5.3.17.

Theorem 5.3.17. *Let τ be a C^1 pw.m.t. for which the second derivatives*

f''_a, $a \in \mathfrak{A}$, exist everywhere in I. Assume that Conditions (C) *and* (BD$^{(2)}$) *hold. Then the invariant density r is a Lipschitz function, and the equation $P_\lambda r = r$ is valid* everywhere *in I.*

Proof. We first show that the Ionescu Tulcea–Marinescu Theorem A2.4 applies to P_λ for the special case of the spaces $X = C^0(I) = C(I)$ and $Y = C^0L(I) = L(I)$ under their usual norms (see Section A2.10(2)).

The fact that P_λ takes $L(I)$ and $C(I)$ into themselves, can be easily verified, using a method similar to that for checking Condition (ITM$_3$) below.

Next, using (5.3.14), we deduce that Condition (ITM$_2$) holds with $H \leqslant C$.

To check Condition (ITM$_3$) choose m and k as in the proof of Lemma 5.3.16. For any $h \in L(I)$ and $t_1, t_2 \in I$, we have

$$P_\lambda^{km} h(t_1) - P_\lambda^{km} h(t_2) = \sum_{a^{(km)} \in \mathfrak{A}^{km}} (|f'_{a^{(km)}}(t_1)| - |f'_{a^{(km)}}(t_2)|) h(f_{a^{(km)}}(t_1))$$
$$+ \sum_{a^{(km)} \in \mathfrak{A}^{km}} |f'_{a^{(km)}}(t_2)| (h(f_{a^{(km)}}(t_1)) - h(f_{a^{(km)}}(t_2))).$$

The first sum above is dominated by $D|h||t_1 - t_2|$, where D is given by (5.3.16). Indeed, the sum can be written as $g(1) - g(0)$ with

$$g(t) = \sum_{a^{(km)} \in \mathfrak{A}^{km}} f'_{a^{(km)}}(tt_1 + (1-t)t_2) h(f_{a^{(km)}}(t_1)) \operatorname{sgn} f'_{a^{(km)}}, \quad t \in I,$$

and the intermediate value formula and Lemma 5.3.16 justify our assertion. Similarly, the second sum is dominated by $q\mathrm{s}(h)|t_1 - t_2|$ with $q = C\alpha^k < 1$. Hence

$$\mathrm{s}(P_\lambda^{km} h) \leqslant q\mathrm{s}(h) + D|h|.$$

Since $|P_\lambda^{km} h| \leqslant C|h|$ we finally get

$$\| P_\lambda^{km} h \|_L \leqslant q \| h \|_L + (C - q + D)|h|, \quad h \in L(I),$$

i.e. Condition (ITM$_3$).

To check Condition (ITM$_4$) we should note that if $A \subset L(I)$ is bounded in $L(I)$, then the set $P_\lambda^k A$ is also bounded in $L(I)$. It remains to use the Arzelà–Ascoli Theorem A1.16.

So far we have proved that $P_\lambda \in \mathscr{L}_{km}(L(I), C(I))$.

Now (compare with the proof of Corollary 5.3.13), setting $h_n = \sum_{j=0}^{n-1} P_\lambda^j 1/n$, we have $h_n \geqslant 0$, and $\int_I h_n \, d\lambda = 1$, $n \in \mathbb{N}^*$. Then the Ionescu Tulcea–Marinescu Theorem A2.4 implies that h_n converges as $n \to \infty$ in both $L(I)$ and $C(I)$ to a probability density which is left invariant by P_λ. Clearly, it should coincide with the invariant density r since r is unique. Thus $r \in L(I)$ as stated. The validity of the equation $P_\lambda r = r$ everywhere in I follows from the fact that both sides are continuous functions in I. □

We shall now state a generalization of Theorem 5.3.17. Assuming that for a fixed natural integer $s \geqslant 2$ the derivatives $f_a^{(s)}$, $a \in \mathfrak{A}$, exist everywhere in I, consider the following condition.

Condition (BD$^{(s)}$)

$$\sup_{t \in I} \left| \sum_{a \in A} f_a^{(s)}(t) \right| < \infty.$$

Theorem 5.3.18. *Let τ be a C^s pw.m.t. for an arbitrarily fixed $s \in \mathbb{N}^*$ such that the derivatives $f_a^{(s+1)}$, $a \in \mathfrak{A}$, exist everywhere in I. Assume that Conditions* (C) *and* (BD$^{(i)}$), $2 \leqslant i \leqslant s+1$, *hold. Then the invariant density r belongs to the space $C^{s-1}L(I)$, and the equation $P_\lambda r = r$ is valid* everywhere *in I.*

The proof is entirely similar to that of Theorem 5.3.17, using the Ionescu Tulcea–Marinescu Theorem A2.4 for the special case of the spaces $X = C^{s-1}(I)$ and $Y = C^{s-1}L(I)$ (see Section A2.10(2)).

Clearly, we first have to prove the obvious generalization of Lemma 5.3.16, which is easy to do. For $s \geqslant 2$, termwise differentiations needed in the proof are justified by dominated convergence ensured by Conditions (BD$^{(i)}$) for $2 \leqslant i \leqslant s+1$.

Remark. The method used to prove Theorems 5.3.17 and 5.3.18 actually implies that r is approached geometrically in $C^{s-1}L(I)$ by the iterated densities

$$\sum_{a^{(n)} \in \mathfrak{A}^n} |f'_{a^{(n)}}(\cdot)| = P_\lambda^n 1, \quad n \in \mathbb{N}^*.$$

Theorem 5.3.18 and the above remark, in conjunction with Theorem 10 of Halfant (1977), lead to Theorem 5.3.19.

Theorem 5.3.19. *Let τ be a pw.m.t. such that the f_a, $a \in \mathfrak{A}$, are analytic in I. Assume that Conditions* (C) *and* (BD$^{(s)}$), *for all $s \geqslant 2$, hold. Then the invariant density r is analytic in I and the equation $P_\lambda r = r$ is valid* everywhere *in I.*

Remark. Theorem 5.3.17 enables us to associate an RSCC $\{(W, \mathscr{W}), (X, \mathscr{X}), u, P\}$ with a C^1 pw.m.t. τ for which Conditions (C) and (BD$^{(2)}$) hold, where

$$(W, \mathscr{W}) = (I, \mathscr{B}_I), \quad (X, \mathscr{X}) = (\mathfrak{A}, \mathscr{P}(\mathfrak{A})),$$

$$u(w, x) = f_x(w), \quad P(w, x) = \frac{r(f_x(w))|f'_x(w)|}{r(w)}, \quad w \in I, \quad x \in \mathfrak{A}.$$

Clearly, the corresponding transition operator U is the Frobenius–Perron operator P_ρ associated with τ under ρ. It is an open problem to find an interpretation of the two sequences of random variables associated with this RSCC in terms of the label sequence $(a_n)_{n\in\mathbb{N}^*}$ defined in Subsection 5.3.1. See also Subsection 5.4.3.

5.3.5 Limit theorems

The results in Subsection 5.3.3 enable us to prove that some type of mixing holds for certain pw.m.t.s considered there. In turn, this enables us to apply known limit theorems to the pw.m.t. setting, which we will do in this subsection. Also, we prove a central limit theorem with optimal convergence rate.

We start with Lemma 5.3.20 (compare with Lemma 5.3.16).

Lemma 5.3.20. *Let τ be a C^1 pw.m.t. Assume that Conditions* (E_m) *for some $m\in\mathbb{N}^*$ and* (BV) *hold. Then*

$$\sup_{n\in\mathbb{N}^*}\sum_{a^{(n)}\in\mathfrak{A}^n}\operatorname{var} f'_{a^{(n)}}<\infty.$$

Proof. Note that $P^n_\lambda(h\chi_{I(a^{(n)})})=|f'_{a^{(n)}}|(h\circ f_{a^{(n)}})$, $n\in\mathbb{N}^*$, $a^{(n)}\in\mathfrak{A}^n$, $h\in L^1_\lambda$, so that by (5.3.11), with $h=1$, we have

$$\sum_{a^{(n)}\in\mathfrak{A}^n}\operatorname{var} f'_{a^{(n)}}=\operatorname{var} g_n,\quad n\in\mathbb{N}^*.$$

Next, for any $n,p\in\mathbb{N}^*$ and $a^{(n+p)}=(a^{(n)},a^{(n+1,n+p)})=(a_1,\ldots,a_n,a_{n+1},\ldots,a_{n+p})\in\mathfrak{A}^{n+p}$, we have

$$|f'_{a^{(n+p)}}|=|f'_{a^{(n)}}(f_{a^{(n+1,n+p)}})|\,|f'_{a^{(n+1,n+p)}}|=P^p_\lambda(|f'_{a^{(n)}}|\chi_{I(a^{(n+1,n+p)})}).$$

Hence, using (5.3.11) again and Proposition 5.3.9, we get

$$\sum_{a^{(n+1,n+p)}\in\mathfrak{A}^p}\operatorname{var} f'_{a^{(n+p)}}=\operatorname{var}(|f'_{a^{(n)}}|g_p)\leqslant q_p\operatorname{var} f'_{a^{(n)}}+Q_p\lambda(I(a^{(n)})).$$

Summing over $a^{(n)}\in\mathfrak{A}^n$ we finally get

$$\operatorname{var} g_{n+p}\leqslant q_p\operatorname{var} g_n+Q_p,\quad n,p\in\mathbb{N}^*. \tag{5.3.17}$$

Now, choosing $k\in\mathbb{N}^*$ as in the proof of Proposition 5.3.10 (i.e. such that $3\alpha^k<1$), and using the notation there, we can write

$$\operatorname{var} g_{n+mk}\leqslant q\operatorname{var} g_n+Q,\quad n\in\mathbb{N}^*.$$

where $q=3\alpha^k$, $0<Q<\infty$. Hence

$$\operatorname{var} g_{nmk}\leqslant q^{n-1}\operatorname{var} g_{mk}+\frac{Q}{1-q},\quad n\in\mathbb{N}^*.$$

Noting that (5.3.17) implies

$$\operatorname{var} g_{nmk+p} \leqslant \left(\max_{1 \leqslant p < mk} q_p \right) \operatorname{var} g_{nmk} + \max_{1 \leqslant p < mk} Q_p, \quad n \in \mathbb{N}^*, \quad 1 \leqslant p < mk,$$

the conclusion

$$\sup_{n \in \mathbb{N}^*} \operatorname{var} g_n < \infty$$

is immediate. □

Corollary 5.3.21. *Under the assumptions of Lemma 5.3.20, for any $h \in BV_\lambda$ we have*

$$\sup_{n \in \mathbb{N}^*} \sum_{a^{(n)} \in \mathfrak{A}^n} \operatorname{var} P_\lambda^n (h \chi_{I(a^{(n)})}) < \infty.$$

Proof. Note that by Lemma 5.3.8 we have

$$|f'_{a^{(n)}}| \leqslant \lambda(I(a^{(n)})) + \operatorname{var} f'_{a^{(n)}}, \quad n \in \mathbb{N}^*, \quad a^{(n)} \in \mathfrak{A}^n.$$

The result now follows since

$$\begin{aligned} \operatorname{var} P_\lambda^n (h \chi_{I(a^{(n)})}) &= \operatorname{var}(|f'_{a^{(n)}}|(h \circ f_{a^{(n)}})) \leqslant |f'_{a^{(n)}}| \operatorname{var}(h \circ f_{a^{(n)}}) + |h| \operatorname{var} f'_{a^{(n)}} \\ &\leqslant (\lambda(I(a^{(n)})) + \operatorname{var} f'_{a^{(n)}}) \operatorname{var} h + |h| \operatorname{var} f'_{a^{(n)}}, \quad n \in \mathbb{N}^*, \quad a^{(n)} \in \mathfrak{A}^n. \end{aligned}$$ □

For $n \leqslant p$, $n, p \in \mathbb{N}^*$, denote by $\mathscr{B}_{[n,p]}$ the σ-algebra generated by the random variables a_u, $n \leqslant u \leqslant p$, defined in Subsection 5.3.1. For $n \in \mathbb{N}^*$ put $\mathscr{B}_{[n,\infty)}$ = the least σ-algebra containing all the $\mathscr{B}_{[n,p]}$, $p \geqslant n$. Note that, under Condition (E_m) for some $m \in \mathbb{N}^*$, by Proposition 5.3.1, $\mathscr{B}_{[1,\infty)}$ coincides with $\mathscr{B}_I$, the σ-algebra of Borel sets in I.

Theorem 5.3.22. *Let τ be a C^1 pw.m.t. Assume that Conditions* (E_m) *for some $m \in \mathbb{N}^*$ and* (BV) *hold. Assume also that the Frobenius–Perron operator P_λ is weak-mixing and let μ be the unique probability such that τ is μ-preserving. Then the sequence $(a_n)_{n \in \mathbb{N}^*}$ is strictly stationary under μ and*

$$\sup_{k \in \mathbb{N}^*} \mathsf{E} \left(\sup_{B \in \mathscr{B}_{[k+n,\infty)}} |\mu(B | \mathscr{B}_{[1,k]}) - \mu(B)| \right) \leqslant c q^n, \quad n \in \mathbb{N}^*, \qquad (5.3.18)$$

with some suitable positive constants $q < 1$ and c.

Proof. The first assertion follows immediately from the fact that τ is μ-preserving.

To prove the second assertion remark that for $B \in \mathscr{B}_{[k+n,\infty)}$, $k, n \in \mathbb{N}^*$, there exists $\tilde{B} \in \mathscr{B}_{[1,\infty)} = \mathscr{B}_I$ such that $B = \tau^{-(k+n)} \tilde{B}$. Then, assuming that

$\mu(I(a^{(k)}))\neq 0$ for arbitrarily fixed $k\in\mathbb{N}^*$ and $a^{(k)}\in\mathfrak{A}^k$, on $I(a^{(k)})$ we have

$$\mu(B|\mathscr{B}_{[1,k]})=\frac{1}{\mu(I(a^{(k)}))}\int_B h\chi_{I(a^{(k)})}\,\mathrm{d}\lambda$$
$$=\frac{1}{\mu(I(a^{(k)}))}\int_{\tilde{B}} P_\lambda^{k+n}(h\chi_{I(a^{(k)})})\,\mathrm{d}\lambda,$$

where $h=\mathrm{d}\mu/\mathrm{d}\lambda\in BV_\lambda$ (see Corollary 5.3.13). Since $\mu(B)=\mu(\tilde{B})=\int_{\tilde{B}} h\mathrm{d}\lambda$, it follows that on $I(a^{(k)})$ we have

$$|\mu(B|\mathscr{B}_{[1,k]})-\mu(B)|\leqslant\int_{\tilde{B}}|P_\lambda^{k+n}(h\chi_{I(a^{(k)})})/\mu(I(a^{(k)}))-h|\,\mathrm{d}\lambda.$$

Hence

$$\sup_{k\in\mathbb{N}^*}\mathbf{E}\left(\sup_{B\in\mathscr{B}_{[k+n,\infty)}}|\mu(B|\mathscr{B}_{[1,k]})-\mu(B)|\right)$$
$$\leqslant\sup_{k\in\mathbb{N}^*}\sum_{a^{(k)}\in\mathfrak{A}^k}\|P_\lambda^{k+n}(h\chi_{I(a^{(k)})})-\mu(I(a^{(k)}))h\|_1.$$

But, for any $k,n\in\mathbb{N}^*$ and $a^{(k)}\in\mathfrak{A}^k$,

$$P_\lambda^{k+n}(h\chi_{I(a^{(k)})})-\mu(I(a^{(k)}))h=P_\lambda^n(P_\lambda^k(h\chi_{I(a^{(k)})})-\mu(I(a^{(k)}))h),$$

so that, by the remark following Corollary 5.3.15 and Lemma 5.3.8, for some $c>0$ we have

$$\|P_\lambda^{k+n}(h\chi_{I(a^{(k)})})-\mu(I(a^{(k)}))h\|_1$$
$$\leqslant\|P_\lambda^{k+n}(h\chi_{I(a^{(k)})})-\mu(I(a^{(k)}))h\|_v\leqslant cq^n\|P_\lambda^k(h\chi_{I(a^{(k)})})-\mu(I(a^{(k)}))h\|_v$$
$$\leqslant 2cq^n\operatorname{var}(P_\lambda^k(h\chi_{I(a^{(k)})})-\mu(I(a^{(k)}))h)$$
$$\leqslant 2cq^n(\operatorname{var}P_\lambda^k(h\chi_{I(a^{(k)})})+\mu(I(a^{(k)}))\operatorname{var}h).$$

Applying Corollary 5.3.21 concludes the proof. □

The mixing property (5.3.18) is called absolute regularity. Since

$$\mu(A\cap B)-\mu(A)\mu(B)=\int_A(\mu(B|\mathscr{B}_{[1,k]})-\mu(B))\mathrm{d}\mu,\quad A\in\mathscr{B}_{[1,k]},\quad B\in\mathscr{B}_I,$$

absolute regularity implies the more usual strong-mixing property, i.e.

$$\sup_{k\in\mathbb{N}^*}\ \sup_{A\in\mathscr{B}_{[1,k]},B\in\mathscr{B}_{[k+n,\infty)}}|\mu(A\cap B)-\mu(A)\mu(B)|\leqslant cq^n,\quad n\in\mathbb{N}^*.$$

This allows us, under the assumptions on τ in Theorem 5.3.22, to state, for some functionals of the sequence $(a_n)_{n\in\mathbb{N}^*}$, various limit theorems which hold for strong-mixing strictly stationary sequences (see, e.g., Philipp & Stout (1975)). To give such a theorem we must first define the concept of bounded p-variation. A function $f:I\to\mathbb{C}$ is said to be of bounded

p-variation, $p \geqslant 1$, iff

$$\operatorname{var}^{(p)} f = \sup \sum_{i=1}^{k-1} |f(t_{i+1}) - f(t_i)|^p < \infty,$$

the supremum being taken over all $t_1 < \cdots < t_k \in I$ and $k \geqslant 2$. Clearly, $\operatorname{var}^{(1)} f = \operatorname{var} f$, and a function of bounded variation is also of bounded p-variation for any $p > 1$, but the converse of the last assertion is in general not true.

Theorem 5.3.23. *Let the assumptions in Theorem 5.3.22 hold. Consider a function $f: I \to \mathbb{R}$ of bounded p-variation for some $p \geqslant 1$, with $\int_I f \,\mathrm{d}\mu = 0$. Define $S_n = \sum_{j=0}^{n-1} f \circ \tau^j$, $n \in \mathbb{N}^*$. Then the series*

$$\sigma^2 = \int_I f^2 \mathrm{d}\mu + 2 \sum_{j \in \mathbb{N}^*} \int_I (f \circ \tau^j) f \,\mathrm{d}\mu$$

converges absolutely, $\int_I S_n^2 \,\mathrm{d}\mu = n\sigma^2 + O(1)$ as $n \to \infty$, and, if $\sigma \neq 0$,

(i) $\sup_{x \in \mathbb{R}} \left| \mu\left(\frac{S_n}{\sigma\sqrt{n}} < x \right) - \Phi(x) \right| = O(n^{-c})$ *as $n \to \infty$ for some $c > 0$;*

(ii) *without changing its distribution we can redefine $(S_n)_{n \in \mathbb{N}^*}$ on a richer probability space together with a Brownian motion process $(\beta_t)_{t \in \mathbb{R}_+}$ on $C(\mathbb{R}_+)$ such that $|S_n/\sigma - \beta_n| = O(n^{1/2-d})$ a.s. as $n \to \infty$ for some $d > 0$.*

For the proof the reader is referred to Hofbauer & Keller (1982, p. 133).

Remarks. 1. Since $\mathscr{B}_I$ is generated by the fundamental intervals, f can be viewed a.e. as a functional of the sequence $(a_n)_{n \in \mathbb{N}^*}$.

2. The order of approximation in (ii) above allows us to obtain the log log law and the functional central limit theorem for the sequence $(f \circ \tau^n)_{n \in \mathbb{N}}$. For details the reader is referred to, e.g. Philipp & Stout (1975).

It is possible to obtain the best rate of convergence in the central limit theorem in the case $p = 1$, i.e. for a function f of bounded variation. More precisely we have Theorem 5.3.24.

Theorem 5.3.24. *Under the assumptions of Theorem 5.3.23 with $p = 1$ and supposing, in addition, that $\operatorname{ess\,inf} \mathrm{d}\mu/\mathrm{d}\lambda > 0$, we have $c = \frac{1}{2}$ in Theorem 5.3.23(i). Moreover, $\sigma > 0$ iff f cannot be written as $f = b \circ \tau - b$ a.e. for some real-valued $b \in BV_\lambda$.*

Proof. The proof of the first assertion is completely analogous to the proof of Theorem 4.2.7. Here, the basic tool is a perturbation $P_{\mu,\theta}$ of the Frobenius–Perron operator P_μ, which is defined as $P_{\mu,\theta} g = P_\mu(g \exp(i\theta f))$, $g \in L_\mu^1$, $\theta \in \mathbb{R}$, while the starting point is the equation $P_{\mu,\theta}^n g = P_\mu^n(g \exp(i\theta S_n))$,

$n \in \mathbb{N}^*$. Hence the characteristic function $\int_I \exp(i\theta S_n)\,d\mu$ of S_n can be written as

$$\int_I \exp(i\theta S_n)\,d\mu = \int_I P_\mu^n(\exp(i\theta S_n))d\mu = \int_I P_{\mu,\theta}^n 1\,d\mu, \quad \theta \in \mathbb{R}, \quad n \in \mathbb{N}^*,$$

(see Section A2.11 for the first equation above). The details can be found in Rousseau–Egèle (1983).

To prove the second assertion, concerning the positivity of σ, we first show that

$$\sigma^2 = \int_I (g - (P_\mu g)\circ\tau)^2\,d\mu, \tag{5.3.19}$$

where $g = f + \sum_{n \in \mathbb{N}^*} P_\mu^n f \in BV_\lambda$. (Note that, since $\int_I f\,d\mu = 0$, we have $P_\mu^n f = T_\lambda^n(fh)/h$, $n \in \mathbb{N}^*$, where $h = d\mu/d\lambda$ – see the proof of Proposition 5.3.14, Corollary 5.3.15, and the remark following the latter. Hence, the series defining g converges in BV_λ.) Using equation (A2.1) we can write

$$\begin{aligned}\sigma^2 &= \sum_{j \in \mathbb{Z}} \int_I (f\circ\tau^{|j|})f\,d\mu = \sum_{j \in \mathbb{Z}} \int_I (P_\mu^{|j|} f)f\,d\mu \\ &= \int_I (2g - f)f\,d\mu = \int_I (g^2 - (P_\mu g)^2)\,d\mu.\end{aligned}$$

(The last equation above holds since $g = f + P_\mu g$.) Now, using (A2.1) again and the fact that τ is μ-preserving, we get

$$\begin{aligned}\int_I (g - (P_\mu g)\circ\tau)^2\,d\mu &= \int_I g^2\,d\mu - 2\int_I g((P_\mu g)\circ\tau)\,d\mu + \int_I ((P_\mu g)\circ\tau)^2\,d\mu \\ &= \int_I g^2\,d\mu - 2\int_I (P_\mu g)^2\,d\mu + \int_I (P_\mu g)^2\,d\mu = \sigma^2,\end{aligned}$$

i.e. (5.3.19) holds.

It follows from (5.3.19) that $\sigma = 0$ iff $g = (P_\mu g)\circ\tau$ a.e. Hence it is immediate that, if $\sigma = 0$, then $f = b\circ\tau - b$ holds a.e. with $b = P_\mu g = g - f \in BV_\lambda$. Conversely, if $f = b\circ\tau - b$ holds a.e. for some real-valued $b \in BV_\lambda$, then $S_n = b\circ\tau^n - b$, $n \in \mathbb{N}^*$, and

$$n^{-1}\int_I S_n^2\,d\mu \leqslant 4n^{-1}\int_I b^2\,d\mu \to 0 \text{ as } n \to \infty, \quad \text{i.e. } \sigma = 0. \qquad \square$$

Remarks. 1. Under the assumptions of Theorem 5.3.23, we obviously have $\operatorname{ess\,inf} d\mu/d\lambda > 0$ if Condition (C) holds. For other cases where $\operatorname{ess\,inf} d\mu/d\lambda > 0$ see Keller (1978) and Kowalski (1979a).

2. In Theorem 5.3.24 the probability μ can be replaced by certain probabilities $\nu \ll \lambda$ for which $d\nu/d\lambda \in BV_\lambda$, but this results in a rate of convergence which, in general, is no longer the optimal one (see Ishitani (1986, p. 164)).

3. For a discussion of approaches to limit theorems in a general context including the special case of pw.m.t.s, see Keller (1988, Ch. 9).

5.3.6 *A few final remarks*

1. The theory in the previous subsections of Section 5.3 has been developed under assumption (5.3.1), which is reminiscent of f-expansions (see Section 5.4), where the assumption is quite natural. In fact, assumption (5.3.1) is not at all essential, and for ensuring the validity of a great number of results (including the basic Theorem 5.3.11) instead of (5.3.1), we can just assume that

$$\inf_{a \in \mathfrak{A}} \lambda(\tau_a(I(a))) > 0, \tag{5.3.1'}$$

which always holds when $\mathfrak{A}$ is a finite set.

When replacing (5.3.1) by (5.3.1′), some notational changes are necessary. The part of the interval $I(a^{(n)})$ should now be played by the non-empty sets of the form $\bigcap_{i=0}^{n-1} \tau^{-i}(I(a_{i+1}))$, $n \in \mathbb{N}^*$, $a^{(n)} = (a_1, \ldots, a_n) \in \mathfrak{A}^n$ (see Subsection 5.3.1). Also, the absolute values of the derivatives $f'_{a^{(n)}}$ have to be replaced by the Jacobians $J_{a^{(n)}}: I \to \mathbb{R}_+$ defined by the equation

$$\lambda\left(\bigcap_{i=0}^{n-1} \tau^{-i}(I(a_{i+1})) \cap \tau^{-n}(A)\right) = \int_A J_{a^{(n)}} \, d\lambda, \quad n \in \mathbb{N}^*, \quad a^{(n)} \in \mathfrak{A}^n,$$

for an arbitrary $A \in \mathscr{B}_I$.

Next, certain results valid under (5.3.1) are no longer valid under (5.3.1′). For example, the results in Subsection 5.3.4 cease to be generally true, as is shown by the case of the transformation τ_α (see Subsection 5.2.5). It should also be mentioned that results claimed to be true without assuming (5.3.1) were later found to be doubtful. See, for example, the additional comments appended to Bowen (1979).

Finally, note that without (5.3.1′) there can be no $\mu \ll \lambda$ such that τ is μ-preserving (see Bugiel (1985)).

2. The basic Theorem 5.3.11 was proved for certain C^1 pw.m.t.s for which the f_a, $a \in \mathfrak{A}$, are functions of bounded variation. The last assumption can be weakened to bounded p-variation for some $p > 1$. The details are to be found in Keller (1985) and in unpublished work by M. Rychlik (see Góra (1984, pp. 81–82). Actually, even the pw.m.t. setting can be abandoned by considering transformations of I with Frobenius–Perron operators which obey the Ionescu Tulcea–Marinescu Theorem A2.4 for suitable Banach spaces (see, e.g., Ishitani (1986)). Of course, this only has a formal value, the main difficulty being to find conditions under which Theorem A2.4 applies.

3. A similar theory can be developed for piecewise many-to-one transformations of the n-dimensional unit interval $[0,1]^n$ or, more generally, of an abstract measure space. The part played by the functions of bounded variation or of bounded p-variation in the one-dimensional case is assumed now by the bounded functions, with oscillation playing the part of variation or p-variation. Some hints are to be found in Gordin (1971b), where, in spite of the fact that there are no proofs, the relevance of the Ionescu Tulcea–Marinescu Theorem A2.4 is obvious.

4. For various other extensions of the conceptual framework of pw.m.t.s the reader is referred to, among others, Hofbauer (1986a), Keller (1988), Kifer (1986b), Lalley (1987), Lasota & Mackey (1985), Pelikan (1984), Pollicott (1984), Thaler (1983), and Walters (1978b).

5.4 *f*-expansions

5.4.1 *Preliminaries*

Let f be a continuous strictly decreasing (increasing) real-valued function defined on $[1,\beta]$, where either $2<\beta\in\mathbb{N}^*$ or $\beta=\infty$ ($[0,\beta]$, where either $1<\beta\in\mathbb{N}^*$ or $\beta=\infty$), such that $f(1)=1$ and $f(\beta)=0$ ($f(0)=0$ and $f(\beta)=1$), with the convention $f(\beta)=\lim_{u\to\beta}f(u)$ for $\beta=\infty$. Denote by f^{-1} the inverse function of f, which is defined on $I=[0,1]$.

Such a function f can be used to represent certain real numbers $t\in I$ as

$$t=f(a_1(t)+f(a_2(t)+\cdots)), \tag{5.4.1}$$

where the 'digits' $a_n(t)$ and the 'remainders'

$$r_n(t)=a_n(t)+f(a_{n+1}(t)+f(a_{n+2}(t)+\cdots))$$

are defined recursively as

$$a_n(t)=[f^{-1}(\{r_{n-1}(t)\})],$$
$$r_0(t)=t,\quad r_n(t)=f^{-1}(\{r_{n-1}(t)\}),\quad n\in\mathbb{N}^*.$$

Clearly, the continued fraction expansion is obtained for $f(u)=1/u, u\geqslant 1$. Representation (5.4.1) is called an f-expansion of t.

The f-expansions were first considered by Kakeya (1924). Unaware of Kakeya's work, Bissinger (1944) and Everett Jr (1946) took up the topic again. Rényi (1957), motivated by the Americans' work, proved that either of the conditions BR and ER given below are sufficient to ensure the validity of (5.4.1). To be precise, let us define recursively $f_1(x_1)=f(x_1)$, $f_2(x_1,x_2)=f_1(x_1+f(x_2))$, $f_{n+1}(x_1,\ldots,x_{n+1})=f_n(x_1,\ldots,x_{n-1},x_n+f(x_{n+1}))$, $n\geqslant 2$, where the x_i, $1\leqslant i\leqslant n+1$, are arbitrary indeterminates. Conditions BR and ER ensure the existence of the digits $a_n(t)$ for any $n\in\mathbb{N}^*$ and the validity of the equation

$$t = \lim_{n\to\infty} f_n(a_1(t), \ldots, a_n(t)) \tag{5.4.2}$$

save possibly a countable t-set $E_f \subset I$.

Condition BR. *f is decreasing, $|f(u_2) - f(u_1)| < u_2 - u_1$ for any $1 \leqslant u_1 < u_2$, and there exists a positive number $\theta < 1$ such that $|f(u_2) - f(u_1)| \leqslant \theta(u_2 - u_1)$ for any $1 + f(2) < u_1 < u_2$.*

Condition ER. *f is increasing and $|f(u_2) - f(u_1)| < u_2 - u_1$ for any $0 \leqslant u_1 < u_2$.*

Let us note that Kakeya's sufficient condition for (5.4.2) to hold a.e. is simply Condition K.

Condition K. *f^{-1} is absolutely continuous and $|\mathrm{d}f^{-1}(t)/\mathrm{d}t| > 1$ a.e. in I.*

Clearly, K (for an increasing absolutely continuous f) implies ER, while K (for a decreasing absolutely continuous f) does not imply BR.

Now, we shall show that the case here enters the general framework of pw.m.t.s studied in Section 5.3. Consider the f-expansion transformation $\tau = \tau_f$ of I defined as $\tau(t) = \{f^{-1}(t)\}$, $t \in I$. (Some caution is necessary in the case where $\beta = \infty$, when either $\tau(0)$ or $\tau(1)$ should be given the value 0.) Then it is easy to see that τ is a pw.m.t. for which

$$\mathfrak{A} = \begin{cases} \{1, 2, \ldots, \beta - 1\}, & \text{if } f \text{ is decreasing and } \beta \in \mathbb{N}^*, \\ \mathbb{N}^*, & \text{if } f \text{ is decreasing and } \beta = \infty, \\ \{0, 1, \ldots, \beta - 1\}, & \text{if } f \text{ is increasing and } \beta \in \mathbb{N}^*, \\ \mathbb{N}, & \text{if } f \text{ is increasing and } \beta = \infty, \end{cases}$$

$I(a) =$ the open interval with endpoints $f(a)$ and $f(a+1)$, $\tau_a(t) = f^{-1}(t) - a$, $t \in I(a)$, $f_a(t) = f(a + t)$, $t \in I$, $a \in \mathfrak{A}$. Also, $f_{a^{(n)}}(t) = f_n(a_1, \ldots, a_{n-1}, a_n + t)$ for $n \in \mathbb{N}^*$ and $a^{(n)} = (a_1, \ldots, a_n) \in \mathfrak{A}^n$, for any $t \in I$, and $I(a^{(n)}) =$ the open interval with endpoints $f_n(a_1, \ldots, a_n)$ and $f_n(a_1, \ldots, a_{n-1}, a_n + 1)$, $n \in \mathbb{N}^*$, $a^{(n)} \in \mathfrak{A}^n$. (Clearly, for $n = 1$, $f_n(a_1, \ldots, a_{n-1}, a_n + t)$ is $f(a_1 + t)$, $t \in I$.)

Note that, in the present case, Conditions (C), (E_m), (A), and (BV) read as follows.

Condition (C)

$$\frac{\operatorname*{ess\,sup}_{t \in I} |\mathrm{d}f_n(a_1, \ldots, a_{n-1}, a_n + t)/\mathrm{d}t|}{\operatorname*{ess\,inf}_{t \in I} |\mathrm{d}f_n(a_1, \ldots, a_{n-1}, a_n + t)/\mathrm{d}t|} \leqslant C, \quad n \in \mathbb{N}^*, \quad a^{(n)} \in \mathfrak{A}^n,$$

where C is a constant $\geqslant 1$.

Condition (E_m)

$$\operatorname*{ess\,sup}_{t\in I} |\mathrm{d}f_m(a_1,\ldots,a_{m-1},a_m+t)/\mathrm{d}t| \leqslant \alpha, \quad a^{(m)}\in\mathfrak{A}^m,$$

where α is a constant <1.

Condition (A)

$$\sup_{a\in\mathfrak{A}} \operatorname*{ess\,sup}_{t\in I} \left|\frac{f''(a+t)}{f'(a+t)}\right| < \infty.$$

Condition (BV)

$$\left(\sum_{a\in\mathfrak{A}} \operatorname{var} f'_a = \sum_{a\in\mathfrak{A}} \operatorname{var}_{[a,a+1]} f'\right) = \operatorname{var} f' < \infty.$$

In particular, for the continued fraction expansion (where $f(u)=1/u$), we have (see (5.2.1′))

$$f_n(a_1,\ldots,a_n+t) = \frac{p_n+tp_{n-1}}{q_n+tq_{n-1}}, \quad n\in\mathbb{N}^*, \quad a^{(n)}\in\mathfrak{A}^n, \quad t\in I,$$

so that Condition (C) holds with $C=\sup_{n\in\mathbb{N}^*}(1+q_{n-1}/q_n)^2=4$. Clearly, Conditions ($\mathrm{E}_m$) for all $m\geqslant 2$, (A), and (BV) all hold, too.

Remark. If β occurring in the description of f at the beginning of this subsection does not belong to $\mathbb{N}^*$, then τ_f does not verify (5.3.1). Such an f leads to a so-called f-expansion with dependent digits. In accordance with the situation described in Subsection 5.3.6, certain f-expansions with dependent digits enjoy properties similar to those of genuine f-expansions (see Mehta (1979), Parry (1964), Rényi (1957), and Walters (1978a)).

It is easy to see that, in terms of τ, the digits a_n and remainders r_n, $n\in\mathbb{N}^*$, can be written as

$$a_1(t)=f^{-1}(t)-\tau(t), \quad a_{n+1}(t)=a_1(\tau^n(t)), \quad t\in I, \quad n\in\mathbb{N}^* \tag{5.4.3}$$

$$r_n(t)=a_1(\tau^{n-1}(t))+\tau^n(t), \quad t\in I, \quad n\in\mathbb{N}^*. \tag{5.4.4}$$

It follows at once from (5.4.3) that the digit sequence associated with f is the label sequence associated with τ_f (see Subsection 5.3.1).

A consequence of the above remark is that Condition (C) is sufficient to ensure the validity of (5.4.2) a.e. in I. Indeed, for any $t\in I$ for which the digits $a_n(t)$ exist for all $n\in\mathbb{N}^*$, we have

$$t=f(a_1(t)+\{r_1(t)\})=f(a_1(t)+f(a_2(t)+\{r_2(t)\}))=\cdots$$

i.e.,

$$t=f_n(a_1(t),\ldots,a_{n-1}(t),a_n(t)+\{r_n(t)\}), \quad n\in\mathbb{N}^*.$$

Thus, for any $n\in\mathbb{N}^*$, such a t belongs to the open interval $I(a^{(n)}(t))$ of rank n with endpoints $f_n(a_1(t),\ldots,a_n(t))$ and $f_n(a_1(t),\ldots,a_{n-1}(t),\ a_n(t)+1)$. But, under Condition (C), by Propositions 5.3.1 and 5.3.2, the length of the largest interval of rank n converges to 0 as $n\to\infty$, which shows that (5.4.2) holds a.e. in I.

5.4.2 *Gauss' problem and mixing properties*

It is easy to show that

$$\lambda(r_0<t)=t,$$

$$\lambda(\{r_n\}<t\,|\,a_1,\ldots,a_n)=\frac{f_n(a_1,\ldots,a_n)-f_n(a_1,\ldots,a_n+t)}{f_n(a_1,\ldots,a_n)-f_n(a_1,\ldots,a_n+1)} \tag{5.4.5}$$

for any $t\in I$ and $n\in\mathbb{N}^*$. The above equations are a generalization of the Brodén–Borel–Lévy formula for the continued fraction expansion (note that $\{r_n\}=f(r_{n+1})$, $n\in\mathbb{N}$). The equations (5.4.5) can only play the part played by the Brodén–Borel–Lévy formula in Subsection 5.2.2, in the case where the fraction in the second equation (5.4.5) can be written as a function of t and the nth iterate of a map from $I\times\mathfrak{A}$ into I. Fortunately, in the case of the general f-expansion, we have at our disposal the Frobenius–Perron operator associated with the f-expansion transformation τ_f. From Subsection 5.3.3, the Frobenius–Perron operator P_λ associated with τ_f under λ is defined as

$$P_\lambda h(t)=\sum_{a\in\mathfrak{A}}|f'(a+t)|h(f(a+t)),\quad t\in I,\quad h\in L^1_\lambda=L^1(I,\mathscr{B}_I,\lambda),$$

under the unique assumption that f is absolutely continuous, which implies the λ-non-singularity of τ_f.

Let us remark that if f is absolutely continuous and Condition (C) holds, then Theorem 5.3.5 applies to the f-expansion transformation τ_f, asserting the existence of a unique probability $\rho\ll\lambda$, such that τ_f is ρ-preserving. Rényi's ρ is in fact equivalent to λ and

$$\frac{1}{C}\leqslant r=\frac{d\rho}{d\lambda}\leqslant C\quad\text{a.e.}$$

Next, the density r verifies the equation $P_\lambda r=r$, i.e.

$$r(t)=\sum_{a\in\mathfrak{A}}|f'(t+a)|r(f(t+a))\quad\text{a.e.} \tag{5.4.6}$$

The fact that τ_f is ρ-preserving implies that the digit sequence $(a_n)_{n\in\mathbb{N}^*}$ is strictly stationary under ρ. (Compare with Proposition 5.2.1 and note that in the case of the continued fraction expansion, where $f(u)=1/u$, Rényi's ρ is Gauss' γ.)

Finally, under Conditions (C) and (BV), the density $r = d\rho/d\lambda \in BV_\lambda$. (See Subsection 5.3.4.)

Further, we shall first give a solution to the obvious generalization to an f-expansion of Gauss' problem.

Consider a probability $\mu \ll \lambda$, and put

$$F_n(w) = F_n(w, \mu) = \mu(\{r_n\} < w). \tag{5.4.7}$$

Note that $F_0(t) = \mu([0, t])$, $t \in I$, and the derivative F'_0 exists a.e. in I, with

$$F_0(t) = \int_0^t F'_0(v)\, dv, \quad t \in I.$$

Proposition 5.4.1. (Solution of Gauss' problem for f-expansions) *Assume that Conditions* (C) *and* (BV) *hold. If the density* $F'_0 = d\mu/d\lambda \in BV_\lambda$ *then there exist two positive constants* $q < 1$ *and* c *such that*

$$\mu(s < \{r_n\} < t) = (1 + \theta_0 q^n)\rho([s, t])$$

for all $s < t$, $s, t \in I$, *and* $n \in \mathbb{N}^*$, *where* $\theta_0 = \theta_0(\mu, n, s, t)$ *with* $|\theta_0| \leqslant c$.

Proof. It follows from (5.4.4) that $\{r_n(t)\} = \tau^n(t) = \tau^n(\{r_0(t)\})$, $n \in \mathbb{N}^*$, $t \in I$, and the probabilistic interpretation of the Frobenius–Perron operator P_λ (see Section A2.11) yields $F'_n = P^n_\lambda F'_0$, $n \in \mathbb{N}^*$. Now, by Corollary 5.3.15,

$$\left\| P^n_\lambda F'_0 - r \int_I F'_0\, d\lambda \right\|_v = \left\| P^n_\lambda F'_0 - r \right\|_v \leqslant c_0 q^n (v(F'_0) + \|F'_0\|_1), \quad n \in \mathbb{N}^*,$$

for suitable positive constants $q < 1$ and c_0. On the other hand, by Lemma 5.3.8, ess sup $|h| \leqslant \|h\|_v$ for any $h \in L^\infty(I, \mathscr{B}_I, \lambda)$. Consequently,

$$\operatorname{ess\,sup} |P^n_\lambda F'_0 - r| \leqslant c_0 q^n (1 + v(F'_0)), \quad n \in \mathbb{N}^*.$$

Hence integrating w.r.t. λ over an arbitrary interval $[s, t] \subset I$ yields the stated result since, by Theorem 5.3.5, we have $\lambda([s, t]) \leqslant C\rho([s, t])$. □

Now, we are prepared to show that in the special case of an f-expansion, stronger mixing properties than that given in Theorem 5.3.22 can be obtained. We use the notation introduced before the statement of that theorem.

Theorem 5.4.2. *Assume that Conditions* (C) *and* (BV) *hold. If the density* $F'_0 = d\mu/d\lambda \in BV_\lambda$, *then for any* $A \in \mathscr{B}_{[1,k]}$, $B \in \mathscr{B}_{[k+n,\infty]}$ *and* $k, n \in \mathbb{N}^*$, *we have*

$$|\mu(A \cap B) - \mu(A)\mu(B)| \leqslant c q^n \mu(B) \tag{5.4.8}$$

with some suitable positive constants $q < 1$ *and* c.

If, instead of (BV), the stronger condition

$$\sup_{k \in \mathbb{N}^*} \sup_{a^{(k)} \in \mathfrak{A}^k} \operatorname{var} f'_{a^{(k)}} / \mu(I(a^{(k)})) < \infty \tag{5.4.9}$$

is assumed, then the digit sequence $(a_n)_{n\in\mathbb{N}^*}$ *is exponentially* ψ*-mixing under* μ.

Proof. For arbitrarily fixed $k\in\mathbb{N}^*$ and $i_1,\ldots,i_k\in\mathfrak{A}$, and for any $n\in\mathbb{N}$ and $t\in I$ put

$$G_n(t)=\mu(\{r_{n+k}\}<t\,|\,a_l=i_l, 1\leqslant l\leqslant k).$$

Note that

$$G_0(t)=\left|\int_{f_{i^{(k)}}(0)}^{f_{i^{(k)}}(t)}\mathrm{d}\mu\right|\Big/\mu(I(i^{(k)})),$$

where $i^{(k)}=(i_1,\ldots,i_k)$ and $f_{i^{(k)}}(t)=f_k(i_1,\ldots,i_{k-1},i_k+t)$.

It follows from (5.4.4) that $\{r_{n+k}\}=\tau^n(\{r_k\})$, $n\in\mathbb{N}$. The probabilistic interpretation of the Frobenius–Perron operator P_λ, yields $G'_n=P_\lambda^n G'_0$, $n\in\mathbb{N}$. As in the preceding proof, by Corollary 5.3.15 and Lemma 5.3.8, we get

$$\operatorname{ess\,sup}|G'_n-r|\leqslant c_0q^n\left(1+\frac{\mathrm{v}((F'_0\circ f_{i^{(k)}})|f'_{i^{(k)}}|)}{\mu(I(i^{(k)}))}\right),\quad n\in\mathbb{N}^*.$$

Using Lemma 5.3.8 again it is easy to see that

$$\begin{aligned}\mathrm{v}((F'_0\circ f_{i^{(k)}})|f'_{i^{(k)}}|)&\leqslant|f'_{i^{(k)}}|\mathrm{v}(F'_0)+\operatorname{ess\,sup}|F'_0|\operatorname{var}f'_{i^{(k)}}\\&\leqslant c_1\mu(I(i^{(k)}))+c_2\operatorname{var}f'_{i^{(k)}}\end{aligned}$$

for suitable positive constants c_1 and c_2 not depending on $i^{(k)}$. Therefore

$$\operatorname{ess\,sup}|G'_n-r|\leqslant c_0q^n\left(1+c_1+\frac{c_2\operatorname{var}f'_{i^{(k)}}}{\mu(I(i^{(k)}))}\right),\quad n\in\mathbb{N}^*.$$

Integrating w.r.t. λ over an arbitrary interval $[s,t]\subset I$ yields

$$\begin{aligned}&|\mu(I(i^{(k)})\cap(s<\{r_{n+k}\}<t))-\mu(I(i^{(k)}))\rho([s,t])|\\&\leqslant c_0q^n((1+c_1)\mu(I(i^{(k)}))+c_2\operatorname{var}f'_{i^{(k)}})\lambda([s,t]),\quad n\in\mathbb{N}^*.\end{aligned}$$

As we have already noted, $\lambda([s,t])\leqslant C\rho([s,t])$, while by Proposition 5.4.1

$$\rho([s,t])=(1+\theta_0q^{n+k})^{-1}\mu(s<\{r_{n+k}\}<t)$$

for all $n\in\mathbb{N}^*$ large enough, say $n\geqslant n_0$, so that we finally get

$$\begin{aligned}&|\mu(I(i^{(k)})\cap(s<\{r_{n+k}\}<t))-\mu(I(i^{(k)}))\mu(s<\{r_{n+k}\}<t))|\\&\leqslant c_3q^n(\mu(I(i^{(k)}))+c_4\operatorname{var}f'_{i^{(k)}})\mu(s<\{r_{n+k}\}<t),\quad n\geqslant n_0,\end{aligned}$$

for suitable positive constants c_3 and c_4 not depending on $i^{(k)}$.

The above inequality proves both assertions of the theorem, since $\sum_{i^{(k)}\in\mathfrak{A}^k}\mu(I(i^{(k)}))=1$, $k\in N^*$, and, by Lemma 5.3.20,

$$\sup_{k\in\mathbb{N}^*}\sum_{i^{(k)}\in\mathfrak{A}^k}\operatorname{var}f'_{i^{(k)}}<\infty.\qquad\square$$

Remarks. 1. The reasoning in the above proof implicitly assumes that $\mu(I(i^{(k)})) \neq 0$. Nevertheless, the final inequality above involving $I(i^{(k)})$ obviously holds even when $\mu(I(i^{(k)})) = 0$.

2. Condition (BV) in the statement of Theorem 5.4.2 can be replaced by a weaker condition involving the p-variation of f' on the intervals $[a, a+1]$, $a \in \mathfrak{A}$ for some $p > 1$. For details see Iosifescu (1987b).

3. Assumption (5.4.9) with $\mu = \rho$ is implied by Conditions (C) and (A), and, more generally, by (C) and the following weaker form of (A): for some $p > 1$ we have

$$\sup_{a \in \mathfrak{A}} \int_0^1 \left| \frac{f''(t+a)}{f'(t+a)} \right|^p dt < \infty.$$

This is stated in Gordin (1968). The reader is invited to provide the proof.

The mixing property (5.4.8) is a less usual one, and was called reversed φ-mixing. It has recently been studied by Peligrad (1983). Her results enable us to state Theorem 5.4.3.

Theorem 5.4.3. *Assume that Conditions* (C) *and* (BV) *hold. Let g be a real-valued function defined on $\mathfrak{A}$ such that $\mathsf{E}(g^2(a_1)) < \infty$, $\mathsf{E}(g(a_1)) = 0$, and $\lim_{n\to\infty} \mathsf{E}(S_n^2) = \infty$, where the expectation is w.r.t. ρ and $S_n = \sum_{i=1}^n g(a_i)$, $n \in N^*$, $S_0 = 0$. Then the limit $\lim_{n\to\infty} \mathsf{E}(S_n^2/n) = \sigma^2$ exists and is positive, and $\rho\boldsymbol{\eta}_n^{-1} \Rightarrow \mathbf{B}$, where $\boldsymbol{\eta}_n$ stands for either of the stochastic processes $\boldsymbol{\eta}_n^C$ and $\boldsymbol{\eta}_n^D$ defined at the beginning of Subsection* 4.1.3.

It is also clear that, under Conditions (C) and (BV), Theorems 5.3.23 and 5.3.24 apply to an f-expansion. (Do not confuse the 'f' here with that in the statements of the two theorems!)

Finally, we remark that the limit theorems valid for the digit sequence $(a_n)_{n\in\mathbb{N}^*}$ associated with the continued fraction expansion (see Subsection 5.2.4) carry over to the digit sequence associated with an f-expansion for which the assumptions ensuring ψ-mixing under ρ hold. The alterations to be made are as follows: Gauss' γ should be replaced by Rényi's ρ, $\mathbb{N}^*$ by $\mathfrak{A}$, τ by τ_f, and

$$\frac{1}{\log 2} \log \frac{1 + v(i_1, \ldots, i_n)}{1 + u(i_1, \ldots, i_n)} \quad \text{by} \quad \rho([u_f(i_1, \ldots, i_n), v_f(i_1, \ldots, i_n)]).$$

Here

$$\begin{Bmatrix} u_f(i_1, \ldots, i_n) \\ v_f(i_1, \ldots, i_n) \end{Bmatrix} = \begin{Bmatrix} \min \\ \max \end{Bmatrix} (f_n(i_1, \ldots, i_n), f_n(i_1, \ldots, i_{n-1}, i_n + 1)).$$

5.4.3 A conjecture

At the end of Subsection 5.3.4 an RSCC $\{(W,\mathscr{W}),(X,\mathscr{X}),u,P\}$ was naturally associated with a C^1 pw.m.t. for which Conditions (C) and (BD$^{(2)}$) hold. This carries over to an f-expansion for which the same assumptions are made. In this latter case the density r is a Lipschitz function, equation (5.4.6) holds for all $t\in I$, and we have

$$(W,\mathscr{W})=(I,\mathscr{B}_I),\qquad (X,\mathscr{X})=(\mathfrak{A},\mathscr{P}(\mathfrak{A})),$$

$$u(w,x)=f(x+w),\quad P(w,x)=\frac{r(f(x+w))|f'(x+w)|}{r(w)},\quad w\in I,\quad x\in X.$$

Clearly, the corresponding transition operator U is the Frobenius–Perron operator P_ρ associated with τ_f under ρ.

It is easy to see that

$$u(w,x^{(n)})=f_n(x_n,\ldots,x_2,x_1+w)$$

for any $w\in W$, $n\in\mathbb{N}^*$ and $x^{(n)}=(x_1,\ldots,x_n)\in X^n$. Therefore, for any arbitrarily given $w_0\in W$, the two sequences $(\xi_n)_{n\in\mathbb{N}^*}$ and $(\zeta_n)_{n\in\mathbb{N}}$ associated with this RSCC (see Theorem 1.1.2) satisfy

$$\mathsf{P}_{w_0}(\xi_1=x)=P(w_0,x)$$

$$\mathsf{P}_{w_0}(\xi_{n+1}=x\,|\,\xi^{(n)})=P(\zeta_n,x),\quad x\in X,$$

$$\zeta_n=f_n(\xi_n,\ldots,\xi_2,\xi_1+w_0),\quad n\in\mathbb{N}^*.$$

It has been conjectured in Iosifescu (1985b) (see also Iosifescu (1988)) that, for $w_0=0$, the ξ-sequence above is equivalent to the digit sequence $(a_n)_{n\in\mathbb{N}^*}$ associated with the f-expansion considered, i.e. in particular,

$$\lambda(a_1=x)=P(0,x),$$

$$\lambda(a_{n+1}=x\,|\,a_i=x_i,\ 1\leqslant i\leqslant n)=P(f_n(x_n,\ldots,x_1),x),\quad x\in X,$$

for all $x_i\in X$, $1\leqslant i\leqslant n$, $n\in\mathbb{N}^*$.

The ζ-sequence appears then to play the part played by the sequence $(s_n)_{n\in\mathbb{N}}$ for the continued fraction expansion.

We have been unable to prove the above conjecture, which is true for both the continued fraction expansion and the D-ary expansion (see the next subsection). Is this blindness or is the conjecture simply not true? And, if it is not true, what relationship does exist between the sequences $(a_n)_{n\in\mathbb{N}^*}$ and $(\xi_n)_{n\in\mathbb{N}^*}$?

5.4.4 The D-ary expansion

Let D be an arbitrary natural integer $\geqslant 2$. Any number $t\in I$ has a D-ary expansion of the form

$$t = \frac{a_1(t)}{D} + \frac{a_2(t)}{D^2} + \cdots = (a_1(t), a_2(t), \ldots), \tag{5.4.10}$$

where the $a_n(t)$, $n \in \mathbb{N}^*$, assume the values $0, 1, \ldots, D-1$. Representation (5.4.10) is not always unique. For example

$$\frac{1}{D} = (1, 0, 0, 0, \ldots) = (0, D-1, D-1, D-1, \ldots).$$

In general, the D-ary representation is not unique exactly for the rational numbers of the form k/D^m, $1 \leqslant k \leqslant D^m$, $m \in \mathbb{N}^*$. Clearly, the set of these numbers is countable.

Actually, the D-ary expansion is the f-expansion for which $f(u) = u/D$, $0 \leqslant D$, i.e. $\beta = D$. Clearly, it satisfies Kakeya's condition and, consequently, Condition ER, too. Next, the associated D-ary transformation τ is defined as $\tau(t) = \{Dt\}$, with $\tau^n(t) = \{D^n t\}$, $t \in I$, $n \in \mathbb{N}^*$, and we have $a_n(t) = [D^n t]$, $r_0(t) = t$, $r_n(t) = D^n t$, $t \in I$, $n \in \mathbb{N}^*$. Since

$$f_n(a_1, \ldots, a_n) = \frac{a_1}{D} + \frac{a_2}{D^2} + \cdots + \frac{a_n}{D^n}$$

and

$$r_n(t) = a_n(t) + \frac{a_{n+1}(t)}{D} + \frac{a_{n+2}(t)}{D^2} + \cdots,$$

$$\{r_n(t)\} = \frac{a_{n+1}(t)}{D} + \frac{a_{n+2}(t)}{D^2} + \cdots, \quad n \in \mathbb{N}^*,$$

it follows from (5.4.5) that, under λ, the r.v.s a_n, $n \in \mathbb{N}^*$, are independent and identically distributed with

$$\lambda(a_n = i) = \frac{1}{D}, \quad i = 0, 1, \ldots, D-1.$$

Remarks. 1. The above result was obtained by Borel (1909). Spătaru (1978a, b) completely solved the problem of describing all the probabilities π on $\mathscr{B}_I$ under which the digits of an f-expansion are independent random variables with prescribed π-distributions.

2. The expression above of f_n shows that for the D-ary expansion, Condition (C) is satisfied with $C = 1$. Hence $\rho = \lambda$.

It is obvious that the independence under λ of the a_n, $n \in \mathbb{N}^*$, enables us to state for them all the classical limit theorems proved under the independence assumption. Nevertheless, even in the present context, both Proposition A3.3 and Theorem A3.7 bring new information to the classical theory. Let h be a measurable real-valued function on I such that

$\int_0^1 |h(t)|\,\mathrm{d}t < \infty$ or $\int_0^1 h^2(t)\,\mathrm{d}t < \infty$ according to whether Proposition A3.3 or Theorem A3.7 is considered. Then, with the notation there, taking $F(a_1, a_2, \ldots) = h((a_1, a_2, \ldots))$, we have

$$\mathsf{E}(f_1) = \int_0^1 h(t)\,\mathrm{d}t (= a),$$

$$\mathsf{E}(f_1 | (a_i)_{1 \leqslant i \leqslant n})(t) = D^n \int_{(i-1)/D^n}^{i/D^n} h(u)\,\mathrm{d}u$$

if $(i-1)/D^n < t < i/D^n$, $1 \leqslant i \leqslant D^n$, $n \in \mathbb{N}^*$, while

$$\sigma^2 = \int_0^1 h^2(t)\,\mathrm{d}t - a^2 + 2 \sum_{n \in \mathbb{N}^*} \left(\int_0^1 h(t) h(\{D^n t\})\,\mathrm{d}t - a^2 \right).$$

It follows from Proposition A3.3 that

$$\lim_{n \to \infty} \frac{1}{n} \sum_{k=1}^{n} h\left(\frac{a_k}{D} + \frac{a_{k+1}}{D^2} + \cdots \right) = \int_0^1 h(t)\,\mathrm{d}t \tag{5.4.11}$$

a.e. in I. This result is known as Raikov's theorem.

It follows from Theorem A3.7 that if

$$\sum_{n \in \mathbb{N}^*} \left(\sum_{i=1}^{D^n} \int_{(i-1)/D^n}^{i/D^n} \left| h(t) - D^n \int_{(i-1)/D^n}^{i/D^n} h(u)\,\mathrm{d}u \right|^2 \mathrm{d}t \right)^{1/2} < \infty$$

and $\sigma > 0$, then

$$\lim_{n \to \infty} \lambda \left(\frac{1}{\sigma \sqrt{n}} \sum_{k=1}^{n} \left(h\left(\frac{a_k}{D} + \frac{a_{k+1}}{D^2} + \cdots \right) - \int_0^1 h(t)\,\mathrm{d}t \right) < x \right) = \Phi(x) \tag{5.4.12}$$

for all $x \in \mathbb{R}$. For the rate of convergence in (5.4.12) see Ladohin & Moskvin (1971) and Misevičius (1972). Note that, by Theorem 5.3.24, if $h \in BV_\lambda$, then, under the assumption that $\sigma > 0$, the rate of convergence in (5.4.12) equals the optimal one $O(n^{-1/2})$.

We close by noticing that for the D-ary expansion the components of the RSCC $\{(W, \mathscr{W}), (X, \mathscr{X}), u, P\}$ considered in Subsection 5.4.3 are as follows

$$(W, \mathscr{W}) = (I, \mathscr{B}_I), \quad X = \{0, 1, \ldots, D-1\}, \quad \mathscr{X} = \mathscr{P}(X),$$

$$u(w, x) = \frac{w + x}{D}, \quad P(w, x) = \frac{1}{D}, \quad w \in I, \quad x \in X.$$

The corresponding transition operator

$$Uh(w) = \frac{1}{D} \sum_{i=0}^{D-1} h\left(\frac{w+i}{D} \right), \quad h \in B(I), \quad w \in I,$$

which is a restriction of the Frobenius–Perron operator associated with the D-ary expansion transformation under λ, is known as the Fortet–Kac operator (see Fortet (1940) and Kac (1938)). These authors were the first to use it to derive the central limit theorem for the D-ary expansion with a Lipschitz h in (5.4.12).

It is clear that for the D-ary expansion the conjecture made in Subsection 5.4.3 is true. Consequently, the associated Markov chain $(\zeta_n)_{n\in\mathbb{N}}$ with $\zeta_0=0$ is equivalent to the sequence $(s_n)_{n\in\mathbb{N}}$, where

$$s_0=0,\quad s_n=(a_n,\ldots,a_1)=\frac{a_n}{D}+\cdots+\frac{a_1}{D^n},\quad n\in\mathbb{N}^*.$$

Similarly, $(\zeta_n)_{n\in\mathbb{N}}$ with $\zeta_0=(i_r,\ldots,i_1)$ is equivalent to the sequence $(s_{n+r})_{n\in\mathbb{N}}$. This enables us to state for the sequences $(s_n)_{n\in\mathbb{N}}$ (under λ) and $(s_{n+r})_{n\in\mathbb{N}}$ (under $D^r\lambda(\cdot\cap[(i_1,\ldots,i_r),(i_1,\ldots,i_r+1)])$, all the limit theorems proved in Section 4.2 for the sequence $(\zeta_n)_{n\in\mathbb{N}}$, with arbitrarily given $r\in\mathbb{N}^*$ and $0\leqslant i_1,\ldots,i_r\leqslant D-1$.

5.5 Strict-sense infinite-order chains

5.5.1 *Preliminaries*

Throughout this book we have used the nomenclature 'infinite-order chain' for the sequence $(\xi_n)_{n\in\mathbb{N}^*}$ of random variables constructed in Theorem 1.1.6. This terminology is not accurate, since by a *strict-sense infinite-order chain* we should mean the RSCC $\{(W,\mathscr{W}),(X,\mathscr{X}),u,({}^tP)_{t\in\mathbb{Z}}\}$, where $(W,\mathscr{W})=(X^{-\mathbb{N}},\mathscr{X}^{-\mathbb{N}})$, and if $w=(\ldots,x_{-n},\ldots,x_{-1},x_0)$, then $u(w,x)=(\ldots,x'_{-n},\ldots,x'_{-1},x'_0)$, where $x'_{-n}=x_{-n+1}$, $n\in\mathbb{N}^*$, $x'_0=x$. For a (strict-sense) infinite-order chain the notation

$$u(w,x^{(n)})=wx^{(n)}=w+x^{(n)},\quad w\in X^{-\mathbb{N}},\quad x^{(n)}\in X^n,\quad n\in\mathbb{N}^*,$$

is traditionally used. The homogeneous case is characterized by the fact that ${}^tP=P$, $t\in\mathbb{Z}$.

Let us note that an element $w=(\ldots,x_{-n},\ldots,x_{-1},x_0)\in X^{-\mathbb{N}}$ can be interpreted as a 'path', to mean the left-infinite sequence of successive states of a system (with state space X): x_0 is the state at the last observation time; x_{-1} the state a time unit before, and so on going back endlessly in time. The path $w+x^{(n)}$ is obtained by simply adding the components of $x^{(n)}$ to the path w. (This explains the special notation we use.) Next, ${}^tP(w,A)$ can be thought of as the probability of being in the set of states $A\in\mathscr{X}$ at time $t+1$, conditional on the path $w\in X^{-\mathbb{N}}$ up to time $t\in\mathbb{Z}$. Bearing this in mind we can assert that an infinite-order chain is a multiple MC whose multiplicity order is infinite.

To conclude these preliminary considerations we shall remark that Theorem 1.1.6 does not agree with the interpretation given above. It is clear that the latter assumes the existence of a *doubly* infinite sequence $(\xi_t)_{t\in\mathbb{Z}}$ of X-valued r.v.s. The existence of such a sequence will be proved in the next subsection.

5.5.2 An existence theorem

Let $(X, \mathscr{X})$ be a measurable space. Define on the measurable space $(X^{\mathbb{Z}}, \mathscr{X}^{\mathbb{Z}})$ a doubly infinite sequence $(\xi_t)_{t\in\mathbb{Z}}$ of X-valued r.v.s by the equations

$$\xi_t(x^{\mathbb{Z}}) = \mathrm{pr}_t x^{\mathbb{Z}} = x_t, \quad t\in\mathbb{Z},$$

if $x^{\mathbb{Z}} = (x_t)_{t\in\mathbb{Z}}$.

For a homogeneous infinite-order chain we shall prove Theorem 5.5.1.

Theorem 5.5.1. *Assume that X is a separable metric space and that $\mathscr{X} = \mathscr{B}_X$. If Conditions* $\mathrm{M}(n_0)$ *for some $n_0\in\mathbb{N}^*$ and* $\mathrm{FLS}(X, 1)$ *hold, then there exists a unique probability P_∞ on $\mathscr{X}^{\mathbb{Z}}$ such that the sequence $(\xi_t)_{t\in\mathbb{Z}}$ is strictly stationary under P_∞, and*

$$\mathsf{P}_\infty((\xi_{n+i})_{1\leqslant i\leqslant r}\in A \mid \xi^{(-\infty,n]}) = P_r(\xi^{(-\infty,n]}, A), \quad \mathsf{P}_\infty - \text{a.s.}, \qquad (5.5.1)$$

for any $n\in\mathbb{Z}$, $r\in\mathbb{N}^*$ and $A\in\mathscr{X}^r$ where $\xi^{(-\infty,n]} = (\xi_t)_{t\leqslant n}$.

Also, for any $n\in\mathbb{Z}$,

$$\varphi(\sigma((\xi_t)_{t\leqslant n}), \sigma((\xi_t)_{t\geqslant s+n})) \leqslant \varepsilon_s, \quad s\in\mathbb{N}^*, \qquad (5.5.2)$$

where the quantities ε_s are those defined in Subsection 4.1.1.

Proof. In what follows, we identify X^{n-m+1} and the Cartesian product $X^{\mathfrak{I}}$ for $\mathfrak{I} = \{m, \ldots, n\} = [m, n]$, $m\leqslant n$, $m, n\in\mathbb{Z}$. With this convention $\mathrm{pr}^{-1}_{[m,n]}(y^{(n-m+1)})$ is the set $\{x^{\mathbb{Z}}: (x_m, \ldots, x_n) = y^{(n-m+1)}\}$.

Under the assumptions of the theorem, the RSCC $\{(X^{-\mathbb{N}}, \mathscr{X}^{-\mathbb{N}}), (X, \mathscr{X}), u, P\}$ is uniformly ergodic (see Theorem 2.2.7). Then the Daniell–Kolmogorov Theorem A1.4 ensures the existence of a unique probability P_∞ on $\mathscr{X}^{\mathbb{Z}}$ satisfying

$$\mathsf{P}_\infty(\mathrm{pr}^{-1}_{[m,n]}(B)) = \mathsf{P}^\infty_{n-m+1}(B)$$

for any $m\leqslant n$, $m, n\in\mathbb{Z}$, and $B\in\mathscr{X}^{n-m+1}$. Clearly, the above equation shows that the sequence $(\xi_t)_{t\in\mathbb{Z}}$ is strictly stationary under P_∞.

Let us prove (5.5.1). For $s\in\mathbb{N}^*$ put $\alpha_s = \sum_{j\geqslant s} b_j$. Using Proposition 2.1.2, for arbitrary $m\leqslant t\leqslant n$, $m, n, t\in\mathbb{Z}$, $r\in\mathbb{N}^*$, $w', w''\in X^{-\mathbb{N}}$, $A\in\mathscr{X}^r$, and $B\in\mathscr{X}^{n-m+1}$, we can write

$$\int_{\mathrm{pr}^{-1}_{[m,n]}(B)} \mathsf{P}_\infty(dx^{\mathbb{Z}}) P_r(\mathrm{pr}_{(-\infty,n]}(x^{\mathbb{Z}}), A)$$

$$= \theta_1 \alpha_{n-t+1} + \int_{\mathrm{pr}^{-1}_{[m,n]}(B)} \mathsf{P}_\infty(dx^{\mathbb{Z}}) P_r(w' + \mathrm{pr}_{[t,n]}(x^{\mathbb{Z}}), A)$$

$$= \theta_1 \alpha_{n-t+1} + \int_B \mathbf{P}^{\infty}_{n-m+1}(dy^{(n-m+1)}) P_r(w' + (y_{t-m+1}, \ldots, y_{n-m+1}), A)$$

$$= \theta_1 \alpha_{n-t+1}$$

$$+ \lim_{q \to \infty} \int_B P^q_{n-m+1}(w'', dy^{(n-m+1)}) P_r(w' + (y_{t-m+1}, \ldots, y_{n-m+1}), A),$$

where $|\theta_1| \leqslant 1$. But for any $q \in \mathbb{N}^*$, by (1.1.3) and (1.1.4) we have

$$\int_B P^{q+1}_{n-m+1}(w'', dy^{(n-m+1)}) P_r(w' + (y_{t-m+1}, \ldots, y_{n-m-1}), A)$$

$$= \int_{X^q \times B} P_{n+q-m+1}(w'', dz^{(n+q-m+1)}) P_r(w' + (z_{t+q-m+1}, \ldots, z_{n+q-m+1}), A)$$

$$= \theta_2 \alpha_{n-t+1} + \int_{X^q \times B} P_{n+q-m+1}(w'', dz^{(n+q-m+1)}) P_r(w'' + z^{(n+q-m+1)}, A)$$

$$= \theta_2 \alpha_{n-t+1} + P^{q+1}_{n+r-m+1}(w'', B \times A),$$

where $|\theta_2| \leqslant 1$. Therefore

$$\int_{\mathrm{pr}^{-1}_{[m,n]}(B)} \mathbf{P}_{\infty}(dx^{\mathbb{Z}}) P_r(\mathrm{pr}_{(-\infty,n]}(x^{\mathbb{Z}}), A)$$

$$= \theta_3 \alpha_{n-t+1} + \lim_{q \to \infty} P^{q+1}_{n+r-m+1}(w'', B \times A)$$

$$= \theta_3 \alpha_{n-t+1} + \mathbf{P}^{\infty}_{n+r-m+1}(B \times A)$$

$$= \theta_3 \alpha_{n-t+1} + \mathbf{P}_{\infty}(\mathrm{pr}^{-1}_{[m,n]}(B) \times \mathrm{pr}^{-1}_{[n+1,n+r]}(A)), \tag{5.5.3}$$

where $|\theta_3| \leqslant 2$. But

$$\mathrm{pr}^{-1}_{[m,n]}(B) = \mathrm{pr}^{-1}_{[m',n]}(X^{m-m'} \times B)$$

for $m' < m$. We may thus let $t \to -\infty$ in (5.5.3), and (5.5.1) is obtained at once.

To prove the uniqueness of $\mathbf{P}_{\infty}$ satisfying (5.5.1) let us remark that this equation allows us to write

$$\int_{X^{\mathbb{Z}}} \mathbf{P}_{\infty}(dx^{\mathbb{Z}}) P_{n+r-m+1}(\mathrm{pr}_{(-\infty,m-r-1]}(x^{\mathbb{Z}}), X^r \times B) = \mathbf{P}_{\infty}(\mathrm{pr}^{-1}_{[m,n]}(B))$$

for any $r \in \mathbb{N}^*$, $m \leqslant n$, $m, n \in \mathbb{Z}$, and $B \in \mathscr{X}^{n-m+1}$. Letting $r \to \infty$, by uniform ergodicity we get

$$\mathbf{P}^{\infty}_{n-m+1}(B) = \mathbf{P}_{\infty}(\mathrm{pr}^{-1}_{[m,n]}(B)),$$

whence the uniqueness of $\mathbf{P}_{\infty}$ satisfying (5.5.1) follows.

Inequality (5.5.2) is an immediate consequence of (5.5.1) and uniform ergodicity. □

5.5.3 The case of a countable state space

In this subsection we shall consider the case where the state space X is countable. Write $P(w, x)$ for $P(w, \{x\})$, $w \in X^{-\mathbb{N}}$, $x \in X$. If we assume that there exists $x_0 \in X$ such that $P(w, x_0) \geqslant \gamma > 0$ for any $w \in X^{-\mathbb{N}}$ and that the series $\sum_{n \in \mathbb{N}^*} b_n$ is convergent where

$$b_n = \frac{1}{2} \sup \sum_{x \in X} |P(w' + x^{(n)}, x) - P(w'' + x^{(n)}, x)|, \quad n \in \mathbb{N}^*,$$

the supremum being taken over all $w', w'' \in X^{-\mathbb{N}}$, $x^{(n)} \in X^n$ (see Section A1.4 and Subsection 2.2.2), then Conditions FLS$(X, 1)$, FLS$(\{x_0\}, 1)$ and $\mathrm{K}_1(\{x_0\}, 0)$ all hold. By Propositions 2.2.6 and 2.2.8, Condition M(1) holds and so do the conclusions of Theorem 5.5.1.

In what follows, we consider chains of infinite order with a countable state space, taken, without any loss of generality, to be the set $\mathbb{N}^*$ of natural integers. The basic device is to associate the path $(\ldots, i_{-n}, \ldots, i_{-1}, i_0)$ with the irrational number which has continued fraction expansion $[i_0, i_1, \ldots, i_{-n}, \ldots]$. It is instructive to compare the results below with the special case just described of Theorem 5.5.1.

Let Y denote the set of irrational numbers in I. For any $y \in Y$ consider a probability $(p_i(y))_{i \in \mathbb{N}^*}$ on $\mathbb{N}^*$. We shall use the notation $(y \equiv y')_n$, $y, y' \in Y$, $n \in \mathbb{N}$, to mean that the first n digits of the continued fraction expansion of y coincide with the first n digits of the continued fraction expansion of y'. Of course, $(y \equiv y')_0$ means that no restriction is imposed on y and y'. It is easy to prove that if $(y \equiv y')_n$, $y, y' \in Y, n \in \mathbb{N}$, then $|y - y'| < (c_n c_{n+1})^{-1}$, where $c_n = 2^{-n-1}((1+\sqrt{5})^{n+1} - (1-\sqrt{5})^{n+1})/\sqrt{5}$, $n \in \mathbb{N}$, i.e. the Fibonacci numbers defined by $c_0 = c_1 = 1$, $c_n = c_{n-1} + c_{n-2}$, $n \geqslant 2$. Let

$$d((p_i(y))_{i \in \mathbb{N}^*}, (p_i(y'))_{i \in \mathbb{N}^*}) = \sum_{i \in \mathbb{N}^*} ((\sigma_i(y) - \max(\sigma_{i-1}(y), \sigma_i(y')))^+$$
$$+ (\sigma_i(y') - \max(\sigma_{i-1}(y'), \sigma_i(y)))^+),$$

where

$$\sigma_0(y) \equiv 0, \quad \sigma_i(y) = \sum_{r \leqslant i} p_r(y), \quad i \in \mathbb{N}^*, \quad y \in Y,$$

and

$$\alpha_n = \sup_{(y \equiv y')_n} d((p_i(y))_{i \in \mathbb{N}^*}, (p_i(y'))_{i \in \mathbb{N}^*}), \quad n \in \mathbb{N}. \tag{5.5.4}$$

Consider Conditions H and U.

Condition H.

$$\sum_{k \in \mathbb{N}} \prod_{n=0}^{k} (1 - \alpha_n) = \infty.$$

Condition U. *The series $\sum_{i \in \mathbb{N}^*} p_i(y)$ converges uniformly w.r.t. $y \in Y$.*

Let us note that Condition U automatically holds (by Proposition A1.15) in the case where $(p_i(y))_{i\in\mathbb{N}^*}$ is a probability for *any* $y\in I$ and the p_i, $i\in\mathbb{N}^*$, are continuous functions on I. (Actually, continuity can be replaced by a weaker assumption, namely that the p_i, $i\in\mathbb{N}^*$, be lower semi-continuous functions on I (see Billingsley (1968, p. 218)).)

We shall use the following theorem, whose proof has been given in Iosifescu & Spătaru (1973).

Theorem 5.5.2. *Let F be a (right-continuous) distribution function, continuous at the rational points, with $F(0)=0$ and $F(1)=1$, which satisfies the functional equation*

$$F(v)=\int_0^1 \sum_{\{i:(i+y)^{-1}\leqslant v\}} p_i(y)\,\mathrm{d}F(y),\ v\in I. \tag{5.5.5}$$

If Conditions H *and* U *hold, then equation* (5.5.5) *has a unique solution of one of the following forms:*

(i)$_i$ *F has a jump of magnitude* 1 *at the point $y_i=(-i+\sqrt{i^2+4})/2$, and this occurs iff $p_i(y_i)=1$, $i=1,2,\ldots$;*†

(ii) *$F(v)=\log(1+v)/\log 2$, $v\in I$, and this occurs iff*

$$p_i(y)=(y+1)/(y+i)\,(y+i+1),\ i\in\mathbb{N}^*,\ y\in Y;$$

(iii) *F is continuous and purely singular, and this occurs iff none of the previous cases apply.*

We are now able to prove another existence theorem for chains of infinite order. As has been already noted, we shall interpret the path $w=(\ldots,i_{-n},\ldots,i_{-1},i_0)$ as the continued fraction expansion (read inversely) of an $y\in Y$. We define $P(w,i)=p_i(y)$ if w and y are associated in this manner (i.e. $y=[i_0,i_{-1},\ldots,i_{-n},\ldots]$). In this context the α_n defined by (5.5.4) will be expressed as

$$\alpha_n=\sup_{(w\equiv w')_n} d((P(w,i))_{i\in\mathbb{N}^*},(P(w',i))_{i\in\mathbb{N}^*}),\quad n\in\mathbb{N}.$$

Here the notation $(w\equiv w')_n$ means that the last n components of the path w coincide with the last n components of the path w'. Next, Condition U amounts to the uniform convergence of the series $\sum_{i\in\mathbb{N}^*}P(w,i)$ w.r.t. $w\in X^{-\mathbb{N}}$.

Theorem 5.5.3. *Assume that Conditions* H *and* U *hold and that $P((\ldots,i,i,i),i)<1$ for any $i\in\mathbb{N}^*$. Then*

† Actually, the sole Condition H is sufficient for the equation $p_i(y_i)=1$ for some $i\in\mathbb{N}^*$ to be equivalent to the existence and uniqueness of an F of this type.

(i) *there exists a strictly stationary doubly infinite sequence* $(\xi_t)_{t\in\mathbb{Z}}$ *of* $\mathbb{N}^*$*-valued random variables on a probability space* $(\Omega, \mathscr{K}, \mathsf{P})$ *such that*

$$\mathsf{P}(\xi_{n+1}=i\,|\,\xi^{(-\infty,n]}) = P((\ldots,\xi_{n-1},\xi_n), i), \ \mathsf{P}\text{-}a.s. \tag{5.5.6}$$

for any $n\in\mathbb{Z}$ and $i\in\mathbb{N}^*$;

(ii) *any other doubly infinite sequence for which equation* (5.5.6) *holds is equivalent to* $(\xi_t)_{t\in\mathbb{Z}}$.

Proof. (i) Using Theorem 5.5.2 and the Daniell–Kolmogorov Theorem A1.4, we can construct, on a suitable probability space $(\Omega, \mathscr{K}, \mathsf{P})$, a strictly stationary Markov sequence $(\eta_t)_{t\in\mathbb{Z}}$, with state space Y and stationary probability given by the solution F of equation (5.5.5), i.e.

$$\mathsf{P}(\eta_t \leqslant v) = F(v), \quad t\in\mathbb{Z}, \quad v\in I,$$

and t.p.f. given by

$$\mathsf{P}\left(\eta_{t+1} = \frac{1}{i+y}\,\middle|\,\eta_t = y\right) = p_i(y), \quad i\in\mathbb{N}^*, \quad y\in Y.$$

Consider the function h defined on Y as $h(y)=$ the first digit in the continued fraction expansion of y. We shall prove that the random variables ξ_t defined as $\xi_t = h(\eta_{t+1})$, $t\in\mathbb{Z}$, satisfy (5.5.6).

First, for any $t\in\mathbb{Z}$ we have

$$\mathsf{P}\left(\eta_{t+1} = \frac{1}{\xi_t+\eta_t}\right) = \sum_{i\in\mathbb{N}^*} \mathsf{P}\left(\eta_{t+1} = \frac{1}{i+\eta_t}\right) = \sum_{i\in\mathbb{N}^*} \int_0^1 p_i(y)\,\mathrm{d}F(y) = 1,$$

whence

$$\mathsf{P}(\eta_t = [\xi_{t-1}, \xi_{t-2}, \ldots]) = \mathsf{P}\left(\eta_s = \frac{1}{\xi_{s-1}+\eta_{s-1}}, \quad s\leqslant t\right) = 1. \tag{5.5.7}$$

Next, it is clear that to prove (5.5.6) it is sufficient to show that, for any $n\in\mathbb{Z}$ and $r, i, i_1, \ldots, i_r \in \mathbb{N}^*$, we have

$$\mathsf{P}(\xi_{n+1}=i, \quad \xi_{n-l+1}=i_l, \quad 1\leqslant l\leqslant r)$$
$$= \int_{\{\xi_{n-l+1}=i_l, 1\leqslant l\leqslant r\}} P((\ldots,\xi_{n-r},\xi_{n-r+1},\ldots,\xi_n), i)\,\mathrm{d}\mathsf{P}. \tag{5.5.8}$$

Put

$$u_r = [i_1, \ldots, i_r], \quad u_r^i = [i, i_1, \ldots, i_r]$$

and, for any $v = [j_1, \ldots, j_k] \in (0,1)$, let

$$a(v) = \min([j_1, \ldots, j_k], [j_1, \ldots, j_k, 1]),$$
$$b(v) = \max([j_1, \ldots, j_k], [j_1, \ldots, j_k, 1]).$$

Then by (5.5.7) we can write

$$\mathsf{P}(\xi_{n+1}=i, \quad \xi_{n-l+1}=i_l, \quad 1\leqslant l\leqslant r)=\mathsf{P}(a(u_r^i)<\eta_{n+2}<b(u_r^i))$$
$$=\int_0^1 \mathsf{P}(a(u_r^i)<\eta_{n+2}<b(u_r^i)|\eta_{n+1}=y)\,\mathrm{d}F(y)=\int_{a(u_r)}^{b(u_r)} p_i(y)\,\mathrm{d}F(y).$$

On the other hand, we have

$$\int_{\{\xi_{n-l+1}=i_l, 1\leqslant l\leqslant r\}} P((\ldots,\xi_{n-r},\xi_{n-r+1},\ldots,\xi_n),i)\,\mathrm{d}\mathsf{P}$$
$$=\int_{\{a(u_r)\leqslant\eta_{n+1}<b(u_r)\}} p_i([i_1,\ldots,i_r,\xi_{n-r},\ldots])\,\mathrm{d}\mathsf{P}=\int_{a(u_r)}^{b(u_r)} p_i(y)\,\mathrm{d}F(y).$$

Therefore (5.5.8) holds so that the proof of (i) is complete.

(ii) It follows from the considerations above that for any $n\in\mathbb{Z}$ and $r, i_1,\ldots,i_r\in\mathbb{N}^*$ we have

$$\mathsf{P}(\xi_{n-l+1}=i_l, 1\leqslant l\leqslant r)=F(b(u_r))-F(a(u_r)). \tag{5.5.9}$$

Therefore the finite dimensional distributions of $(\xi_t)_{t\in\mathbb{Z}}$ are completely determined by F. Thus the truth of our assertion follows from the uniqueness of F (see Theorem 5.5.2). □

Remark. It is well known that (5.5.6) implies that

$$\lim_{r\to\infty} \mathsf{P}(\xi_1=i|\xi_{-l}, \quad 0\leqslant l\leqslant r)=P((\ldots,\xi_{-1},\xi_0),i), \quad \mathsf{P}\text{-a.s.}$$

for any $i\in\mathbb{N}^*$. It is easy to prove that if for some $i\in\mathbb{N}^*$, p_i is continuous on I, then

$$\lim_{r\to\infty} \mathsf{P}(\xi_1=i|\xi_{-l}=i_{-l}, \quad 0\leqslant l\leqslant r)=P((\ldots,i_{-1},i_0),i)$$

for any $(\ldots,i_{-1},i_0)\in X^{-\mathbb{N}}$, provided that the left side is defined for any $r\in\mathbb{N}^*$.

This result shows that in the case of Theorem 5.5.2 (ii) the probability $\mathsf{P}(\cdot|\xi_{-l}=i_{-l}, l\in\mathbb{N})$ on $\sigma((\xi_n)_{n\in\mathbb{N}^*})$ coincides for $w=[i_0, i_{-1},\ldots,i_{-n},\ldots]$ with the probability P_w constructed in the existence Theorem 1.1.2, for the RSCC associated with the continued fraction expansion (see the Remark preceding Proposition 5.2.1).

We conclude this subsection by stating a theorem which establishes a certain type of mixing for the sequence $(\xi_t)_{t\in\mathbb{Z}}$. The proof of this theorem can be found in Iosifescu & Spătaru (1973).

Theorem 5.5.4. *Under the assumptions of Theorem* 5.5.3, *we have*

$$\lim_{n\to\infty}\frac{1}{n}\sum_{m=1}^{n} \mathsf{P}(\xi_{m-l}=i_l, 1\leqslant l\leqslant r|A)=\mathsf{P}(\xi_{-l}=i_l, 1\leqslant l\leqslant r)$$

uniformly w.r.t. $r, i_1,\ldots,i_r\in\mathbb{N}^*$ *and* $A\in\sigma((\xi_s)_{s\leqslant 0})$ *for which* $\mathsf{P}(A)\neq 0$.

5.5.4 The case of a finite state space

By analogy with the case of a countable state space, the case of a finite state space taken, without any loss of generality, to be the set $\mathbb{N}_D = \{0, 1, \ldots, D-1\}$ for a natural integer $D \geqslant 2$, can be treated by the device which consists of associating the path $w = (\ldots, i_{-n}, \ldots, i_{-1}, i_0)$ with the number whose D-ary expansion is $(i_0, i_{-1}, \ldots, i_{-n}, \ldots)$.

In what follows we shall only indicate the changes to be made in the treatment of the countable state space case to suit the present case.

For any $t \in I$ consider a probability $(p_i(t))_{i \in \mathbb{N}_D}$ on $\mathbb{N}_D$. Use the notation $(t \equiv t')_n$, $t, t' \in I$, $n \in \mathbb{N}$, to mean that the first n digits of the D-ary expansion of t coincide with the first n digits of the D-ary expansion of t'.[†] Of course $(t \equiv t')_0$ means that no restriction is imposed on t and t'. Let

$$d((p_i(t))_{i \in \mathbb{N}_D}, (p_i(t'))_{i \in \mathbb{N}_D}) = \sum_{i \in \mathbb{N}_D} ((\sigma_i(t) - \max(\sigma_{i-1}(t), \sigma_i(t')))^+ + (\sigma_i(t') - \max(\sigma_{i-1}(t'), \sigma_i(t)))^+),$$

where

$$\sigma_{-1}(t) \equiv 0, \quad \sigma_i(t) = \sum_{r \leqslant i} p_r(t), \quad i \in \mathbb{N}_D,$$

and

$$\alpha'_n = \sup_{(t \equiv t')_n} d((p_i(t))_{i \in \mathbb{N}_D}, (p_i(t'))_{i \in \mathbb{N}_D}), \quad n \in \mathbb{N}. \tag{5.5.4'}$$

Consider Conditions H′ and U′.

Condition H′

$$\sum_{k \in \mathbb{N}} \prod_{n=0}^{k} (1 - \alpha'_n) = \infty.$$

Condition U′. *There exists $i_0 \in \mathbb{N}_D$ such that $p_{i_0}(t) \geqslant \delta > 0$ for any $t \in I$.*

Theorem 5.5.2′. *Let F be a (right-continuous) distribution function with $F(0-) = 0$ and $F(1) = 1$, which satisfies the functional equation*

$$F(t) = \int_0^1 \sum_{\{i : i + v \leqslant Dt\}} p_i(v)\, dF(v), \quad t \in I. \tag{5.5.5'}$$

[†] Concerning the D-ary expansion in ambiguous cases we adopt the following convention. The D-ary expansion of 1 is taken as $(D-1, D-1, D-1, \ldots)$ while any other expansion terminating in $(D-1)$ is replaced by the corresponding one terminating in 0s. Thus in the decimal system we take $1 = 0.999\ldots$ while e.g. $\frac{1}{2} = 0.5$ (rather than $0.4999\ldots$).

If Conditions H′ *and* U′ *hold, then equation* (5.5.5′) *has a unique solution of one of the following forms:*

(i′)$_i$ *F has a jump of magnitude* 1 *at the point* $t_i = i/(D-1)$, *and this occurs iff* $p_i(t_i) = 1$, $i = 0, 1, \ldots, D-1$;

(ii′) $F(t) = t$, $t \in I$. *This occurs iff* $p_i(t) = 1/D$, $i \in \mathbb{N}_D$, $t \in I$;

(iii′) *F is continuous and purely singular. This occurs iff none of the previous cases apply.*

As has been already noted, we shall interpret the path $w = (\ldots, i_{-n}, \ldots, i_{-1}, i_0)$ as the D-ary expansion (read inversely) of a $t \in I$. We shall not consider the paths for which just a finite number of components are different from $D-1$, except for the path $(\ldots, D-1, D-1, D-1)$ for which all the components are equal to $D-1$ (see the convention concerning the ambiguous cases of D-ary representation). Let us denote by W_0 the set of the paths discarded. (Clearly, W_0 is a countable set.) We define $P(w, i) = p_i(t)$ if t and w are associated in this manner (i.e., $t = (i_0, i_{-1}, \ldots, i_{-n}, \ldots)$). In this context, the α'_n defined by (5.5.4′) will be expressed as

$$\alpha'_n = \sup_{\substack{(w \equiv w')_n \\ w, w' \notin W_0}} d((P(w, i))_{i \in \mathbb{N}_D}, (P(w', i))_{i \in \mathbb{N}_D}), \quad n \in \mathbb{N}.$$

Here the notation $(w \equiv w')_n$ means that the last n components of the path w coincide with the last n components of the path w'. Then Condition U′ amounts to the existence of an $i_0 \in \mathbb{N}_D$ such that $P(w, i_0) \geq \delta > 0$ for any $w \notin W_0$.

Theorem 5.5.3′. *Assume that Conditions* H′ *and* U′ *hold and that* $P((\ldots, i, i, i), i) < 1$ *for any* $i \in \mathbb{N}_D$. *Then*

(i′) *there exists a strictly stationary doubly infinite sequence* $(\xi_t)_{t \in \mathbb{Z}}$ *of* $\mathbb{N}_D$*-valued random variables on a probability space* $(\Omega, \mathscr{K}, \mathbf{P})$ *such that*

$$\mathbf{P}(\xi_{n+1} = i \mid \xi^{(-\infty, n]}) = P((\ldots, \xi_{n-1}, \xi_n), i), \ \mathbf{P}\text{-a.s.} \qquad (5.5.6')$$

for any $n \in \mathbb{Z}$ *and* $i \in \mathbb{N}_D$;

(ii′) *any other doubly infinite sequence for which equation* (5.5.6′) *holds is equivalent to* $(\xi_t)_{t \in \mathbb{Z}}$.

Remark. Equation (5.5.6′) implies that

$$\lim_{r \to \infty} \mathbf{P}(\xi_1 = i \mid \xi_{-l}, 0 \leq l \leq r) = P((\ldots, \xi_{-1}, \xi_0), i), \ \mathbf{P}\text{-a.s.}$$

for any $i \in \mathbb{N}_D$. If, for some $i \in \mathbb{N}_D$, p_i is continuous on I, then

$$\lim_{r \to \infty} \mathbf{P}(\xi_1 = i \mid \xi_{-l} = i_{-l}, 0 \leq l \leq r) = P((\ldots, i_{-1}, i_0), i)$$

for any $(\ldots, i_{-1}, i_0) \notin W_0$, provided that the left side is defined for any $r \in \mathbb{N}^*$.

We also have Theorem 5.5.4′.

Theorem 5.5.4′. *Under the assumptions of Theorem 5.5.3′, we have*

$$\lim_{n\to\infty} \frac{1}{n} \sum_{m=1}^{n} \mathsf{P}(\xi_{m-l}=i_l,\ 1\leqslant l\leqslant r\,|\,A) = \mathsf{P}(\xi_{-l}=i_l,\ 1\leqslant l\leqslant r)$$

uniformly w.r.t. $r\in\mathbb{N}^*$, $i_1,\ldots,i_r\in\mathbb{N}_D$, *and* $A\in\sigma((\xi_s)_{s\leqslant 0})$ *for which* $\mathsf{P}(A)\neq 0$.

Problems

1. (*The Onicescu–Mihoc urn*) Consider an initial urn U_0 containing $a_j^{(0)}=a_j$ balls of colour j, $1\leqslant j\leqslant m+1$, and denote by $a_j^{(n)}$, $1\leqslant j\leqslant m+1$, the structure of the urn U_n, $n\in\mathbb{N}^*$, given by the following rule: if the structure of the urn U_{n-1} was $a_j^{(n-1)}$, $1\leqslant j\leqslant m+1$, and on trial n (which is a drawing from U_{n-1}) a ball of colour i was drawn, then $a_j^{(n)}=a_j^{(n-1)}+d_{ij}^{(n-1)}$, $1\leqslant j\leqslant m+1$, where the $d_{ij}^{(n-1)}$, $1\leqslant i$, $j\leqslant m+1$, $n\in\mathbb{N}^*$, are non-negative integers. Let $\xi_n=$ the colour of the ball drawn on trial n, $n\in\mathbb{N}^*$. Show that $(\xi_n)_{n\in\mathbb{N}^*}$ is an OM chain and determine its transition maps. Discuss the special cases: (i) $m=1$, $d_{11}^{(n)}=d_{22}^{(n)}=\alpha_1$, $d_{12}^{(n)}=d_{21}^{(n)}=\alpha_2$ (Friedman's urn); (ii) $m=1$, $d_{11}^{(n)}=\alpha_1$, $d_{21}^{(n)}=\alpha_2$, $d_{12}^{(n)}=d_{22}^{(n)}=0$ (Thurstone's urn); (iii) $m=1$, $d_{11}^{(n)}=a_1^{(n)}/(1+\beta_1)$, $d_{21}^{(n)}=a_1^{(n)}/(1+\beta_2)$, $d_{12}^{(n)}=d_{22}^{(n)}=0$ (Luce's urn).

2. Consider a finite state homogeneous OM chain. Prove that if there exist $\mathbf{q}\in\Delta'$ and $0<a<1$ such that
$$d(\boldsymbol{\varphi}_i(\mathbf{p}),\mathbf{q})\leqslant a\,d(\mathbf{p},\mathbf{q})$$
for any $\mathbf{p}\in\Delta'$ and $i\in X$, then uniform ergodicity holds and the limit probabilities are given by $P^{\infty}_{i_1\cdots i_r}=P_{i_1\cdots i_r}(\mathbf{q})$, $r\in\mathbb{N}^*$, $i_1,\ldots,i_r\in X$, the rate of convergence being $O(a^n)$. (Fortet (1938))
(*Hint*: The condition assumed implies $d(\boldsymbol{\varphi}_{i_1\cdots i_r}(\mathbf{p}),\mathbf{q})\leqslant a^r d(\mathbf{p},\mathbf{q})$ for all $\mathbf{p}\in\Delta'$, $r\in\mathbb{N}^*$ and $i_1,\ldots,i_r\in X$.)

3. Prove that for an alternate linear homogeneous OM chain for which
$$\alpha=\frac{b}{1-a}=\frac{d}{1-c},$$
we have
$$P^n(p)=\alpha+\sum_{r=1}^{n} c_r(p-\alpha)^r[\alpha_1,\ldots,\alpha_r]_{n-r},\quad n\in\mathbb{N}^*,\quad p\in[0,1],$$
where
$$c_r=\begin{cases}1, & \text{if } r=1\\ \prod_{l=1}^{r-1}(a^l-c^l), & \text{if } r\geqslant 2,\end{cases}$$

$$\alpha_l = \alpha a^l + (1-\alpha)c^l, \quad l \in \mathbb{N}^*,$$

$$[\alpha_1, \ldots, \alpha_r]_s = \sum_{\substack{m_1+\cdots+m_r=s \\ m_u \in \mathbb{N}, 1 \leqslant u \leqslant r}} \alpha_1^{m_1} \cdots \alpha_r^{m_r}, \quad r \in \mathbb{N}^*, \quad s \in \mathbb{N}.$$

(Ismail & Theodorescu (1980), Rickert & Theodorescu (1983))

4. Consider an alternate linear homogeneous OM chain for which $a, c < 1$, $a+b<1$, $d>0$. Prove that the existence of a unique value $p_0 \in [0,1]$ such that $P^n(p_0) = p_0$, $n \in \mathbb{N}^*$, is equivalent to the fact that either $a = c$ or $b/(1-a) = d(1-c)$. (Rickert & Theodorescu (1983).)

5. Prove that for an alternate linear homogeneous OM chain the probabilities $P^n(p)$ are given by

$$P^n(p) = \sum_{r=0}^{n} c_{n,r} p^r, \quad n \in \mathbb{N}^*, \quad p \in [0,1],$$

where $c_{1,0} = 0$, $c_{1,1} = 1$ and, for $n \in \mathbb{N}^*$,

$$c_{n+1,r} = \begin{cases} \sum_{l=0}^{m} c_{n,l} d^l, & \text{if } r = 0 \\ \sum_{l=0}^{n-r+1} \binom{r+l-1}{r-1} c_{n,r+l-1}(a^{r-1}b^l - c^{r-1}d^l) & \\ \quad + \sum_{l=0}^{n-r} \binom{r+l}{r} c_{n,r+l} c^r d^l, & \text{if } 1 \leqslant r \leqslant n \\ c_{n,n}(a^n - c^n), & \text{if } r = n+1. \end{cases}$$

In the special case where $a = c$ we have (with the convention $0^\circ = 1$)

$$P^n(p) = (a+b-d)^{n-1} p + \frac{d}{1-a-b+d}[1-(a+b-d)^{n-1}], \quad n \in \mathbb{N}^*, \quad p \in [0,1],$$

if $a+b-d \neq 1$ and

$$P^n(p) = p, \quad n \in \mathbb{N}^*, \quad p \in [0,1],$$

if $a+b-d = 1$.

6. For an alternate linear homogeneous OM chain derive the asymptotic behaviour of the probability $P^n(p)$ as $n \to \infty$ in the following cases

   22° $0 \leqslant a \leqslant 1, \quad b = 1-a, \quad c = 0, \quad d = 1$

   23° $a = 0, \quad b = d = 1, \quad c = -1$

   24° $0 \leqslant a \leqslant 1, \quad b = 1-a, \quad 0 < c < 1, \quad d = 1-c$

   25° $0 \leqslant a < 1, \quad b = 1-a, \quad |c| < 1, \quad 0 < d \leqslant 1, \quad c+d < 1$

   26° $-1 < c < 0, \quad b = d = 1, \quad -1 < a < 0$

   27° $|a| < 1, \quad b \leqslant 1, \quad a+b < 1, \quad 0 \leqslant c < 1, \quad d = 1-c$

28° $|a|<1,\quad b<1,\quad a+b<1,\quad |c|<1,\quad d=1$

29° $-1<a<0,\quad b=1,\quad |c|<1,\quad d<1,\quad c+d<1$

30° $|a|<1,\quad 0<b<1,\quad 0<a+b<1,\quad c=-1,\quad d=1$

31° $a=d=1,\quad b=0,\quad c=-1$

32° $0\leqslant a<1,\quad 0<b\leqslant 1,\quad c=1,\quad d=0$

33° $a=1,\quad b=0,\quad -1<c<0,\quad d=1$

34° $|a|<1,\quad 0<b\leqslant 1,\quad 0<a+b<1,\quad c=1,\quad d=0$

35° $a=-1,\quad b=d=1,\quad c=0$

36° $0<a<1,\quad b=1-a,\quad c=-1,\quad d=1$

37° $a=-1,\quad b=1,\quad 0<c<1,\quad d=1-c$

38° $-1<a<0,\quad b=d=1,\quad c=-1$

39° $a=-1,\quad b=d=1,\quad -1<c<0.$

(*Hint*: See the remark preceding the list of cases 1°–21° in Subsection 5.1.3.)

7. (*Generalized OM chains*) A generalized (homogeneous) OM chain is a GRSCC $\{(W,\mathscr{W}),(X,\mathscr{X}),\Pi,P\}$ for which

$$(W,\mathscr{W})=(\Delta',\mathscr{B}_{\Delta'}),\quad X=\{1,\ldots,m+1\},\quad \mathscr{X}=\mathscr{P}(X),$$

$$\Pi(i,\mathbf{p},A)=\frac{\lambda^m(A\cap S_{r_i}(\boldsymbol{\varphi}_i(\mathbf{p}))}{\lambda^m(S_{r_i}(\boldsymbol{\varphi}_i(\mathbf{p})))},\quad i\in X,\quad \mathbf{p}\in\Delta',\quad A\in\mathscr{B}_{\Delta'},$$

where r_i, $i\in X$, are given positive numbers, λ^m is the m-dimensional Lebesgue measure, $S_r(a)$ is the (open) sphere of centre a and radius r in $\mathbb{R}^m$, and φ_i, $i\in X$, are $(\mathscr{B}_{\Delta'},\mathscr{B}_{\Delta'})$-measurable maps from Δ' into itself. Prove that

   (i) if the maps $\boldsymbol{\varphi}_i$, $i\in X$, are continuous, then the associated MC is continuous.

   (ii) if the maps $\boldsymbol{\varphi}_i$, $i\in X$, satisfy the assumptions of Theorem 5.1.6, then the associated MC is regular w.r.t. $C(\Delta')$. (Grigorescu (1977a))

(*Hint*: Use Theorem 3.3.31.)

8. Consider a linear homogeneous OM chain $(\xi_n)_{n\in\mathbb{N}^*}$ with arbitrary state space $(X,\mathscr{X})$ and transition maps as in Subsection 5.1.4. Put

$$\beta(x,A)=\int_A \Lambda(x,\mathrm{d}y)\alpha(y),\quad x\in X,\quad A\in\mathscr{X}$$

$$\mathfrak{M}_1=\{p:p\in\mathfrak{M},\int_X \alpha(x)p(\mathrm{d}x)<1\},\quad X_1=\{x:x\in X,\alpha(x)<1\}.$$

Call $p \in \mathfrak{M}$ a stationary probability for $(\xi_n)_{n\in\mathbb{N}^*}$ iff this sequence is strictly stationary under $\mathbf{P}_p$. Prove the following assertions.

(i) For any $p \in \mathfrak{M} - \mathfrak{M}_1$, the sequence $(\xi_n)_{n\in\mathbb{N}^*}$ is a sequence of independent identically distributed random variables under $\mathbf{P}_p$.

(ii) Let $p \in \mathfrak{M}_1$. Then the following statements are equivalent:

(a) p is a stationary probability for $(\xi_n)_{n\in\mathbb{N}^*}$;

(b) $P_2^2(p, A) = P_2(p, A)$ for any $A \in \mathscr{X}^2$;

(c) $P_1^2(p, A) = p(A)$ for any $A \in \mathscr{X}$, and, for any $A \in \mathscr{X}$, there exists $B(A) \in \mathscr{X}$ such that $p(B(A)) = 1$ and $\beta(x', A) = \beta(x'', A)$ for any $x', x'' \in B(A) \cap X_1$.

(iii) Let $p \in \mathfrak{M}_1$ and assume that $\beta(x', A) = \beta(x'', A)$ for any $A \in \mathscr{X}$ and $x', x'' \in X_1$. Then the following statements are equivalent

(a) p is a stationary probability for $(\xi_n)_{n\in\mathbb{N}^*}$;

(b) $P_1^2(p, A) = p(A)$ for any $A \in \mathscr{X}$;

(c) p satisfies the equation

$$(1 - \beta)p(A) = \int_X p(\mathrm{d}x)(1 - \alpha(x))\Lambda(x, A), \quad A \in \mathscr{X},$$

where β is the common value of the $\beta(x, X)$, $x \in X$.

(iv) Assume that X is a compact topological space, α is a continuous function, $\Lambda(\cdot, A)$ is a continuous function for any $A \in \mathscr{X}$, $\beta(x', A) = \beta(x'', A)$ for any $x', x'' \in X$ and $A \in \mathscr{X}$, and $\beta(x, X) = \beta < 1$ for any $x \in X$. Then the set of stationary probabilities for $(\xi_n)_{n\in\mathbb{N}^*}$ is non-empty and any element of it is a convex combination of a finite number of linearly independent stationary probabilities. (Bert, Boudreau & Theodorescu (1985), Bert, Herkenrath & Theodorescu (1985), Junge & Theodorescu (1983), Pruscha & Theodorescu (1983))

9. Consider a C^1 pw.m.t. satisfying the assumptions of Theorem 5.3.17. Show that the transition operator U associated with the RSCC defined in the Remark concluding Subsection 5.3.4 is regular w.r.t. $L(I)$, and $\mathbf{Q}^\infty = \rho$. (Iosifescu (1974, 1988))

10. Prove that Kakeya's Condition K for an f-expansion is sufficient for (5.4.2) to hold a.e.

11. Consider an f-expansion for which Conditions (C) and (BV) hold. Show that the Frobenius–Perron operator P_ρ associated with the f-expansion transformation τ_f under Rényi's ρ satisfies

$$\operatorname{ess\,sup}\left| P_\rho h - \int_0^1 h \mathrm{d}\rho \right| \leqslant \left\| P_\rho^n h - \int_0^1 h \mathrm{d}\rho \right\|_{\mathrm{v}} \leqslant c q^n \mathrm{v}(h),$$

for any $n \in \mathbb{N}^*$ and $h \in BV_\lambda$, with suitable positive constants $q < 1$ and c. (Gordin (1968), Iosifescu (1987a))

12. Derive from (5.4.11) *Borel's theorem: the asymptotic relative frequency of any digit* $0, 1, \ldots, D-1$ *in the D-ary expansion of* $t \in I$ *equals* $1/D$ *a.e.*

13. (*Open problem*) Does Theorem 5.5.1 remain valid if instead of Condition FLS$(X, 1)$ we assume Condition FLS(A_0, v) for some $A_0 \in \mathscr{X}$, $A_0 \neq X$, and $v \in \mathbb{N}^*$?

Remark. Such a result was stated without proof in Iosifescu (1966d). The proof given in Iosifescu & Theodorescu (1969, pp. 188–190) and reproduced in the Romanian version of this book is not correct.

14. (*Partially observed Markov chains*) Consider a homogeneous MC $(\theta_n)_{n \in \mathbb{N}}$ with state space Θ (a denumerable set), initial distribution $\mathbf{p}_0 = (\mathsf{P}(\theta_0 = j))_{j \in \Theta}$, and transition probabilities $p_{ij} = \mathsf{P}(\theta_{n+1} = j \mid \theta_n = i)$, $i, j \in \Theta$. Let $\mathfrak{X}$ be a denumerable set. For any $j \in \Theta$, consider a probability $q(j) = (q_x(j))_{x \in \mathfrak{X}}$ on $\mathfrak{X}$. For any $n \in \mathbb{N}^*$, let x_n be an $\mathfrak{X}$-valued random variable having the following properties:
    (i) $\mathsf{P}(x_n = x \mid \theta_n, x_1, \ldots, x_{n-1}) = q_x(\theta_n), \quad n \geqslant 2,$
    $\mathsf{P}(x_1 = x \mid \theta_1) = q_x(\theta_1), \quad x \in \mathfrak{X};$
    (ii) for any $n \in \mathbb{N}^*$, the random variables $(x_1, \ldots, x_n)$ and θ_{n+1} are conditionally independent given θ_n.

Let

$$p_0^j = \mathsf{P}(\theta_0 = j),$$
$$p_n^j = \mathsf{P}(\theta_n = j \mid x_1, \ldots, x_n), \quad n \in \mathbb{N}^*, \quad j \in \Theta.$$

Prove the following assertions.

(a) The scheme described above is a special case of Example 11 in Section 1.2 with

$$p(\theta^{(n)}, x^{(n)}; \theta, x) = p_{ij} q_x(j)$$

if $\theta_n = i$, $\theta = j$, for any $n \in \mathbb{N}^*$, $\theta_0, \ldots, \theta_{n-1} \in \Theta$, and $x^{(n)} \in \mathfrak{X}^n$.

(b) We have

$$p_{n+1}^j = \frac{\sum_{i \in \Theta} p_n^i p_{ij} q_{x_{n+1}}(j)}{\sum_{j \in \Theta} \sum_{i \in \Theta} p_n^i p_{ij} q_{x_{n+1}}(j)}, \quad n \in \mathbb{N}, \quad j \in \Theta.$$

(c) The sequence $(\mathbf{p}_n)_{n \in \mathbb{N}}$, where $\mathbf{p}_n = (p_n^j)_{j \in \Theta}$, is equivalent to the MC associated with the RSCC $\{(W, \mathscr{W}), (X, \mathscr{X}), u, P\}$, where

$$W = \left\{ \mathbf{p} = (p^j)_{j \in \Theta} : p^j \geqslant 0, \sum_{j \in \Theta} p^j = 1 \right\}, \quad \mathscr{W} = \mathscr{B}_W, \quad X = \mathfrak{X}, \quad \mathscr{X} = \mathscr{P}(\mathfrak{X}),$$

$$u(\mathbf{p}, x) = \left(\frac{\sum_{i\in\Theta} p^i p_{ij} q_x(j)}{\sum_{j\in\Theta}\sum_{i\in\Theta} p^i p_{ij} q_x(j)} \right)_{j\in\Theta}$$

$$P(\mathbf{p}, x) = \sum_{i,j\in\Theta} p^i p_{ij} q_x(j), \quad \mathbf{p}\in W, \quad x\in\mathfrak{X}.$$

(Iosifescu & Mandl (1966))

(d) Setting $\mathbf{M}(x) = (m_{ij}(x))_{i,j\in\Theta}$, $x\in\mathfrak{X}$, where $m_{ij}(x) = p_{ij}q_x(j)$, we have

$$\mathbf{p}_n = \frac{\mathbf{p}_0\mathbf{M}(x_1)\cdots\mathbf{M}(x_n)}{|\mathbf{p}_0\mathbf{M}(x_1)\cdots\mathbf{M}(x_n)|}, \quad n\in\mathbb{N},$$

with $|\mathbf{u}| = \sum_{j\in\Theta} |u_j|$, $\mathbf{u} = (u_j)_{j\in\Theta}$.

(e) If the sets Θ and $\mathfrak{X}$ are finite, $p_{ij} > 0$, $i, j\in\Theta$, and for any $x\in\mathfrak{X}$ there exists $j = j(x)\in\Theta$ such that $q_x(j) > 0$, then the sequence $(\mathbf{p}_n)_{n\in\mathbb{N}}$ converges in $\mathsf{P}_{\mathbf{p}_0}$-distribution to a limit independent of $\mathbf{p}_0$. (Kaijser (1973))

(f) If, in particular, $\mathfrak{X}$ is a class of subsets of Θ, which constitute a partition of Θ, and we define $x_n = x$ iff $\theta_n\in x$, $n\in\mathbb{N}^*$, $x\in\mathfrak{X}$, then properties (i) and (ii) above hold as well as the condition concerning $q_x(j)(=\chi_{\{x\}}(j))$ from (e). (In this special case the sequence $(x_n)_{n\in\mathbb{N}^*}$ is said to be a grouped (lumped) Markov chain. See Iosifescu (1980b, p. 166).)

15. (*Grouped Markov chains as chains of infinite order*) Let $\mathbf{P} = (p_{ij})_{1\leqslant i,j\leqslant s}$ be a positive stochastic matrix, and let $(\pi_j)_{1\leqslant j\leqslant s}$ be the unique stationary probability satisfying $\pi_j = \sum_{i=1}^{s} \pi_i p_{ij}$, $1\leqslant j\leqslant s$. Let $(\mu_t)_{t\in\mathbb{Z}}$ be a strictly stationary Markov sequence with state space $\{1,\ldots,s\}$, transition matrix $\mathbf{P}$, and stationary probability $(\pi_j)_{1\leqslant j\leqslant s}$. Let $A_0\cup\cdots\cup A_{D-1}$ be a partition of the set $\{1,\ldots,s\}$. Consider the random variables ξ_t and η_t defined as

$$\xi_t = i \quad \text{iff} \quad \mu_t\in A_i, \quad 0\leqslant i\leqslant D-1,$$

$$\eta_t = \sum_{j\in\mathbb{N}^*} \frac{\xi_{t-j}}{D^j}, \quad t\in\mathbb{Z}.$$

Let

$$H_j(u) = \mathsf{P}(\eta_t\leqslant u\,|\,\mu_t = j), \quad u\in[0,1],$$

$$\theta_j(z) = \int_0^1 e^{zu}\,dH_j(u), \quad 1\leqslant j\leqslant s,$$

$$\mathbf{P}(z) = (p_{ij}\pi_j e^{z\nu(j)}/\pi_i)_{1\leqslant i,j\leqslant s},$$

for any complex number z, where $\nu(j) = i$ iff $j\in A_i$, $0\leqslant i\leqslant D-1$.

Prove the following assertions.

(i) $(\xi_t)_{t\in\mathbb{Z}}$ is a φ-mixing strictly stationary sequence.

(ii) $(\eta_t)_{t\in\mathbb{Z}}$ is a strictly stationary Markov sequence, and the distribution function G of η_t is continuous and strictly increasing on $[0,1]$.

(iii) $\theta_j(z)$ is the sum of the elements of row j of the convergent matrix product $\Pi_{n\in\mathbb{N}^*}\mathbf{P}(z/D^n)$, $1\leqslant j\leqslant s$. (This enables us to determine the distribution functions H_j, $1\leqslant j\leqslant s$.)

(iv) For any $r\in\mathbb{N}^*$ and any $0\leqslant i_1,\ldots,i_r,\ i\leqslant D-1$, we have

$$\mathsf{P}(\xi_t=i\,|\,\xi_{t-1}=i_1,\ldots,\xi_{t-r}=i_r)=\sum_{j\in A_i}\pi_j\frac{H_j(u_2)-H_j(u_1)}{G(u_2)-G(u_1)},$$

where

$$u_1=\sum_{l=1}^{r}\frac{i_l}{D^l},\quad u_2=u_1+\frac{1}{D^r},$$

$$G(u)=\sum_{j=1}^{s}\pi_jH_j(u),\quad u\in[0,1].$$

(v) For any $r\in\mathbb{N}^*$ and any $0\leqslant i_1,\ldots,i_r,\ i\leqslant D-1$ we have

$$\mathsf{P}(\xi_t=i\,|\,\xi_{t-r}=i_r, r\in\mathbb{N}^*)=\frac{\mathrm{d}}{\mathrm{d}G(y)}\sum_{j\in A_i}\pi_jH_j(y)\big|_{y=(i_1,i_2,\ldots)},$$

the derivative occurring in the formula above being understood to be a right-hand one. (Harris (1955))

Appendix 1

Spaces, measures and functions

A1.1

Let X be an arbitrary non-empty set. A non-empty collection $\mathscr{X}$ of subsets of X is said to be an algebra (in X) iff it is closed under the formation of complements and finite unions. Clearly, $\varnothing$ and X both belong to $\mathscr{X}$, and $\mathscr{X}$ is also closed under the formation of finite intersections. An algebra which is closed under the formation of countable unions, is said to be a σ-algebra. Clearly, a σ-algebra is also closed under the formation of countable intersections. A non-empty collection $\mathscr{C}$ of subsets of X is said to be a monotone class iff it contains the limit of any monotone sequence of its elements. An algebra is a σ-algebra iff it is a monotone class. The intersection of an arbitrary non-empty family of σ-algebras in X, is still a σ-algebra. It should be noted that the union of a strictly increasing sequence of σ-algebras in X is *not* a σ-algebra (Broughton & Huff (1977)). For any non-empty collection $\mathscr{C}$ of subsets of X, the σ-algebra generated by $\mathscr{C}$, denoted $\sigma(\mathscr{C})$, is defined as the smallest σ-algebra in X which contains $\mathscr{C}$. (Clearly, $\sigma(\mathscr{C})$ is the intersection of all σ-algebras in X which contain $\mathscr{C}$.)

A pair $(X, \mathscr{X})$, consisting of a non-empty set X and a σ-algebra $\mathscr{X}$ in X, is called a measurable space.

Measurable spaces are usually constructed by singling out a collection $\mathscr{C}$ of interesting subsets of a set X and then considering the σ-algebra $\sigma(\mathscr{C})$. Thus, in the special case where $X = \mathbb{R}$ (the real line), the usual σ-algebra in $\mathbb{R}$ is $\mathscr{B}$ = the collection of Borel sets in $\mathbb{R}$ = the σ-algebra generated by the open intervals in $\mathbb{R}$. More generally, if X is a subset M of the real line, the usual σ-algebra in M is $\mathscr{B}_M = \mathscr{B} \cap M = \{A \cap M : A \in \mathscr{B}\}$. In the special case where X is a denumerable set the usual σ-algebra in X is $\mathscr{P}(X)$, the collection of all subsets of X. Clearly, $\mathscr{P}(X)$ is generated by the elements of X: $\mathscr{P}(X) = \sigma(\{x\}, x \in X)$.

A1.2

Let $(X_i, \mathscr{X}_i)$, $i \in \mathfrak{I}$, be an arbitrary non-empty family of measurable spaces. If Λ is a finite subset of $\mathfrak{I}$, a set $\prod_{i \in \mathfrak{I}} A_i$, where $A_i = X_i$, $i \in \mathfrak{I} - \Lambda$, $A_i \in \mathscr{X}_i$,

$i \in \Lambda$, is called a (measurable) rectangle with sides A_i, $i \in \Lambda$, in $\prod_{i \in \mathfrak{J}} X_i$.[†] The σ-algebra generated by the collection of all rectangles in $\prod_{i \in \mathfrak{J}} X_i$ is called the product σ-algebra of the σ-algebras $\mathscr{X}_i$, $i \in \mathfrak{J}$, and is denoted $\bigotimes_{i \in \mathfrak{J}} \mathscr{X}_i$. The measurable space $(\prod_{i \in \mathfrak{J}} X_i, \bigotimes_{i \in \mathfrak{J}} \mathscr{X}_i)$ is called the product measurable space of the given measurable spaces $(X_i, \mathscr{X}_i)$, $i \in \mathfrak{J}$. In the special case where all the measurable spaces $(X_i, \mathscr{X}_i)$ are identical, $(X_i, \mathscr{X}_i) = (X, \mathscr{X})$, $i \in \mathfrak{J}$, their product measurable space is denoted $(X^{\mathfrak{J}}, \mathscr{X}^{\mathfrak{J}})$; and if $\mathfrak{J} = \{1, \ldots, n\}$, $n \in \mathbb{N}^*$, then we simply write $(X^n, \mathscr{X}^n)$ for $(X^{\mathfrak{J}}, \mathscr{X}^{\mathfrak{J}})$. In particular, the elements of $\mathscr{B}^{\mathfrak{J}}$ are called the Borel subsets of $\mathbb{R}^{\mathfrak{J}}$.

For any non-empty (finite or infinite) subset Λ of $\mathfrak{J}$, the Λ-projection $\mathrm{pr}_\Lambda : \prod_{i \in \mathfrak{J}} X_i \to \prod_{i \in \Lambda} X_i$ is defined as $\mathrm{pr}_\Lambda((x_i)_{i \in \mathfrak{J}}) = (x_i)_{i \in \Lambda}$. Note that the rectangle with sides $A_i \in \mathscr{X}_i$, $i \in \Lambda$, can be written as $\mathrm{pr}_\Lambda^{-1}(\prod_{i \in \Lambda} A_i)$ (see Section A1.3).

A1.3

Let $(X, \mathscr{X})$ and $(Y, \mathscr{Y})$ be two measurable spaces. A map (function, transformation, etc.) $f: X \to Y$ from X into Y is said to be $(\mathscr{X}, \mathscr{Y})$-measurable iff the inverse image $f^{-1}(A) = \{x : f(x) \in A\}$ of every set $A \in \mathscr{Y}$ is in $\mathscr{X}$. Setting $f^{-1}(\mathscr{Y}) = \{f^{-1}(A) : A \in \mathscr{Y}\}$, the above condition can be compactly written as $f^{-1}(\mathscr{Y}) \subset \mathscr{X}$. Since the collection $\{A \in \mathscr{Y} : f^{-1}(A) \in \mathscr{X}\}$ is a σ-algebra in Y, the latter condition is equivalent to $f^{-1}(\mathscr{C}) \subset \mathscr{X}$, for any collection $\mathscr{C}$ of subsets of Y which generates $\mathscr{Y}$ ($\sigma(\mathscr{C}) = \mathscr{Y}$). In the special case $(Y, \mathscr{Y}) = (\mathbb{R}^n, \mathscr{B}^n)$, $n \in \mathbb{N}^*$ the phrase 'an $(\mathscr{X}, \mathscr{B}^n)$-measurable map' is usually simplified to 'an $\mathscr{X}$-measurable map'.

Let $(Z, \mathscr{Z})$ be a third measurable space and consider a map $g: Y \to Z$ from Y into Z. If f is $(\mathscr{X}, \mathscr{Y})$-measurable and g is $(\mathscr{Y}, \mathscr{Z})$-measurable, then the composition $g \circ f : X \to Z$ from X into Z defined as $(g \circ f)(x) = g(f(x))$, $x \in X$, is $(\mathscr{X}, \mathscr{Z})$-measurable. In simple language, a measurable function of a measurable function is still measurable.

The concept of an $(\mathscr{X}, \mathscr{Y})$-measurable map is identical to that of a Y-valued random variable (r.v.) on X. In particular, if $(Y, \mathscr{Y}) = (\mathbb{R}, \mathscr{B})$, the phrase 'an $(\mathscr{X}, \mathscr{B})$-measurable map' has the same meaning as the phrase 'a (real) random variable on X'; and if $(Y, \mathscr{Y}) = (\mathbb{R}^n, \mathscr{B}^n)$, $n \in \mathbb{N}^*$, $n \geqslant 2$, the phrase 'an $(\mathscr{X}, \mathscr{B}^n)$-measurable map' has the same meaning as the phrase 'a (real) n-dimensional random vector on X'. A complex random variable is a complex-valued map for which both the real and imaginary parts are random variables. By the composition property above, the usual arithmetic

[†] In the case where $\mathfrak{J}$ is a finite set, the rectangles in $\prod_{i \in \mathfrak{J}} X_i$ are all the Cartesian products $\prod_{i \in \mathfrak{J}} A_i$, $A_i \in \mathscr{X}_i$, $i \in \mathfrak{J}$. In the case where $\mathfrak{J} = \mathbb{N}^*$, the rectangles in $\prod_{i \in \mathfrak{J}} X_i$ are all the Cartesian products $(\prod_{i=1}^r A_i) \times X \times X \times X \times \cdots$, $A_i \in \mathscr{X}_i$, $1 \leqslant i \leqslant r$, $r \in \mathbb{N}^*$.

operations applied to real or complex r.v.s. lead to new real or complex r.v.s.

The Λ-projection map pr_Λ (see Section A1.2) is a $\prod_{i\in\Lambda} X_i$-valued random variable on $\prod_{i\in\mathfrak{I}} X_i$.

Let $(X, \mathscr{X})$ be a measurable space, $((Y_i, \mathscr{Y}_i))_{i\in\mathfrak{I}}$ a family of measurable spaces, and, for any $i\in\mathfrak{I}$, let $f_i: X \to Y_i$ be a Y_i-valued r.v. on X. Consider all the σ-algebras $\mathscr{S}$ in X having the property that f_i is $(\mathscr{S}, \mathscr{Y}_i)$-measurable for all $i\in\mathfrak{I}$. (Clearly, in particular, $\mathscr{X}$ itself enjoys that property.) The intersection of all these σ-algebras (i.e., the smallest σ-algebra enjoying that property), denoted $\sigma((f_i)_{i\in\mathfrak{I}})$, is called the σ-algebra generated by the family $(f_i)_{i\in\mathfrak{I}}$.

In particular, we write $\sigma(f)$ for the case of a family consisting of a single element f.

A1.4

Let $(X, \mathscr{X})$ be a measurable space. A set function $\mu: \mathscr{X} \to \bar{\mathbb{R}}_+$ (which is thus allowed to take the value ∞), for which $\mu(\varnothing) = 0$, is said to be a measure on $\mathscr{X}$ iff it is completely additive (i.e. for any sequence $(A_i)_{i\in\mathbb{N}^*}$ of pairwise disjoint elements of $\mathscr{X}$ we have $\mu(\bigcup_{i\in\mathbb{N}^*} A_i) = \sum_{i\in\mathbb{N}^*}\mu(A_i)$). For a measure for which $\mu(A) < \infty$ for some $A\in\mathscr{X}$, $A \neq \varnothing$, complete additivity is equivalent to finite additivity (i.e., for any finite collection $A_1, \ldots, A_n$ of pairwise disjoint elements of $\mathscr{X}$, we have $\mu(\bigcup_{i=1}^n A_i) = \sum_{i=1}^n \mu(A_i)$) in conjunction with continuity at $\varnothing$ (i.e., for any decreasing sequence $A_1 \supset A_2 \supset \cdots$ of elements of $\mathscr{X}$ with $\bigcap_{i\in\mathbb{N}^*} A_i = \varnothing$ we have $\lim_{n\to\infty}\mu(A_n) = 0$).

A measure μ is called finite iff $\mu(X) < \infty$ and σ-finite iff there exists a sequence $(A_i)_{i\in\mathbb{N}^*}$ of elements of $\mathscr{X}$ such that $X = \bigcup_{i\in\mathbb{N}^*} A_i$ and $\mu(A_i) < \infty$ for all $i\in\mathbb{N}^*$. (Note that for a finite or a σ-finite measure the equation $\mu(\varnothing) = 0$ is implied by finite additivity.)

In the special case where X is a denumerable set, a measure μ on $\mathscr{P}(X)$ is defined by simply giving the values $\mu(\{x\})$ for the elements $x\in X$.

A triple $(X, \mathscr{X}, \mu)$ consisting of a measurable space $(X, \mathscr{X})$ and a measure μ on $\mathscr{X}$ is called a measure space. A measure space $(X, \mathscr{X}, \mu)$ is called finite, or σ-finite, according to whether the measure μ is finite, or σ-finite.

A set function $\mu: \mathscr{X} \to \mathbb{R}_- \cup \bar{\mathbb{R}}_+$ (which is thus allowed to take on the value ∞ but not the value $-\infty$), for which $\mu(\varnothing) = 0$, is said to be a signed measure on $\mathscr{X}$ iff it is completely additive (in this requirement it is implicit that, if $(A_i)_{i\in\mathbb{N}^*}$ is a sequence of pairwise disjoint elements of $\mathscr{X}$, then the series $\sum_{i\in\mathbb{N}^*}\mu(A_i)$ is either convergent or divergent to ∞, i.e. that the term $\sum_{i\in\mathbb{N}^*}\mu(A_i)$ makes sense).

The adjectives 'finite' and 'σ-finite' are used for a signed measure just as for a measure. If μ is a signed measure on $\mathscr{X}$, then there exist two disjoint sets X^+ and X^- in $\mathscr{X}$, whose union is X, such that

$$\mu(A \cap X^+) \geqslant 0, \quad \mu(A \cap X^-) \leqslant 0$$

for all $A \in \mathscr{X}$. (For a very simple proof see Doos (1980).) The decomposition $X = X^+ \cup X^-$ is called a Hahn decomposition of X w.r.t. μ. In spite of the fact that a Hahn decomposition is not necessarily unique, the equations

$$\mu^+(A) = \mu(A \cap X^+), \quad \mu^-(A) = -\mu(A \cap X^-)$$

unambiguously define two measures μ^+ and μ^- on $\mathscr{X}$, the last one being finite, and $\mu = \mu^+ - \mu^-$ (the Jordan decomposition of μ). If μ is finite or σ-finite then so is μ^+. The set function $|\mu| = \mu^+ + \mu^-$ is a measure on $\mathscr{X}$, called the total variation measure associated with μ. The quantity $|\mu|(X) = \mu^+(X) + \mu^-(X) = \mu(X^+) - \mu(X^-)$ is called the total variation of μ, and is denoted $\operatorname{var} \mu$ or $\|\mu\|$. We have

$$\operatorname{var} \mu = \sup \sum_{i=1}^{n} |\mu(A_i)|,$$

where the supremum is taken over all finite partitions $X = \bigcup_{i=1}^{n} A_i$, $A_i \in \mathscr{X}$, $A_i \cap A_j = \varnothing$, $i \neq j$, $1 \leqslant i$, $j \leqslant n$, $n \in \mathbb{N}^*$. In the special case where X is denumerable with elements x_i, $i \in \mathfrak{J}$, and μ is a signed measure on $\mathscr{P}(X)$ for which $\mu(X) = 0$, we have

$$\operatorname{var} \mu = 2 \sup_{A \subset X} \mu(A) = \sum_{i \in \mathfrak{J}} |\mu(\{x_i\})|.$$

A1.5

Let $(X, \mathscr{X})$ be a measurable space. A probability on $\mathscr{X}$ is a measure P on $\mathscr{X}$ satisfying $\mathsf{P}(X) = 1$. In particular, complete additivity for a probability is equivalent to finite additivity in conjunction with continuity at $\varnothing$ (see Section A1.4).

A measure space $(X, \mathscr{X}, \mathsf{P})$, where P is a probability, is called a probability space. (The traditional notation for a probability space is $(\Omega, \mathscr{K}, \mathsf{P})$. The points $\omega \in \Omega$ are interpreted as the possible outcomes (elementary events) of a random experiment, and the sets $A \in \mathscr{K}$ as the (random) events associated with it (these are the subsets of Ω arising as the truth sets of certain statements concerning the experiment).) We say that $A \in \mathscr{X}$ occurs P-almost surely, and write A P-a.s., iff $\mathsf{P}(A) = 1$.

Let $(Y, \mathscr{Y})$ be a measurable space and let f be a Y-valued r.v. on X. The P-distribution of f is the probability $\mathsf{P}f^{-1}$ defined as $(\mathsf{P}f^{-1})(A) = \mathsf{P}(f^{-1}(A))$, $A \in \mathscr{Y}$. Let g be another Y-valued r.v. on a probability space $(X', \mathscr{X}', \mathsf{P}')$. The r.v.s f and g are said to be equivalent iff $\mathsf{P}f^{-1} = \mathsf{P}'g^{-1}$. They are said to be versions of each other iff $(X, \mathscr{X}, \mathsf{P}) = (X', \mathscr{X}, \mathsf{P}')$ and $f = g$ P-a.s. More generally, for any $i \in \mathfrak{I}$, let f_i and g_i be Y-valued r.v.s on probability spaces $(X, \mathscr{X}, \mathbf{P})$ and $(X', \mathscr{X}', \mathbf{P}')$, respectively. The collections $(f_i)_{i \in \mathfrak{I}}$ and $(g_i)_{i \in \mathfrak{I}}$ of r.v.s are said to be equivalent iff for any finite set $\Lambda \subset \mathfrak{I}$ the Y^Λ-valued r.v.s $(f_i)_{i \in \Lambda}$ and $(g_i)_{i \in \Lambda}$ are equivalent.

A collection $(f_i)_{i \in \mathfrak{I}}$ of Y-valued r.v.s on the same probability space $(X, \mathscr{X}, \mathbf{P})$, where $\mathfrak{I}$ is a finite or infinite interval of the real line, is called a Y-valued (continuous parameter) stochastic process on X.

A1.6

In a certain sense, taking the limit does not alter the collection of probabilities on a σ-algebra $\mathscr{X}$. More precisely, we have Theorem A1.1.

Theorem A1.1. (Vitali–Hahn–Saks) *Let* $(\mathbf{P}_n)_{n \in \mathbb{N}^*}$ *be a sequence of probabilities on a* σ*-algebra* $\mathscr{X}$. *If the limit* $\mathbf{P}(A) = \lim_{n \to \infty} \mathbf{P}_n(A)$ *exists for any* $A \in \mathscr{X}$, *then* $\mathbf{P}$ *is a probability on* $\mathscr{X}$.

A1.7

Let $(X_i, \mathscr{X}_i, \mathbf{P}_i)$, $i \in \mathfrak{I}$, be an arbitrary family of probability spaces. On the collection of rectangles in $\prod_{i \in \mathfrak{I}} X_i$ (see Section A1.2), we define the set function $\mathbf{P}_{\mathfrak{I}}$ by

$$\mathbf{P}_{\mathfrak{I}}\left(\mathrm{pr}_\Lambda^{-1}\left(\prod_{i \in \Lambda} A_i\right)\right) = \prod_{i \in \Lambda} \mathbf{P}_i(A_i). \tag{A1.1}$$

Theorem A1.2. (Andersen–Jessen) *There exists a unique probability* $\mathbf{P}_{\mathfrak{I}}$ *on* $\bigotimes_{i \in \mathfrak{I}} \mathscr{X}_i$ *which satisfies equation* (A1.1).

The probability $\mathbf{P}_{\mathfrak{I}}$ is called the product probability of the probabilities $\mathbf{P}_i$, $i \in \mathfrak{I}$, while the probability space $(\prod_{i \in \mathfrak{I}} X_i, \bigotimes_{i \in \mathfrak{I}} \mathscr{X}_i, \mathbf{P}_{\mathfrak{I}})$ is called the product probability space of the probability spaces $(X_i, \mathscr{X}_i, \mathbf{P}_i)$, $i \in \mathfrak{I}$. We write $\mathbf{P}_{\mathfrak{I}} = \prod_{i \in \mathfrak{I}} \mathbf{P}_i$, and, in particular, $\mathbf{P}_{\mathfrak{I}} = \mathbf{P}_1 \times \cdots \times \mathbf{P}_n$ if $\mathfrak{I} = \{1, \ldots, n\}$, $n \in \mathbb{N}^*$.

Let $(X, \mathscr{X})$ and $(Y, \mathscr{Y})$ be two measurable spaces. A real valued function P defined on $X \times \mathscr{Y}$ is said to be a transition probability function (t.p.f.) from $(X, \mathscr{X})$ to $(Y, \mathscr{Y})$ iff $P(x, \cdot)$ is a probability on $\mathscr{Y}$ for any $x \in A$, and $P(\cdot, A)$ is a r.v. on X for any $A \in \mathscr{Y}$. In the special case where $(X, \mathscr{X}) = (Y, \mathscr{Y})$, we speak of a t.p.f. on $(X, \mathscr{X})$.

In the case where the index set $\mathfrak{I} = \mathbb{N}^*$, the Andersen–Jessen theorem can be generalized as follows. Assume that, for any $n \geqslant 2$, a t.p.f. P_n from $(\prod_{i=1}^{n-1} X_i, \bigotimes_{i=1}^{n-1} \mathscr{X}_i)$ to $(X_n, \mathscr{X}_n)$ is given. Let $\mathbf{P}_1$ be a probability on $\mathscr{X}_1$.

Define on the collection of rectangles in $\prod_{i \in \mathbb{N}^*} X_i$ the set function $\mathbf{P}_{\mathbb{N}^*}$ as

$$\mathbf{P}_{\mathbb{N}^*}(\mathrm{pr}^{-1}_{\{1,\ldots,n\}}(\textstyle\prod_{i=1}^n A_i))$$
$$= \begin{cases} \mathbf{P}_1(A_1), & \text{if } n=1, \\ \displaystyle\int_{A_1} \mathbf{P}_1(\mathrm{d}x_1) \int_{A_1} P_2(x_1, \mathrm{d}x_2) \cdots \int_{A_n} P_n((x_1, \ldots, x_{n-1}), \mathrm{d}x_n), & \text{if } n>1 \end{cases} \tag{A1.2}$$

Theorem A1.3. (Ionescu Tulcea) *There is a unique probability* $\mathbf{P}_{\mathbb{N}^*}$ *on* $\bigotimes_{i \in \mathbb{N}^*} \mathscr{X}_i$ *satisfying equation* (A1.2).

A1.8

Let $(X_i, \mathscr{X}_i)$, $i \in \mathfrak{I}$, be an arbitrary family of measurable spaces and $\mathbf{P}_{\mathfrak{I}}$ a probability on $\bigotimes_{i \in \mathfrak{I}} \mathscr{X}_i$. For any non-empty subset Λ of $\mathfrak{I}$, define the projection $\mathbf{P}_\Lambda$ of $\mathbf{P}_{\mathfrak{I}}$ on $\bigotimes_{i \in \Lambda} \mathscr{X}_i$ as $\mathbf{P}_\Lambda = \mathbf{P}_{\mathfrak{I}} \mathrm{pr}_\Lambda^{-1}$, i.e.

$$\mathbf{P}_\Lambda(A) = \mathbf{P}_{\mathfrak{I}}(\mathrm{pr}_\Lambda^{-1}(A)), \quad A \in \bigotimes_{i \in \Lambda} \mathscr{X}_i.$$

The projections $\mathbf{P}_\Lambda$, $\Lambda \subset \mathfrak{I}$, are consistent. This means that, if $\Lambda' \cap \Lambda'' = \Lambda \neq \varnothing$, then the projections of $\mathbf{P}_{\Lambda'}$ and $\mathbf{P}_{\Lambda''}$ on $\bigotimes_{i \in \Lambda} \mathscr{X}_i$ coincide with $\mathbf{P}_\Lambda$. A converse is also true. More precisely, we have Theorem A1.4.

Theorem A1.4. (Daniell–Kolmogorov) *Let X_i be a separable metric space and $\mathscr{X}_i$ the σ-algebra of Borel subsets of X_i, $i \in \mathfrak{I}$ (see Sections* A1.12 *and* A1.13). *Then, given a collection of probabilities* $\mathbf{P}_\Lambda$ *on* $\bigotimes_{i \in \Lambda} \mathscr{X}_i$, *where* Λ *varies over all finite subsets of* $\mathfrak{I}$, *which are consistent in the sense defined above, there exists a unique probability* $\mathbf{P}_{\mathfrak{I}}$ *on* $\bigotimes_{i \in \mathfrak{I}} \mathscr{X}_i$ *such that any* $\mathbf{P}_\Lambda$ *is the projection of* $\mathbf{P}_{\mathfrak{I}}$ *on* $\bigotimes_{i \in \Lambda} X_i$.

A1.9

In Sections A1.9 to A1.11, we shall consider only real (or complex) r.v.s on a probability space $(\Omega, \mathscr{K}, \mathbf{P})$ unless otherwise stated. We do not distinguish between two r.v.s f and g which are versions of each other, i.e., such that $\mathbf{P}(f \neq g) = \mathbf{P}(\{\omega: f(\omega) \neq g(\omega)\}) = 0$. A r.v. is thus determined $\mathbf{P}$-a.s. = $\mathbf{P}$-almost surely (except for random events of probability 0). We mention that the construction below of the integral with respect to $\mathbf{P}$ holds when $\mathbf{P}$ is replaced by either a finite measure or, more generally, a finite signed measure.

A real r.v. f is called simple iff it can be written as $f = \sum_{i=1}^n a_i \chi_{A_i}$, with

$a_i \in \mathbb{R}$ and pairwise disjoint $A_i \in \mathscr{K}$, $1 \leqslant i \leqslant n$. The integral of a simple r.v. f, denoted $\int_\Omega f(\omega)\mathbf{P}(d\omega)$, or, simply, $\int_\Omega f d\mathbf{P}$, is defined as

$$\int_\Omega f d\mathbf{P} = \sum_{i=1}^{n} a_i \mathbf{P}(A_i),$$

its value being independent of the representation of f (which is not unique). Now, for any non-negative r.v. f there exists a non-decreasing sequence $(f_n)_{n\in\mathbb{N}^*}$ of simple r.v.s which converges pointwise to f. We can, for example, take

$$f_n(\omega) = \begin{cases} (i-1)/2^n, & \text{if } \ (i-1)/2^n \leqslant f(\omega) < i/2^n, \quad 1 \leqslant i \leqslant n2^n, \\ n, & \text{if } \ f(\omega) \geqslant n. \end{cases}$$

The integral of a non-negative r.v. is defined as

$$\int_\Omega f d\mathbf{P} = \lim_{n\to\infty} \int_\Omega f_n d\mathbf{P},$$

its value being independent of the chosen sequence $(f_n)_{n\in\mathbb{N}^*}$. A non-negative r.v. f is called **P**-integrable iff $\int_\Omega f d\mathbf{P} < \infty$, and if $\int_\Omega f d\mathbf{P} = \infty$ then f is said to have an infinite integral. Finally, for an arbitrary real r.v. f we can write $f = f^+ - f^-$, where $f^+(\omega) = \max\{f(\omega), 0\} \geqslant 0$ and $f^-(\omega) = \max\{-f(\omega), 0\} \geqslant 0, \omega \in \Omega$; f is called **P**-integrable (integrable for short, when **P** need not be emphasized) iff both f^+ and f^- are **P**-integrable, and the integral of f is defined as

$$\int_\Omega f d\mathbf{P} = \int_\Omega f^+ d\mathbf{P} - \int_\Omega f^- d\mathbf{P}.$$

(If one of f^+ and f^- is integrable and the other has an infinite integral, then f is said to have an infinite integral.) A complex r.v. $f = \operatorname{Re} f + i \operatorname{Im} f$ is called **P**-integrable iff both $\operatorname{Re} f$ and $\operatorname{Im} f$ are **P**-integrable, and the integral of f is defined as

$$\int_\Omega f d\mathbf{P} = \int_\Omega \operatorname{Re} f d\mathbf{P} + i \int_\Omega \operatorname{Im} f d\mathbf{P}.$$

It should be emphasized that f is integrable iff $|f|$ is integrable. Also, assuming $f = g$ **P**-a.s., one of f and g is integrable iff the other is, and $\int_\Omega f d\mathbf{P} = \int_\Omega g \, d\mathbf{P}$.

If f is **P**-integrable, then $\int_A f d\mathbf{P}$ denotes $\int_\Omega f \chi_A d\mathbf{P}$ for any $A \in \mathscr{K}$.

Proposition A1.5. (Change of variable formula) *Let $(X, \mathscr{X})$ be a measurable space, h an X-valued r.v. on Ω and f a real r.v. on X. Then f is $\mathbf{P}h^{-1}$-integrable iff $f \circ h$ is **P**-integrable, and for any $A \in \mathscr{X}$ we have*

$$\int_{h^{-1}(A)} (f \circ h)(\omega) \mathbf{P}(d\omega) = \int_A f(x) (\mathbf{P}h^{-1})(dx).$$

The mean (value) $\mathsf{E}(f)$ of a r.v. f is defined as the P-integral of f whenever f is P-integrable. By Proposition A1.5, we thus have

$$\mathsf{E}(f) = \int_{\Omega} f(\omega)\mathsf{P}(d\omega) = \int_{\mathsf{R}} x(\mathsf{P}f^{-1})(dx).$$

A real r.v. is said to have an infinite mean iff the integral defining its mean value is infinite.

Let f_n, $n \in \mathbb{N}^*$, and f be r.v.s on Ω. The sequence $(f_n)_{n\in\mathbb{N}^*}$ is said to converge in P-probability to f ($f_n \xrightarrow{\mathsf{P}} f$) iff $\lim_{n\to\infty} \mathsf{P}(|f_n - f| \geq \varepsilon) = 0$ for any $\varepsilon > 0$. The sequence $(f_n)_{n\in\mathbb{N}^*}$ is said to converge P-almost surely to f ($f_n \to f$ P-a.s.) iff $\mathsf{P}(\lim_{n\to\infty} f_n = f) = 1$.

Theorem A1.6. (Beppo Levi) *Let $f = \sum_{n\in\mathbb{N}^*} f_n$ be the sum of a sequence $(f_n)_{n\in\mathbb{N}^*}$ of non-negative P-integrable r.v.s. If $a = \sum_{n\in\mathbb{N}^*} \mathsf{E}(f_n) < \infty$, then f is P-integrable and $\mathsf{E}(f) = a$.*

Proposition A1.7. (Fatou's lemma) *If $f_n \geq 0$, $n \in \mathbb{N}^*$, then $\mathsf{E}(\liminf_{n\to\infty} f_n) \leq \liminf_{n\to\infty} \mathsf{E}(f_n)$.*

Theorem A1.8. (Lebesgue's dominated convergence theorem) *If $f_n \to f$ P-a.s. and $|f_n| \leq g$, $n \in \mathbb{N}^*$, while $\mathsf{E}(g) < \infty$, then $\lim_{n\to\infty} \mathsf{E}(f_n) = \mathsf{E}(f)$.*

A1.10

Let P and Q be two probabilities on the σ-algebra $\mathscr{K}$. The probability P is said to be absolutely continuous w.r.t. Q ($\mathsf{P} \ll \mathsf{Q}$) iff $A \in \mathscr{K}$ and $\mathsf{Q}(A) = 0$ imply $\mathsf{P}(A) = 0$. The probabilities P and Q are said to be equivalent, ($\mathsf{P} \equiv \mathsf{Q}$) iff $\mathsf{P} \ll \mathsf{Q}$ and $\mathsf{Q} \ll \mathsf{P}$.

Theorem A1.9. (Radon–Nikodym) *We have $\mathsf{P} \ll \mathsf{Q}$ iff there exists a non-negative r.v. g such that $\mathsf{P}(A) = \int_A g(\omega)\mathsf{Q}(d\omega)$, $A \in \mathscr{K}$. The r.v. g is unique P-a.s. For any P-integrable r.v. f we have*

$$\int_{\Omega} f(\omega)\mathsf{P}(d\omega) = \int_{\Omega} f(\omega)\, g(\omega)\mathsf{Q}(d\omega).$$

The r.v. g occurring in the Radon–Nikodym theorem is called the density of P w.r.t. Q, and we write $d\mathsf{P} = g\, d\mathsf{Q}$.

A1.11

Let $\mathscr{L} \subset \mathscr{K}$ be a σ-algebra in Ω and f a r.v. with finite mean. The conditional mean of f given $\mathscr{L}$ is defined as any $\mathscr{L}$-measurable r.v.

$g = \mathbf{E}(f \mid \mathscr{L})$ which satisfies the equation

$$\int_A g(\omega)\mathbf{P}(d\omega) = \int_A f(\omega)\mathbf{P}(d\omega)$$

for any $A \in \mathscr{L}$. The existence of the conditional mean $\mathbf{E}(f \mid \mathscr{L})$ is ensured by the Radon–Nikodym theorem. Clearly, $\mathbf{E}(f \mid \mathscr{L})$ is unique $\mathbf{P}$-a.s. In particular, $\mathbf{E}(\chi_A \mid \mathscr{L}) = \mathbf{P}(A \mid \mathscr{L})$ is called the conditional probability of A given $\mathscr{L}$.

Let $((Y_i, \mathscr{Y}_i))_{i \in \mathfrak{I}}$ be a family of measurable spaces and, for any $i \in \mathfrak{I}$, let $f_i : \Omega \to Y_i$ be a Y_i-valued r.v. on Ω. The conditional mean $\mathbf{E}(f \mid (f_i)_{i \in \mathfrak{I}})$ of f given $(f_i)_{i \in \mathfrak{I}}$ is defined as the conditional mean $\mathbf{E}(f \mid \sigma((f_i)_{i \in \mathfrak{I}})$ (see Section A1.3).

Let $\mathscr{M} \subset \mathscr{L}$ be a σ-algebra in Ω. In the statement below, equalities and inequalities hold $\mathbf{P}$-a.s.

Proposition A1.10. (i) $f \geqslant 0$ *implies* $\mathbf{E}(f \mid \mathscr{L}) \geqslant 0$;

(ii) $\mathbf{E}(f \mid \mathscr{L})$ *is linear in* f;

(iii) $\mathbf{E}(f \mid \mathscr{L}) = f$ *if* f *is* $\mathscr{L}$*- measurable;*

(iv) $\mathbf{E}(fg \mid \mathscr{L}) = f\mathbf{E}(g \mid \mathscr{L})$ *if* f *is* $\mathscr{L}$*-measurable;*

(v) $\mathbf{E}(f) = \mathbf{E}(\mathbf{E}(f \mid \mathscr{L}))$;

(vi) $\mathbf{E}(\mathbf{E}(f \mid \mathscr{L}) \mid \mathscr{M}) = \mathbf{E}(f \mid \mathscr{M})$;

(vii) $\mathbf{E}(f \mid \mathscr{L}) = \mathbf{E}(f)$ *if* $\sigma(f)$ *and* $\mathscr{L}$ *are independent* (*i.e.*, $\mathbf{P}(A \cap B) = \mathbf{P}(A)\mathbf{P}(B)$ *for all* $A \in \sigma(f)$ *and* $B \in \mathscr{L}$).

A sequence $(f_n)_{n \in \mathbb{N}^*}$ of r.v.s is said to be a martingale w.r.t. a non-decreasing sequence $(\mathscr{K}_n)_{n \in \mathbb{N}^*}$ of σ-algebras included in $\mathscr{K}$ iff $\mathbf{E}(f_{n+1} \mid \mathscr{K}_n) = f_n$ $\mathbf{P}$-a.s. for all $n \in \mathbb{N}^*$.

Theorem A1.11. (Martingale convergence theorem) *If* $\sup_{n \in \mathbb{N}^*} \mathbf{E}(|f_n|) < \infty$, *then the martingale* $(f_n)_{n \in \mathbb{N}^*}$ *converges* $\mathbf{P}$*-a.s. to a r.v.* f *with finite mean.*

A1.12

A set X is said to be a metric space if a metric d is defined in X. A metric is a non-negative function defined on $X \times X$ such that $d(x, y) = 0$ iff $x = y$, $d(x, y) = d(y, x)$ for all $x, y \in X$ (the symmetry property) and $d(x, z) \leqslant d(x, y) + d(y, z)$ for all $x, y, z \in X$ (the triangle property). When we wish to emphasize the metric of the space, we write (X, d). The distance $d(x, A)$ from a point $x \in X$ to a subset A of X is defined as

$$d(x, A) = \inf\{d(x, y) : y \in A\}.$$

A sequence $(x_n)_{n \in \mathbb{N}^*}$ of elements of X is said to be convergent to $x \in X$

($x_n \to x$) iff $d(x_n, x) \to 0$ as $n \to \infty$. A set $A \subset X$ is said to be closed iff $x_n \in A$, $n \in \mathbb{N}^*$, and $x_n \to x$ imply that $x \in A$. A set $A \subset X$ is said to be open iff the set $A^c = X - A$ is closed. Equivalently, an open set can be defined as follows. Let the (open) sphere $S_r(x)$ of centre x and radius r be the set of all points $y \in X$ such that $d(y, x) < r$. A set $A \subset X$ is said to be open iff for any $x \in A$ there exists a sphere $S_r(x) \subset A$.

The collection of open (closed) sets in a metric space X is closed under the formation of arbitrary unions (intersections) and under the formation of finite intersections (unions). The empty set $\varnothing$ and the total space X are simultaneously open and closed. The collection of all open sets in X constitutes the topology of the metric space X.

A sequence $(x_n)_{n \in \mathbb{N}^*}$ of elements of X is said to be a Cauchy sequence iff $d(x_m, x_n) \to 0$ as $m, n \to \infty$. A metric space is said to be complete iff any Cauchy sequence is convergent.

A metric space X is said to be separable iff there exists a denumerable set $D \subset X$ which is dense in X. This means that for any $x \in X$ there exists a sequence $(x_n)_{n \in \mathbb{N}^*} \subset D$ such that $x_n \to x$. A metric space X is said to be compact iff any sequence of elements of X contains a convergent subsequence. A metric space X is said to be totally bounded iff, for any $\varepsilon > 0$, there exists a finite collection of spheres of radii $\leqslant \varepsilon$ which cover X (i.e. any point of X belongs to at least one of the spheres). A metric space is compact iff it is totally bounded and complete. A compact metric space is separable and bounded (i.e. $\sup_{x,y \in X} d(x, y) < \infty$).

To any subset A of a metric space X there are assigned an open set A° (the interior of A) and a closed set $\bar{A}$ (the closure of A) as follows. The set A° is the largest open set contained in A, which is the union of all open sets contained in A, while $\bar{A}$ is the smallest closed set containing A, which is the intersection of all closed sets containing A. A set A is closed (open) iff $A = \bar{A}$ ($A = A^\circ$). The set $\partial A = \bar{A} - A^\circ$ is called the boundary of A. A point $x \in X$ is said to be an accumulation point of a subset A of the metric space X, iff any open set containing x also contains points $y \in A$, $y \neq x$. The set of all accumulation points of a set A is called the derived set A' of A. Clearly, $\bar{A} = A \cup A' = \partial A \cup A^\circ$.

A subset A of the metric space X is said to be compact iff any sequence of elements of A contains a subsequence convergent to an element of A; and A is said to be relatively compact if $\bar{A}$ is compact. In a compact metric space a set is compact iff it is closed.

An open covering of a subset A of a metric space X is a collection of open subsets of X such that any point of A belongs to at least one of the subsets of the collection.

Theorem A1.12. (Borel–Lebesgue) *Any open coverning of a compact subset of a metric space contains a finite subcovering.*

In some cases the weaker result below is still useful.

Theorem A1.13. (Lindelöf) *Any open covering of a separable metric space contains a countable subcovering.*

Many definitions and results concerning metric spaces do not use the concept of a metric, but just that of an open (closed) set. In particular, in many instances, a metric d in X can be replaced by any equivalent metric d' (i.e., such that $a \leqslant d(x, y)/d'(x, y) \leqslant b$ for all $x, y \in X$, $x \neq y$, where a and b are two given positive numbers); a subset $A \subset X$ is open (closed) w.r.t. d iff it is open (closed) w.r.t. d'. This has led naturally to the concept of a topological space.

A non-empty set is said to be a topological space iff there exists a non-empty collection $\mathcal{O}$ of subsets of X which contains $\varnothing$ and X, and is closed under the formation of arbitrary unions and finite intersections. The collection $\mathcal{O}$ is called the topology of X, while the elements of $\mathcal{O}$ are called the open subsets in X. The dual collection of the complements of the open sets is the collection of closed sets in X. When we wish to emphasize the topology of the space, we write $(X, \mathcal{O})$.

A topological space $(X, \mathcal{O})$ is said to be a Hausdorff space (or a separated space) iff for any pair of points $x, y \in X$, $x \neq y$, there exist two sets $A_x \in \mathcal{O}$ and $A_y \in \mathcal{O}$ such that $x \in A_x$, $y \in A_y$, and $A_x \cap A_y = \varnothing$. (While it is obvious that any metric space is a Hausdorff space, not every topological space is a Hausdorff space.) Compactness is defined for a topological space by the open covering property (see Theorem A1.12): a set (in particular, the whole space) is said to be compact iff any open covering of it contains a finite subcovering.

A1.13

The elements of the σ-algebra $\mathscr{B}_X$ generated by the topology $\mathcal{O}$ of the topological space $(X, \mathcal{O})$ are called the Borel sets in X. Let $(Y, \mathcal{O}')$ be another topological space. A map $f: X \to Y$ from X into Y is said to be continuous iff for any open (closed) subset A of Y the inverse image $f^{-1}(A)$ is open (closed) in X. It follows that a continuous map from X into Y is $(\mathscr{B}_X, \mathscr{B}_Y)$-measurable, thus it is a Y-valued r.v. on X.

Proposition A1.14. *A real-valued continuous function on a compact topo-*

logical space is bounded and takes the values of its supremum and infimum over the space.

Proposition A1.15. (Dini's theorem) *Let $(f_n)_{n\in\mathbb{N}^*}$ be a monotone sequence of real-valued continuous functions defined on a compact topological space X. If $\lim_{n\to\infty} f_n(x) = f(x)$ for all $x\in X$, where f is a continuous function, then the convergence is uniform.*

A1.14

Let (X, d') be a compact metric space and let $C(X)$ denote the collection of all complex-valued continuous functions on X. Then

$$d(f, g) = \sup_{x\in X} |f(x) - g(x)|$$

is a metric in $C(X)$, called the uniform metric.

Theorem A.16. (Arzelà–Ascoli) *A subset A of the metric space $(C(X), d)$ is relatively compact iff A is bounded and equicontinuous, i.e.,*

$$\sup_{f\in A,\, x\in X} |f(x)| < \infty$$

and

$$\lim_{\delta\to 0} \sup_{f\in A} \sup_{d'(x,y)<\delta} |f(x) - f(y)| = 0.$$

A1.15

The collection $\mathcal{O}_u$ consisting of the real line, the empty set and the intervals $(-\infty, a), a\in\mathbb{R}$, is a topology in $\mathbb{R}$, which is called the upper topology. Let $(X, \mathcal{O})$ be an arbitrary topological space. A real-valued function f on X is said to be upper semi-continuous iff it is continuous as a map from $(X, \mathcal{O})$ into $(\mathbb{R}, \mathcal{O}_u)$.

The upper semi-continuous functions enjoy some of the properties of the continuous functions, as shown below.

Proposition A1.17. *Let X be a compact topological space and let $f: X\to\mathbb{R}$ be an upper semi-continuous function. Then f is bounded from above and takes the value of its supremum over the space; also, the set $\{x: f(x) = a\}$ is closed for any $a\in\mathbb{R}$.*

Proposition A1.18. *Let $(f_n)_{n\in\mathbb{N}^*}$ be a non-increasing sequence of upper semi-continuous functions defined on a compact topological space X. Then* *the limit* $f(x)=\lim_{n\to\infty} f_n(x)$, $x\in X$, *is upper semi-continuous. If* $f(x)=0$, $x\in X$, *then the convergence is uniform.*

A1.16

Let X be a metric space and $\mathscr{B}_X$ the σ-algebra of Borel sets in X. Any finite measure μ on $\mathscr{B}_X$ is regular. This means that, for any $A\in\mathscr{B}_X$ and $\varepsilon>0$, there exist a closed set $F\subset X$ and an open set $G\subset X$ such that $F\subset A\subset G$ and $\mu(G-F)<\varepsilon$. (In an arbitrary topological space this property does not, generally, hold.)

Proposition A1.19. *Let X be a separable metric space and μ a measure on $\mathscr{B}_X$. Then there exists a unique closed set $S\subset X$ such that*

(i) $\mu(S)=\mu(X)$;

(ii) *if $F\subset X$ is a closed set for which $\mu(F)=\mu(X)$, then $F\supset S$.*

The set S, whose existence is guaranteed by Proposition A1.19, is called the support of the measure μ, and is denoted $\operatorname{supp}\mu$. In a separable metric space the support of a measure can be also characterized as follows: $\operatorname{supp}\mu$ is the set of all the points $x\in X$ such that $\mu(A)>0$ for any open set $A\subset X$ containing x.

A1.17

For any $n\in\mathbb{N}^*$ let $\mathbf{P}_n$ be a probability on the σ-algebra $\mathscr{B}_X$ of Borel sets of the metric space X. The sequence $(\mathbf{P}_n)_{n\in\mathbb{N}^*}$ is said to converge weakly to a probability $\mathbf{P}$ on $\mathscr{B}_X$, and we write $\mathbf{P}_n\Rightarrow\mathbf{P}$, iff

$$\lim_{n\to\infty}\int_X h(x)\mathbf{P}_n(\mathrm{d}x)=\int_X h(x)\mathbf{P}(\mathrm{d}x) \tag{A1.3}$$

for any $h\in C_r(X)=$ the collection of all real-valued bounded continuous functions on X.

Theorem A1.20. (Alexandrov–Prohorov) *The five assertions below are equivalent:*

(i) $\mathbf{P}_n\Rightarrow\mathbf{P}$;

(ii) (A1.3) *holds for any real-valued bounded uniformly continuous function h on X;*

(iii) $\limsup_{n\to\infty}\mathbf{P}_n(F)\leqslant\mathbf{P}(F)$ *for any closed set $F\subset X$;*

(iv) $\liminf_{n\to\infty}\mathbf{P}_n(G)\geqslant\mathbf{P}(G)$ *for any open set $G\subset X$;*

(v) $\lim_{n\to\infty}\mathbf{P}_n(A)=\mathbf{P}(A)$ *for any $A\in\mathscr{B}_X$ with $\mathbf{P}(\partial A)=0$.*

Let (X, d) and (Y, d') be two metric spaces. Consider a Y-valued r.v. f on X. The set D_f of the discontinuity points of f can be written as $D_f = \bigcup_\varepsilon \bigcap_\delta A_{\varepsilon,\delta}$, where ε and δ vary over the positive rational numbers, and $A_{\varepsilon,\delta}$ is the (open) set of all points $x \in X$ for which there exist $x', x'' \in X$ such that $d(x, x') < \delta$, $d(x, x'') < \delta$ and $d'(f(x'), f(x'')) \geqslant \varepsilon$. Clearly, $D_f \in \mathscr{B}_X$.

Proposition A1.21. *If $\mathsf{P}_n \Rightarrow \mathsf{P}$ and $\mathsf{P}(D_f) = 0$, then $\mathsf{P}_n f^{-1} \Rightarrow \mathsf{P} f^{-1}$. i.e., the P_n-distribution of f converges weakly to the P-distribution of f.*

In particular, the above result holds for a continuous f, for which, clearly, $D_f = \varnothing$.

Let $(f_n)_{n \in \mathbb{N}^*}$ be a sequence of Y-valued r.v.s on X. The sequence $(f_n)_{n \in \mathbb{N}^*}$ is said to converge in P-distribution to a probability Q on $\mathscr{B}_Y$, and we write $f_n \overset{\mathsf{P}}{\Rightarrow} \mathsf{Q}$, iff the P-distribution of f_n converges weakly to Q, i.e. $\mathsf{P} f_n^{-1} \Rightarrow \mathsf{Q}$. Let $(g_n)_{n \in \mathbb{N}^*}$ be another sequence of Y-valued r.v.s on X.

Proposition A1.22. *If (Y, d') is a separable metric space, then the distance $d': Y \times Y \to R$ is $\mathscr{B}_Y^2$-measurable. If $f_n \overset{\mathsf{P}}{\Rightarrow} \mathsf{Q}$ and the sequence $(d'(f_n, g_n))_{n \in \mathbb{N}^*}$ converges in P-probability to 0, then $g_n \overset{\mathsf{P}}{\Rightarrow} \mathsf{Q}$.*

A1.18

Let $C = C_r([0, 1])$ be the metric space of real-valued continuous functions on $[0, 1]$ with the uniform metric

$$d(\mathbf{x}, \mathbf{y}) = \sup_{t \in [0,1]} |\mathbf{x}(t) - \mathbf{y}(t)|, \quad \mathbf{x}, \mathbf{y} \in C.$$

The space C is complete and separable. The σ-algebra $\mathscr{B}_C$ of Borel sets in C coincides with the σ-algebra $\mathscr{B}^{[0,1]} \cap C$. Hence f is a C-valued r.v. iff $\xi(t) = \mathrm{pr}_t \circ f$ is a real r.v. for any $t \in [0, 1]$. Thus, to any stochastic process $(\xi(t))_{t \in [0,1]}$ with continuous trajectories there corresponds a C-valued r.v. $f = \xi(\cdot)$, and, conversely, to any C-valued r.v. f there corresponds a stochastic process $(\mathrm{pr}_t \circ f)_{t \in [0,1]}$ with continuous trajectories.

Of paramount importance is the probability B on $\mathscr{B}_C$ known as the Wiener measure, for which

$$\mathsf{B}(\mathbf{x} : \mathbf{x}(0) = 0) = 1,$$

$$\mathsf{B}(\mathbf{x} : \mathbf{x}(t_i) - \mathbf{x}(t_{i-1}) \leqslant a_i,\ 1 \leqslant i \leqslant k) = \prod_{i=1}^{k} \frac{1}{\sqrt{2\pi(t_i - t_{i-1})}} \int_{-\infty}^{a_i} \mathrm{e}^{-u^2/2(t_i - t_{i-1})}\, du$$

for all $k \in \mathbb{N}^*$, $0 \leqslant t_0 < t_1 \cdots < t_k \leqslant 1$, $a_i \in \mathbb{R}$, $1 \leqslant i \leqslant k$. Hence, under Wiener measure, the real-valued stochastic process $\boldsymbol{\beta} = (\mathrm{pr}_t)_{t\in[0,1]}$ on C has continuous trajectories, and its increments $\mathrm{pr}_t - \mathrm{pr}_s$ over disjoint intervals $[s,t] \subset [0,1]$ are independent and normally distributed with mean 0 and variance $t-s$. The process $\boldsymbol{\beta}$ is called the one-dimensional standard Brownian motion process.

A1.19

Let $D = D([0,1])(\supset C_r([0,1]))$ be the metric space of real-valued functions on $[0,1]$, which are right continuous and have left limits, with the Skorohod metric d_0 to be defined below.

Let $\mathfrak{L}$ denote the collection of all continuous functions $l: [0,1] \to [0,1]$, which are strictly increasing, with $l(0)=0$, $l(1)=1$, and put $s_0(l) = \sup_{s\neq t}|\log[(l(t)-l(s))/(t-s)]|$ for any $l\in\mathfrak{L}$. For $\mathbf{x}, \mathbf{y}\in D$, the distance $d_0(\mathbf{x},\mathbf{y})(\leqslant d(\mathbf{x},\mathbf{y}))$ is defined as the infimum of all $\varepsilon > 0$ for which there exists $l\in\mathfrak{L}$ such that $s_0(l)\leqslant\varepsilon$ and $\sup_{t\in[0,1]}|\mathbf{x}(t)-\mathbf{y}(l(t))|\leqslant\varepsilon$. Under this metric (which is equivalent to the uniform metric in D, defined similarly to that in C), D is complete and separable. (The latter property does not hold under the uniform metric.) Under d_0, the σ-algebra $\mathscr{B}_D$ of Borel sets in D coincides with the σ-agebra $\mathscr{B}^{[0,1]}\cap D$. While in C the projections pr_t, $t\in[0,1]$, are continuous, thus they are real r.v.s on C, in D the projection pr_t is continuous at $\mathbf{x}\in D$ iff the function $\mathbf{x}$ is continuous at t. In any case, pr_t is a real random variable on D for any $t\in[0,1]$. Hence f is a D-valued r.v. iff $\xi(t) = \mathrm{pr}_t\circ f$ is a real r.v. for any $t\in[0,1]$. Thus, to any stochastic process $(\xi(t))_{t\in[0,1]}$ whose trajectories are right continuous and have left limits, there corresponds a D-valued r.v. $f = \xi(\cdot)$ and, conversely, to any D-valued r.v. f there corresponds a stochastic process $(\mathrm{pr}_t\circ f)_{t\in[0,1]}$ whose trajectories are right continuous and have left limits.

Wiener measure $\mathbf{B}$ can be immediately extended from $\mathscr{B}_C$ to $\mathscr{B}_D$ as the topologies induced in C by the metrics d_0 and d are identical. Hence $A\cap C\in\mathscr{B}_C$ for any $A\in\mathscr{B}_D$. This allows us to define $\mathbf{B}(A) = \mathbf{B}(A\cap C)$, $A\in\mathbf{B}_D$, and, clearly, C is the support of $\mathbf{B}$ in D.

General references

Bergström (1982), Billingsley (1968), Grauert & Lieb (1968), Halmos (1950), Kelley (1955), Loève (1977), Mukherjea & Pothoven (1984), Natanson (1961).

Appendix 2
Notions of functional analysis

A2.1

A set X is said to be a linear vector space, a linear space or a vector space over a field $\mathbb{K}$ (more often than not, $\mathbb{K}$ is either the field $\mathbb{R}$ of real numbers, or the field $\mathbb{C}$ of complex numbers) iff there exist two maps, the first one from $X \times X$ into X, written as $(x_1, x_2) \to x_1 + x_2$ and called addition in X, and the second from $\mathbb{K} \times X$ into X, written as $(\alpha, x) \to \alpha x$ and called multiplication of the elements in X by scalars in $\mathbb{K}$, satisfying the properties below:

(i) the set X is an Abelian group w.r.t. addition;
(ii) for any $x, x_1, x_2 \in X$ and $\alpha, \beta \in \mathbb{K}$:
 (a) $1x = x$ (here 1 denotes the multiplicative identity in $\mathbb{K}$);
 (b) $\alpha(\beta x) = (\alpha\beta)x$;
 (c) $(\alpha + \beta)x = \alpha x + \beta x$;
 (d) $\alpha(x_1 + x_2) = \alpha x_1 + \alpha x_2$.

The elements in X are called vectors, while the elements in $\mathbb{K}$ are called scalars. A sum $\sum_{i=1}^{n} \alpha_i x_i$, $x_i \in X$, $\alpha_i \in \mathbb{K}$, $1 \leqslant i \leqslant n$, is said to be a linear combination of the vectors $x_1, \ldots, x_n$. The vectors $x_1, \ldots, x_n$ are said to be linearly independent iff the equation $\sum_{i=1}^{n} \alpha_i x_i = 0$ (here 0 stands for the zero element of the additive group X) implies that $\alpha_i = 0$, $1 \leqslant i \leqslant n$; they are said to be linearly dependent iff they are not linearly independent.

It is easy to verify that a necessary and sufficient condition for the vectors $x_1, \ldots, x_n$ to be linearly dependent is that one of them is a linear combination of the others.

A (non-empty) set $A \subset X$ is said to be linearly independent iff any $n \in \mathbb{N}^*$ distinct vectors $x_1, \ldots, x_n$ of A are linearly independent. A linear space is said to be infinite dimensional iff there exists an infinite linearly independent subset of X; otherwise, the space is said to be finite dimensional. A linearly independent set, which is maximal in the sense of the inclusion of sets, is called a basis or a Hamel basis for X. Any two bases for X have the same cardinal. In the case of a finite-dimensional linear space X, a set E is a basis for X iff it is linearly independent and any element of X is a linear combination of the elements of E.

Let $A \subset X$. If A is also a vector space over $\mathbb{K}$ with respect to the same operations, then A is called a linear subspace of X (or, simply, a subspace).

A2.2

Let X and Y be linear spaces over the same field $\mathbb{K}$. A one-to-one map J from X onto Y is said to be an isomorphism of X and Y iff $J(\alpha x_1 + \beta x_2) = \alpha J x_1 + \beta J x_2$, $x_1, x_2 \in X$, $\alpha, \beta \in \mathbb{K}$. It is easy to see that, if J is an isomorphism of X and Y, then the map $J^{-1}: Y \to X$ does exist and it is an isomorphism of Y and X.

A map $U: X \to Y$ is said to be an operator from X into Y; X and $UX = \{Ux: x \in X\}$ are said to be the domain and range of U, respectively. An operator U is said to be additive iff $U(x_1 + x_2) = Ux_1 + Ux_2, x_1, x_2 \in X$; it is said to be homogeneous iff $U(\alpha x) = \alpha U x$, $x \in X$, $\alpha \in \mathbb{K}$. An operator U is said to be linear iff it is additive and homogeneous.

Let X, Y, Z be linear spaces over the same field $\mathbb{K}$, and let $U: X \to Y$ and $V: Y \to Z$ be operators. The operator $W: X \to Z$, given by the formula $Wx = V(Ux)$, $x \in X$, is said to be the product of the operators V and U. In particular, if $X = Y = Z$, one can define the nth iterate (power) U^n of an operator U by the equation $U^n = UU^{n-1}$, $n \in \mathbb{N}^*$, with the convention $U^0 = I$, where I, defined as $Ix = x$, $x \in X$, is called the identity operator. The collection of all linear operators from X into Y is a linear space over the field $\mathbb{K}$ if equipped with the addition of two operators, given by the equation $(U + V)x = Ux + Vx, x \in X$, and the multiplication of an operator by a scalar of $\mathbb{K}$, given by the equation $(\alpha U)x = \alpha U x$, $x \in X$.

Let X be a linear space over the field $\mathbb{K}$. A map $p: X \to \mathbb{K}$ is called a functional on X. Since a functional is a special case of an operator, it is natural to say that a functional is linear iff it is additive ($p(x_1 + x_2) = p(x_1) + p(x_2)$, $x_1, x_2 \in X$) and homogeneous ($p(\alpha x) = \alpha p(x)$, $x \in X, \alpha \in \mathbb{K}$). The collection of all linear functionals on the linear space X is a linear space over the field $\mathbb{K}$ if equipped with the addition of two functionals, given by the equation $(p + q)(x) = p(x) + q(x)$, $x \in X$, and the multiplication of a functional by a scalar of $\mathbb{K}$, given by the equation $(\alpha p)(x) = \alpha p(x)$, $x \in X$, $\alpha \in \mathbb{K}$. This linear space is called the dual of X and is denoted by X^*.

A2.3

From now on let the field $\mathbb{K}$ be either the field $\mathbb{R}$ of real numbers or the field $\mathbb{C}$ of complex numbers and let X be a linear space over $\mathbb{K}$. A real non-negative functional on X, denoted by $\|\cdot\|_X = \|\cdot\|$,[†] is called a norm iff it satisfies the following conditions:

(i) the equation $\|x\| = 0$ implies $x = 0$;

(ii) $\|x_1 + x_2\| \leqslant \|x_1\| + \|x_2\|$ for all $x_1, x_2 \in X$ (this is called the triangle inequality);

(iii) $\|\alpha x\| = |\alpha| \, \|x\|$ for all $\alpha \in \mathbb{K}$ and $x \in X$.

It is obvious from the definition that $\|x\| = 0$ iff $x = 0$. A linear space X with norm $\|\cdot\|$ is said to be a linear normed space (or, in short, a normed space) and is denoted $(X, \|\cdot\|)$ when its norm should be emphasized.

In a normed space $(X, \|\cdot\|)$ a metric d can be defined by the equation $d(x_1, x_2) = \|x_1 - x_2\|$, $x_1, x_2 \in X$, and hence such a space can be viewed as a metric space. The metric topology in a normed linear space is called its norm or strong topology. It is easy to prove that, in this topology, the maps $(x_1, x_2) \to x_1 + x_2$, $(\alpha, x) \to \alpha x$, and $x \to \|x\|$ are all continuous.

A set $A \subset X$ is said to be bounded iff the set $\{\|x\| : x \in A\}$ is bounded (in $\mathbb{R}$).

Let $(X, \|\cdot\|_X)$ and $(Y, \|\cdot\|_Y)$ be two normed spaces, and let $J: X \to Y$ be an isomorphism; J is said to be isometric iff $\|Jx\|_Y = \|x\|_X$ for all $x \in X$.

A complete normed space is called a Banach space. A Banach algebra is a Banach space $(X, \|\cdot\|)$ for which there exists one more map from $X \times X$ into X, called multiplication and written as $(x_1, x_2) \to x_1 x_2$, $x_1, x_2 \in X$, satisfying for all $x_1, x_2, x_3 \in X$ and $\alpha, \beta \in \mathbb{K}$, the following properties:

(i) $(x_1 x_2) x_3 = x_1 (x_2 x_3)$ (associativity of multiplication);

(ii) $x_1(x_2 + x_3) = x_1 x_2 + x_1 x_3$ and $(x_1 + x_2) x_3 = x_1 x_3 + x_2 x_3$ (distributivity of multiplication w.r.t. addition);

(iii) $(\alpha x_1)(\beta x_2) = (\alpha\beta)(x_1 x_2)$;

(iv) $\|x_1 x_2\| \leqslant \|x_1\| \, \|x_2\|$.

A Banach algebra is said to be with unit iff there exists an element $1_X \in X$ such that $1_X x = x 1_X = x$ for all $x \in X$, and it is said to be commutative iff $x_1 x_2 = x_2 x_1$ for all $x_1, x_2 \in X$.

A2.4

In this section some special Banach spaces, which are often mentioned in this book, are described.

Let $(W, \mathscr{W})$ be an arbitrary measurable space. We denote by $B(W, \mathscr{W})$ the collection of all bounded $\mathscr{W}$-measurable complex-valued functions

† Different notation e.g. $|\cdot|$ or $|\!|\!|\cdot|\!|\!|$, will be also used

defined on W. This is a commutative Banach algebra with unit under the supremum norm $|f| = \sup_{w\in W}|f(w)|$, $f\in B(W,\mathscr{W})$.

If $(W,\mathscr{O})$ is a topological space and $\mathscr{W}$ is the σ-algebra of the Borel subsets of W (see Section A1.13), then we shall simply write $B(W)$ instead of $B(W,\mathscr{W})$. Under the same conditions we denote by $C(W)(C_r(W))$ the collection of all bounded, continuous complex-(real-)valued functions defined on W. Both $C(W)$ and $C_r(W)$ are commutative Banach algebras with unit under the supremum norm defined above.

If $(W,\mathscr{W})$ is an arbitrary measurable space, let $\text{ba}(W,\mathscr{W})$ denote the collection of all finitely additive finite complex-valued set functions defined on $\mathscr{W}$, and let $\text{ca}(W,\mathscr{W})$ denote the collection of all completely additive finite complex-valued set functions defined on $\mathscr{W}$. (Finite additivity and complete additivity for complex-valued set functions are defined as for measures and signed measures – see Section A1.4.) Both $\text{ba}(W,\mathscr{W})$ and $\text{ca}(W,\mathscr{W})$ are Banach spaces under the total variation norm $\|\mu\| = \text{var}\,\mu$ defined as $\text{var}\,\mu = \sup\sum_{i=1}^{n}|\mu(A_i)|$, where the supremum is taken over all finite partitions $W=\bigcup_{i=1}^{n}A_i$, $A_i\in\mathscr{W}$, $A_i\cap A_j=\varnothing$, $i\neq j$, $1\leqslant i,j\leqslant n$, $n\in\mathbb{N}^*$.

If $(W,\mathscr{O})$ is a topological space and $\mathscr{W}$ is the σ-algebra of the Borel subsets of W, then we shall simply write $\text{ca}(W)$ for $\text{ca}(W,\mathscr{W})$. Under the same conditions, the space $\text{rca}(W)\subset\text{ca}(W)$ of all regular completely additive complex-valued set functions is also a Banach space under the same norm of total variation. ($\mu\in\text{ca}(W)$ is said to be regular iff for any Borel set $A\subset W$ and $\varepsilon>0$ there exist a closed set $F\subset W$ and an open set $G\subset W$ with $F\subset A\subset G$ and $|\mu(G-F)|<\varepsilon$ – see Section A1.16.) If the topology $\mathscr{O}$ is given by a metric, then $\text{rca}(W)=\text{ca}(W)$.

Let $(\Omega,\mathscr{K},\mathbf{P})$ be a probability space. For any real number $p\geqslant 1$ we denote by $L^p(\Omega,\mathscr{K},\mathbf{P})$ the collection of all complex-valued r.v.s f on $(\Omega,\mathscr{K},\mathbf{P})$ for which $\int_\Omega|f|^p\,d\mathbf{P}<\infty$. As mentioned in Section A1.9, we do not distinguish between two r.v.s f and g which are versions of each other. Therefore $L^p(\Omega,\mathscr{K},\mathbf{P})$ is a collection of classes of $\mathbf{P}$-indistinguishable r.v.s as defined previously. Clearly, $L^p(\Omega,\mathscr{K},\mathbf{P})\subset L^{p'}(\Omega,\mathscr{K},\mathbf{P})$ if $p\geqslant p'$. Next, $L^p(\Omega,\mathscr{K},\mathbf{P})$ is a Banach space under the norm $\|f\|_p=(\int_\Omega|f|^p d\mathbf{P})^{1/p}$.

For a real r.v. f on $(\Omega,\mathscr{K},\mathbf{P})$, the $\mathbf{P}$-essential supremum $\text{ess sup}\,f$ and $\mathbf{P}$-essential infimum $\text{ess inf}\,f$ are defined as

$$\text{ess sup}\,f=\inf\{a\colon a\in\mathbb{R}, \mathbf{P}(f>a)=0\},\quad \text{ess inf}\,f=-\,\text{ess sup}(-f).$$

A complex-valued r.v. is said to be essentially bounded iff $\text{ess sup}\,|f|<\infty$.

Denote by $L^\infty(\Omega,\mathscr{K},\mathbf{P})$ the collection of all classes of essentially bounded indistinguishable r.v.s on $(\Omega,\mathscr{K},\mathbf{P})$; $L^\infty(\Omega,\mathscr{K},\mathbf{P})$ is a commutative Banach algebra with unit under the norm $\|f\|_\infty=\text{ess sup}\,|f|$. Clearly, $L^\infty(\Omega,\mathscr{K},\mathbf{P})\subset L^p(\Omega,\mathscr{K},\mathbf{P})$ for any $p\geqslant 1$.

A2.5

A normed space X can also be given another convergence, called the weak convergence, by using its dual X^*.

A sequence $(x_n)_{n\in\mathbb{N}^*}\subset X$ is said to converge weakly to $x\in X$ iff $\lim_{n\to\infty} x^*(x_n)=x^*(x)$ for all $x^*\in X^*$. A set $A\subset X$ is said to be weakly sequentially closed iff, together with any weakly convergent sequence included in it, it also contains its limit. A set $A\subset X$ is said to be weakly sequentially open iff its complement is weakly sequentially closed. A set $A\subset X$ is said to be weakly sequentially compact iff any covering of it with weakly sequentially open sets contains a finite subcovering.

In the special case of the space $C(W)$, where W is a compact metric space, the weak convergence of a sequence $(f_n)_{n\in\mathbb{N}^*}\subset C(W)$ to a function $f\in C(W)$ is equivalent to $\lim_{n\to\infty} f_n(w)=f(w)$ for all $w\in W$.

A2.6

Let $(X,\|\cdot\|_X)$ and $(Y,\|\cdot\|_Y)$ be normed spaces. A linear operator U from X into Y is said to be bounded iff there is a constant $M>0$ such that $\|Ux\|_Y\leqslant M\|x\|_X$ for all $x\in X$. The norm $|||U|||$ of a bounded linear operator U from X into Y is defined as

$$|||U|||=\sup_{\|x\|_X=1}\|Ux\|_Y=\sup_{0\neq x\in X}\frac{\|Ux\|_Y}{\|x\|_X}.$$

We shall denote by $\mathscr{L}(X,Y)$ the collection of all bounded linear operators from X into Y. It is easy to prove that $(\mathscr{L}(X,Y),|||\cdot|||)$ is a normed space. It is a Banach space iff Y is a Banach space. In the case $X=Y$, $\mathscr{L}(X,Y)$ will be denoted $\mathscr{L}(X)$, and we shall write $\|\cdot\|_X=\|\cdot\|$ instead of $|||\cdot|||$. If $U\in\mathscr{L}(X)$, U is said to be an operator on X. If X is a Banach space, then $\mathscr{L}(X)$ is a (non-commutative) Banach algebra with unit, the multiplication of two operators being defined as their product (see Section A2.2).

A linear operator U *in* (as opposed to 'on') a Banach space $(X,\|\cdot\|)$ is a linear map from a linear subspace A of X into X, i.e. $U(x_1+x_2)=Ux_1+Ux_2$, $x_1,x_2\in A$, and $U(\alpha x)=\alpha Ux$, $\alpha\in\mathbb{K}$, $x\in A$. The norm of such a U is defined as

$$\|U\|_A=\sup_{0\neq x\in A}\frac{\|Ux\|}{\|x\|},$$

and U is said to be bounded iff $\|U\|_A<\infty$. If U is bounded, then there exists a unique linear map U' from the (strong) closure $\bar{A}$ of A into X such that $U'x=Ux$ for any $x\in A$ and $\|U'\|_{\bar{A}}=\|U\|_A$; U' is called the continuous extension of U to $\bar{A}$.

Let X and Y be normed spaces and let $U \in \mathscr{L}(X, Y)$. The adjoint U^* of U is defined as the linear operator from Y^* into X^* given by $U^* y^* = y^*(U), y^* \in Y^*$.

Let $(X, \|\cdot\|)$ and Y be Banach spaces. Denote by $S = \{x \in X : \|x\| \leqslant 1\}$ the closed unit sphere in X. An operator $U \in \mathscr{L}(X, Y)$ is said to be compact (weakly compact) iff the image US of S is a compact (weakly sequentially compact) set in Y. An operator $U \in \mathscr{L}(X, Y)$, with finite-dimensional range, is compact. An operator $U \in \mathscr{L}(X, Y)$ is compact iff its adjoint is compact (Schauder's theorem). An operator $U \in \mathscr{L}(X)$ is said to be quasi-compact (quasi-weakly compact) iff there exist $m \in \mathbb{N}^*$ and a compact (weakly compact) operator $T \in \mathscr{L}(X)$ such that $\|U^m - T\| < 1$. An immediate consequence is that an operator $U \in \mathscr{L}(X)$, for which there exists a compact iterate U^m, $m \in \mathbb{N}^*$, is quasi-compact. If $(W, \mathcal{O})$ is a compact Hausdorff space, then the product of two weakly compact operators on $C(W)$ is compact; in particular, if U is a weakly compact operator on $C(W)$, then U^2 is compact.

A2.7

The dual space X^* of a normed space $(X, \|\cdot\|)$ is simply $\mathscr{L}(X, \mathbb{K})$ (see Section A2.6), which is a Banach space under the norm

$$|||x^*||| = \sup_{\|x\|=1} |x^*(x)| = \sup_{0 \neq x \in X} \frac{|x^*(x)|}{\|x\|}, \quad x^* \in X^*.$$

For certain Banach spaces the dual spaces can be explicitly described. Such is the case of the space $C(W)$, as shown in Theorem A2.1.

Theorem A2.1. (Riesz representation theorem) *If $(W, \mathcal{O})$ is a compact Hausdorff space then there is an isometric isomorphism J of* rca(W) *and $C^*(W)$ given by the equation*

$$(J\mu)(f) = (J\mu, f) = \int_W f(w)\mu(\mathrm{d}w), \quad f \in C(W), \quad \mu \in \mathrm{rca}(W).$$

A2.8

Let $(X, \|\cdot\|)$ be a Banach space and let $U \in \mathscr{L}(X)$. For any $\sigma \in \mathbb{C}$ let $E(\sigma) = \{x \in X : Ux = \sigma x\}$. The set $E(\sigma)$ is a non-empty (obviously, $0 \in E(\sigma)$) linear subspace of X. If $E(\sigma) \neq \{0\}$, then σ is called an eigenvalue of U, while $E(\sigma)$ is called the eigenspace corresponding to the eigenvalue σ. The multiplicity of an eigenvalue σ is the dimension of $E(\sigma)$ (= the cardinal of any basis of $E(\sigma)$).

Theorem A2.2. (Mean ergodic theorem) *Let $U \in \mathscr{L}(X)$ and assume that*

(i) *there exists a constant $H > 0$ such that $\| U^n \| \leqslant H$, $n \in \mathbb{N}^*$.*

(ii) *for any $x \in X$, the sequence $((1/n)\sum_{j=1}^{n} U^j x)_{n \in \mathbb{N}^*}$ contains a subsequence which converges weakly to a point $\bar{x} \in X$. Then*

(a)
$$\lim_{n \to \infty} \left\| \frac{1}{n} \sum_{j=1}^{n} U^j x - \bar{x} \right\| = 0.$$

(b) *if we put $\bar{x} = U_1 x$, $x \in X$, then $U_1 \in \mathscr{L}(X)$ and*

$$\| U_1 \| \leqslant H, \quad E(1) = U_1(X),$$
$$U_1^2 = U_1 = UU_1 = U_1 U.$$

Theorem A2.3. (Uniform ergodic theorem) *Let $U \in \mathscr{L}(X)$ be compact (quasi-compact) and assume there exists a constant $H > 0$ such that $\| U^n \| \leqslant H$, $n \in \mathbb{N}^*$. Then*

(i) *the set E of eigenvalues of U of modulus 1 is finite and each of them is of finite multiplicity.*

(ii) *there are compact operators $U_\sigma \in \mathscr{L}(X)$, $\sigma \in E$, such that*

$$\| U_\sigma \| \leqslant H, E(\sigma) = U_\sigma X, \quad U_\sigma^2 = U_\sigma, \quad UU_\sigma = U_\sigma U = \sigma U_\sigma, \quad U_\sigma U_{\sigma'} = 0,$$
$$\sigma \neq \sigma', \quad \sigma, \sigma' \in E.$$

(iii) *the operator $T = U - \sum_{\sigma \in E} \sigma U_\sigma$ is compact (quasi-compact), has spectral radius*

$$r(T) = \lim_{n \to \infty} \| T^n \|^{1/n} < 1$$

and $U_\sigma T = TU_\sigma = 0$ for any $\sigma \in E$.

(iv) *$U^n = \sum_{\sigma \in E} \sigma^n U_\sigma + T^n$ for all $n \in \mathbb{N}^*$.*

Remark. Let $U \in \mathscr{L}(X)$ be an operator for which some power U^m, $m \in \mathbb{N}^*$, is compact. Then, by the Riesz–Schauder theory, the set of eigenvalues of U is denumerable, having no point of accumulation except possibly $\sigma = 0$; any non-zero eigenvalue of U is of finite multiplicity.

A2.9

Throughout this section we consider two complex Banach spaces $(X, |\cdot|)$ and $(Y, \|\cdot\|)$ such that Y is a linear subspace of X and Condition (ITM_1) below holds.

Condition (ITM_1). *If $y_n \in Y$, $n \in \mathbb{N}^*$, $\sup_{n \in \mathbb{N}^*} \| y_n \| = c < \infty$, and $\lim_{n \to \infty} | y_n - y | = 0$ for some $y \in X$, then $y \in Y$ and $\| y \| \leqslant c$.*

Denote by $\mathscr{L}_k(Y,X)$, $k\in\mathbb{N}^*$, the collection of all linear operators U on Y which are bounded w.r.t. both $\|\cdot\|$ and $|\cdot|_Y$, where the latter is the restriction of $|\cdot|$ to Y, and in addition satisfy the following conditions.

Condition (ITM$_2$).

$$H=\sup_{n\in\mathbb{N}}|U^n|_Y<\infty.$$

Condition (ITM$_3$). *There exist two positive constants $q<1$ and Q such that*

$$\|U^k y\|\leqslant q\|y\|+Q|y|,\quad y\in Y.$$

Condition (ITM$_4$). *If A is a bounded subset of $(Y,\|\cdot\|)$ then U^kA has compact closure in $(X,|\cdot|)$.*

Notice that Conditions (ITM$_2$) and (ITM$_3$) imply that $J=\sup_{n\in\mathbb{N}}\|U^n\|<\infty$.

For any complex number σ set

$$E(\sigma)=\{y\in Y\colon Uy=\sigma y\},$$

so that σ is an eigenvalue of U iff $E(\sigma)\neq\{0\}$. If $|\sigma|=1$ let

$$U_\sigma^n=\frac{1}{n}\sum_{j=0}^{n-1}\sigma^{-j}U^j,\quad n\in\mathbb{N}^*.$$

Theorem A2.4. (Ionescu Tulcea–Marinescu) *Let $U\in\mathscr{L}_k(Y,X)$ for some $k\in\mathbb{N}^*$. Then*

(i) *the set E of eigenvalues of U of modulus 1 is finite and each of them is of finite multiplicity;*

(ii) *for any σ of modulus 1 and any $y\in Y$ there exists a $U_\sigma y\in Y$ to which $U_\sigma^n y$ converges in $(X,|\cdot|)$ as $n\to\infty$ (i.e. $\lim_{n\to\infty}|U_\sigma^n y-U_\sigma y|=0$). The operators $U_\sigma\in\mathscr{L}(y)$, $\sigma\in E$, are compact and $|U_\sigma|_Y\leqslant H$, $\|U_\sigma\|\leqslant J$, $E(\sigma)=U_\sigma Y$, $U_\sigma^2=U_\sigma$, $UU_\sigma=U_\sigma U=\sigma U_\sigma$, and $U_\sigma U_{\sigma'}=0$, $\sigma\neq\sigma'$, $\sigma,\sigma'\in E$.*

(iii) *the operator $T=U-\sum_{\sigma\in E}\sigma U_\sigma$ belongs to $\mathscr{L}_k(Y,X)$, has spectral radius*

$$r_Y(T)=\lim_{n\to\infty}\|T^n\|^{1/n}<1,$$

and $U_\sigma T=TU_\sigma=0$ for any $\sigma\in E$.

(iv) *$U^n=\sum_{\sigma\in E}\sigma^n U_\sigma+T^n$ for all $n\in\mathbb{N}^*$.*

The conclusions of Theorem A2.4 can be expanded in the important special case where Y is dense in X, i.e. the closure of Y in the norm $|\cdot|$ coincides with X.

For any bounded linear operator U on $(Y,|\cdot|_Y)$ let U' be the unique continuous extension of U to X (see Section A2.6), which is a bounded linear operator on X with $|U'| = |U|_Y$. For any complex number σ, set

$$E'(\sigma) = \{x \in X: U'x = \sigma x\},$$

so that σ is an eigenvalue of U' iff $E'(\sigma) \neq \{0\}$. If $|\sigma| = 1$ let

$$U'^n_\sigma = \frac{1}{n} \sum_{j=0}^{n-1} \sigma^{-j} U'^j, \quad n \in \mathbb{N}^*.$$

We have Theorem A2.4′.

Theorem A2.4′. *Let $U \in \mathscr{L}_k(Y, X)$ for some $k \in \mathbb{N}^*$. Assume that Y is dense in X. Then*

(i′) *the set of eigenvalues of U' of modulus 1 coincides with E and $E'(\sigma) = E(\sigma)$ for any $\sigma \in E$. Thus, the corresponding eigenspaces of U' are are contained in Y;*

(ii′) *for any σ of modulus 1 and any $x \in X$ there exists a $U'_\sigma x \in X$ to which $U'^n_\sigma x$ converges as $n \to \infty$ and $U'_\sigma X = E(\sigma)$. The operators $U'_\sigma \in \mathscr{L}(X)$ are compact, $|U'_\sigma| \leqslant H$ and $\sup_{|x|=1} \| U'_\sigma x \| < \infty$, while $U'_\sigma U'_{\sigma'} = 0, \sigma \neq \sigma'$, $U'^2_\sigma = U'_\sigma$.*

(iii′) *the operator $T' = U' - \sum_{\sigma \in E} \sigma U'_\sigma$ has the following properties: $\sup_{n \in \mathbb{N}^*} |T'^n| < \infty$, $\lim_{n \to \infty} T'^n x = 0$ for any $x \in X$, and $U'_\sigma T' = T' U'_\sigma = 0$ for any $\sigma \in E$.*

(iv′) *$U'^n = \sum_{\sigma \in E} \sigma^n U'_\sigma + T'^n$ for all $n \in \mathbb{N}^*$.*

It is obvious that in the context considered of a Y dense in X, the relationship between U and U' can be reversed, i.e. we can start from an arbitrary bounded linear operator U' on X; U is the *restriction* of U' to Y and belongs to $\mathscr{L}_k(X, Y)$ for some $k \in \mathbb{N}^*$. The conclusions of both Theorems A2.4 and A2.4′ hold for U and U' in this relationship.

A2.10

We now give a few examples of Banach spaces X and Y satisfying Condition (ITM_1) in Section A2.9 while Y is dense in X.

(1) Let (W, d) be a metric space. For any function $f \in C(W)$ (see Section A2.4) set

$$s(f) = \sup_{w_1 \neq w_2 \in W} \frac{|f(w_1) - f(w_2)|}{d(w_1, w_2)}, \quad \|f\|_L = |f| + s(f).$$

Let $L(W) = \{f: f \in C(W), \|f\|_L < \infty\}$ be the collection of all bounded Lipschitz complex-valued functions on W. Then $L(W)$ is a Banach space under the norm $\|\cdot\|_L$, and $(X, |\cdot|) = (C(W), |\cdot|)$ and $(Y, \|\cdot\|) = (L(W), \|\cdot\|_L)$ satisfy Condition (ITM_1). This pair of Banach spaces is of paramount importance in the compact Markov chain theory (see Section 3.2).

(2) Let I denote the interval $[0, 1]$. For any complex-valued function f

on I set

$$|f| = \sup_{t \in I} |f(t)|, \quad s(f) = \sup_{t_1 \neq t_2 \in I} \frac{|f(t_1) - f(t_2)|}{|t_1 - t_2|}$$

For any $p \in \mathbb{N}$, the collection $C^p(I)$ of all complex-valued functions on I which have continuous derivatives of order up to and including p is a Banach space under the norm

$$|f|_p = \sum_{i=0}^{p} |f^{(i)}|/i!$$

with $f^{(0)} = f$, $f^{(i)} = i$th derivative of f, $i \neq 0$. Clearly, $(C^0(I), |\cdot|_0)$ is $(C(I), |\cdot|)$ (see Section A2.4). A function belonging to $C^p(I)$ is called a function of class C^p on I. Next, the linear subspace $C^pL(I)$ of $C^p(I)$, consisting of the elements $f \in C^p(I)$ for which $s(f^{(p)}) < \infty$, is also a Banach space under the norm

$$\|f\|_{L,p} = |f|_p + s(f^{(p)})/p!$$

Clearly, $(C^0L(I), \|\cdot\|_{L,0})$ is $(L(I), \|\cdot\|_L)$. The pair $(X, |\cdot|) = (C^p(I), |\cdot|_p)$, $(Y, \|\cdot\|) = (C^pL(I), \|\cdot\|_{L,p})$ satisfies Condition (ITM$_1$) for any $p \in \mathbb{N}$.

(3) The variation $\mathrm{var}_A f$ over $A \subset I$ of a complex-valued function f on I is defined as $\sup \sum_{i=1}^{k-1} |f(t_i) - f(t_{i+1})|$, the supremum being taken over all $t_1 < \cdots < t_k \in A$ and $k \geq 2$.

We write simply var f for $\mathrm{var}_I f$. If var $f < \infty$, then f is called a function of bounded variation. Let λ denote Lebesgue measure. A variation $\mathrm{v}(f)$ for $f \in L^\infty(I, \mathscr{B}_I, \lambda)$, the collection of all classes of λ-essentially bounded measurable complex-valued λ-indistinguishable functions on I, is defined as $\mathrm{v}(f) = \inf \mathrm{var}\, f$, the infimum being taken over all versions of f (i.e. all measurable $\tilde{f}: I \to \mathbb{C}$ such that $\tilde{f} = f$ λ-a.s.) It can be shown that

$$\mathrm{v}(f) = \lim_{0 < a \to 0} \frac{1}{a} \int_0^1 |f(u+a) - f(u)| \, \mathrm{d}u,$$

where, for $x > 1$, we define $f(x) = f(1)$.

Clearly, in general, $\mathrm{v}(f) \leq \mathrm{var}\, f$. However, if f is a continuous function, then $\mathrm{v}(f) = \mathrm{var}\, f$. This is a special instance of the following much more general result (see Stadje (1985)). If $\mathrm{v}(f) < \infty$, then, for any $t \in I$, the limit

$$\tilde{f}(t) = \lim_{0 < a \to 0} \frac{1}{a} \int_t^{t+a} f(u) \, \mathrm{d}u$$

exists. The function $\tilde{f}$ is a right-continuous version of f and var $\tilde{f} = \mathrm{v}(f)$. The set $BV(I, \mathscr{B}_I, \lambda) = \{f : f \in L^\infty(I, \mathscr{B}_I, \lambda),\ f$ has a version of bounded variation$\}$ is a Banach space under the norm $\|f\|_{\mathrm{v}} = \mathrm{v}(f) + \|f\|_1$, where $\|\cdot\|_1$ is the usual L^1-norm $\|f\|_1 = \int_I |f| \, \mathrm{d}\lambda$ (see Section A2.4). The pair

$(L^1(I, \mathscr{B}_I, \lambda), \|\cdot\|_1), (BV(I, \mathscr{B}_I, \lambda), \|\cdot\|_v)$ satisfies Condition (ITM_1). This pair of Banach spaces, as well as the pairs in the preceding example, are important in the study of piecewise monotonic transformations (see Section 5.3).

Notice that, except for $L^1(I, \mathscr{B}_I, \lambda)$, as we have already mentioned, all the Banach spaces considered in the above examples are, in fact, commutative Banach algebras with unit (see Section A2.3.) For $BV(I, \mathscr{B}_I, \lambda)$, which is not a simple case, see Răuţu & Zbăganu (1989).

A2.11

Let $(X, \mathscr{X}, \mu)$ be a probability space. An X-valued r.v. on X, i.e. an $(\mathscr{X}, \mathscr{X})$-measurable map from X into itself, will be called a transformation of X. A transformation τ of X is said to be μ-non-singular iff $\mu(\tau^{-1}(A)) = 0$ for all $A \in \mathscr{X}$ for which $\mu(A) = 0$; it is said to be measure-preserving iff $\mu\tau^{-1} = \mu$, i.e. $\mu(\tau^{-1}(A)) = \mu(A)$ for all $A \in \mathscr{X}$ – see Section A1.5. (When the probability μ should be emphasized we shall say that τ is μ-preserving.) Clearly, any μ-preserving transformation is μ-non-singular. A pair (τ, μ), where τ is a μ-preserving transformation of X, is called an endomorphism of X.

The Frobenius–Perron operator $P_\mu = P$ associated with a μ-non-singular transformation τ is defined as the linear bounded operator on $L^1(X, \mathscr{X}, \mu) = L^1$ (see Section A2.4), which takes $f \in L^1$ into $Pf \in L^1$ with

$$\int_A Pf \, d\mu = \int_{\tau^{-1}(A)} f \, d\mu, \quad A \in \mathscr{X},$$

or, equivalently

$$\int_X gPf \, d\mu = \int_X (g \circ \tau) f \, d\mu \tag{A2.1}$$

for all $f \in L^1$ and $g \in L^\infty(X, \mathscr{X}, \mu) = L^\infty$. Actually, P so defined takes $L^p(X, \mathscr{X}, \mu)$ into itself for any $p \geqslant 1$ and $p = \infty$. The probabilistic interpretation of P is immediate: if an X-valued r.v. ξ on X has μ-density h (i.e. $\mu(\xi \in A) = \int_A h \, d\mu$, $A \in \mathscr{X}$, with $h \geqslant 0$ and $\int_X h \, d\mu = 1$), then $\tau \circ \xi$ has μ-density Ph. The following properties hold:

(a) P is positive i.e. $Pf \geqslant 0$ if $f \geqslant 0$;
(b) P preserves integrals, i.e. $\int_X Pf \, d\mu = \int_X f \, d\mu$, $f \in L^1$;
(c) $\|P\|_p \leqslant 1$ for all $p \geqslant 1$ and $p = \infty$;
(d) for all $n \in \mathbb{N}^*$ the nth power P^n of P is the Frobenius–Perron operator associated with the nth iterate τ^n of τ;
(e) $\overline{Pf} = P\bar{f}$ for all $f \in L^1$;

(f) $P((g\circ\tau)f) = gPf$ for all $f\in L^1$ and $g\in L^\infty$;
(g) $Pf = f$ iff (τ, ν) is an endomorphism of X, where ν is defined as $\nu(A) = \int_A f \, d\mu$, $A\in\mathscr{X}$. In particular, $P1 = 1$ iff τ is μ-preserving.

A μ-non-singular transformation τ of X is said to be ergodic (or metrically transitive, or indecomposable) w.r.t. μ iff the sets $A\in\mathscr{X}$ with $\tau^{-1}A = A$ satisfy either $\mu(A) = 0$ or $\mu(A) = 1$.

Proposition A2.5. *If a transformation τ of X is ergodic w.r.t. μ then there exists at most one μ-density f such that $Pf = f$. Conversely, if there is a unique μ-density f such that $f > 0$ μ-a.s. satisfying $Pf = f$, then τ is ergodic w.r.t. μ.*

Theorem A2.6. (Birkhoff's individual ergodic theorem) *Let τ be a μ-preserving transformation of X. Then for any $f\in L^1(X, \mathscr{X}, \mu)$ there exists $f^*\in L^1(X, \mathscr{X}, \mu)$ such that*

$$\lim_{n\to\infty} \frac{1}{n} \sum_{k=0}^{n-1} f\circ\tau^k = f^* \quad \mu\text{-a.s.}$$

and $f^*\circ\tau = f^*$ μ-a.s. Moreover, $\int_X f^* \, d\mu = \int_X f \, d\mu$, and if in addition τ is ergodic w.r.t. μ, then f^* is a constant equal to $\int_X f \, d\mu$.

An endomorphism (τ, μ) of X is said to be exact iff, putting $\mathscr{X}_n = \{\tau^{-n}A, A\in\mathscr{X}\}$, $n\in\mathbb{N}^*$, the tail σ-algebra $\bigcap_{n\in N^*}\mathscr{X}_n$ is μ-trivial, i.e. it contains only sets A for which either $\mu(A) = 0$ or $\mu(A) = 1$. If the endomorphism (τ, μ) is exact then τ is ergodic w.r.t. μ; also, for any $A\in\mathscr{X}$ for which $\mu(A) > 0$ and $\tau^n(A)\in\mathscr{X}$, $n\in\mathbb{N}^*$, we have $\lim_{n\to\infty}\mu(\tau^n(A)) = 1$.

Proposition A2.7. (M. Lin) *Let τ be a μ-preserving transformation of X for which $\tau(A)\in\mathscr{X}$ for all $A\in\mathscr{X}$. Then the endomorphism (τ, μ) is exact iff*

$$\lim_{n\to\infty} \left\| P^n f - \int_X f \, d\mu \right\|_1 = 0$$

for all $f\in L^1(X, \mathscr{X}, \mu)$.

General references

Cornfeld, Fomin & Sinai (1982), Dunford & Schwartz (1958), Lasota & Mackey (1985), Mukherjea & Pothoven (1986), Natanson (1961), Norman (1972).

Appendix 3
Mixing and Markovian dependence

A3.1

Let $(\Omega, \mathscr{K}, \mathsf{P})$ be a probability space and consider two σ-algebras $\mathscr{K}_1$ and $\mathscr{K}_2$ included in the σ-algebra $\mathscr{K}$. The φ-dependence coefficient $\varphi(\mathscr{K}_1, \mathscr{K}_2)$ of $\mathscr{K}_1$ and $\mathscr{K}_2$ is defined as

$$\varphi(\mathscr{K}_1, \mathscr{K}_2) = \sup |\mathsf{P}(B|A) - \mathsf{P}(B)|,$$

where the supremum is taken over all $A \in \mathscr{K}_1$ for which $\mathsf{P}(A) \neq 0$ and all $B \in \mathscr{K}_2$. The ψ-dependence coefficient $\psi(\mathscr{K}_1, \mathscr{K}_2)$ of $\mathscr{K}_1$ and $\mathscr{K}_2$ is defined as

$$\psi(\mathscr{K}_1, \mathscr{K}_2) = \sup \left| \frac{\mathsf{P}(B|A)}{\mathsf{P}(B)} - 1 \right|,$$

where the supremum is taken over all $A \in \mathscr{K}_1$ and $B \in \mathscr{K}_2$ for which $\mathsf{P}(A)\mathsf{P}(B) \neq 0$.

Both coefficients φ and ψ do not increase (decrease) when either $\mathscr{K}_1$ or $\mathscr{K}_2$, or both, are replaced by smaller (larger) σ-algebras. Clearly, $\varphi(\mathscr{K}_1, \mathscr{K}_2) \leqslant \psi(\mathscr{K}_1, \mathscr{K}_2)$ and $0 \leqslant \varphi(\mathscr{K}_1, \mathscr{K}_2) \leqslant 1, 0 \leqslant \psi(\mathscr{K}_1, \mathscr{K}_2) \leqslant \infty$, where the bounds can be attained (the lower bound is attained when $\mathscr{K}_1$ and $\mathscr{K}_2$ are independent, i.e. $\mathsf{P}(A \cap B) = \mathsf{P}(A)\mathsf{P}(B)$ for all $A \in \mathscr{K}_1$ and $B \in \mathscr{K}_2$, the upper one when $\mathscr{K}_1 = \mathscr{K}_2$ and there exist in $\mathscr{K}_1$ sets of arbitrarily small probability). Finally, let us note that, when defining $\varphi(\mathscr{K}_1, \mathscr{K}_2)$ and $\psi(\mathscr{K}_1, \mathscr{K}_2)$, the range of variation of A and B can be restricted to any two algebras which generate $\mathscr{K}_1$ and $\mathscr{K}_2$, respectively.

Let $(X, \mathscr{X})$ be a measurable space and consider a doubly infinite sequence $(\xi_n)_{n \in \mathbb{Z}}$ of X-valued r.v.s on Ω. (The usual case of an infinite sequence $(\xi_n)_{n \in \mathbb{N}^*}$ is obtained by taking $\xi_k \equiv 0$ for $k \leqslant 0$.) The φ-mixing coefficients $\varphi(n)$ and the ψ-mixing coefficients $\psi(n)$, $n \in \mathbb{N}^*$, of $(\xi_n)_{n \in \mathbb{Z}}$ are defined as (see Section A1.3)

$$\varphi(n) = \sup_{l \in \mathbb{Z}} \varphi(\sigma((\xi_i)_{i \leqslant l}), \sigma((\xi_i)_{i \geqslant l+n})),$$

$$\psi(n) = \sup_{l \in \mathbb{Z}} \psi(\sigma((\xi_i)_{i \leqslant l}), \sigma((\xi_i)_{i \geqslant l+n})).$$

Clearly, both $\varphi(\cdot)$ and $\psi(\cdot)$ are non-increasing and $\varphi(n) \leqslant \psi(n)$, $n \in \mathbb{N}^*$.

In the case of a strictly stationary sequence we get the same values $\varphi(n)$ and $\psi(n)$ when $\sigma((\xi_i)_{i\leqslant l})$ is replaced by $\sigma((\xi_i)_{1\leqslant i\leqslant l})$ and the supremum is taken over $l \in \mathbb{N}^*$. This means that a strictly stationary sequence and a doubly infinite strictly stationary sequence with the same finite-dimensional distributions (the latter, in general, being defined on a new probability space) have the same mixing coefficients $\varphi(n)$ and $\psi(n)$, $n \in \mathbb{N}^*$.

A sequence $(\xi_n)_{n\in\mathbb{Z}}$ is called φ-mixing or ψ-mixing depending on whether $\varphi(n) \to 0$ or $\psi(n) \to 0$ as $n \to \infty$. Clearly, a ψ-mixing sequence is φ-mixing, too.

Proposition A3.1. (0–1 law: Cohn (1965b)) *If for a sequence $(\xi_n)_{n\in\mathbb{N}^*}$ there exists $n_0 \in \mathbb{N}^*$ such that $\varphi(n_0) < \frac{1}{2}$, then its tail σ-algebra $\bigcap_{n\in\mathbb{N}^*}\sigma((\xi_i)_{i\geqslant n})$ is trivial, i.e. it contains only sets of probability* 0 *or* 1.

Corollary A3.2. *The tail σ-algebra of a φ-mixing sequence is trivial.*

Proposition A3.3. (Law of large numbers: Iosifescu (1974)) *Let $(\xi_n)_{n\in\mathbb{N}^*}$ be a φ-mixing strictly stationary sequence of X-valued r.v.s on Ω. Consider a real r.v. F on $X^{\mathbb{N}^*}$ and set $f_n = F(\xi_n, \xi_{n+1}, \ldots)$, $n \in \mathbb{N}^*$. If $\mathsf{E}(|f_1|) < \infty$, then $(\sum_{i=1}^n f_i)/n \to \mathsf{E}(f_1)$ P-a.s.*

Proposition A3.4. (Doob–Ibragimov–Ueno: Iosifescu & Theodorescu (1969, pp. 10–12)) *Let f_i be $\mathscr{K}_i$-measurable $(f_i^{-1}(\mathscr{B}) \subset \mathscr{K}_i)$, $i = 1, 2$, real r.v.s on Ω. If $\mathsf{E}(|f_1|^u) < \infty$ and $\mathsf{E}(|f_2|^v) < \infty$, with $1/u + 1/v = 1$, $u, v > 1$, then*

$$|\mathsf{E}(f_1 f_2) - \mathsf{E}(f_1)\mathsf{E}(f_2)| \leqslant 2\varphi^{1/u}(\mathscr{K}_1, \mathscr{K}_2)\mathsf{E}^{1/u}(|f_1|^u)\mathsf{E}^{1/v}(|f_2|^v).$$

If $\mathsf{E}(|f_1|) < \infty$ and f_2 is bounded, then

$$|\mathsf{E}(f_1 f_2) - \mathsf{E}(f_1)\mathsf{E}(f_2)| \leqslant \varphi(\mathscr{K}_1, \mathscr{K}_2)\mathsf{E}(|f_1|) \operatorname{osc} f_2.$$

Theorem A3.5. (Central limit theorem with remainder: Iosifescu (1968)) *Let $(\xi_n)_{n\in\mathbb{Z}^*}$ be a φ-mixing strictly stationary sequence of X-valued r.v.s on Ω. Consider a real r.v. F on X^k for some $k \in \mathbb{N}^*$, and set $f_n = F(\xi_n, \ldots, \xi_{n+k-1})$, $n \in \mathbb{N}^*$. Assume that $\mathsf{E}(|f_1|^{2+\delta}) < \infty$ for some $\delta > 0$ and that $\sum_{n\in\mathbb{N}^*}\varphi^{(1+\delta)/(2+\delta)}(n) < \infty$. Then the series $\sigma^2 = \mathsf{E}(f_1^2) - \mathsf{E}^2(f_1) + 2\sum_{n\in\mathbb{N}^*}(\mathsf{E}(f_1 f_{n+1}) - \mathsf{E}^2(f_1))$ is absolutely convergent and $\sigma^2 \geqslant 0$. If $\sigma > 0$, then there exist two positive constants $v < 1$ and c such that*

$$\left|\mathsf{P}\left(\frac{\sum_{r=1}^n f_r - n\mathsf{E}(f_1)}{\sigma\sqrt{n}} < a\right) - \Phi(a)\right| \leqslant cn^{-v}$$

for all $a \in \mathbb{R}$ and $n \in \mathbb{N}^*$.

Let $(\xi_n)_{n\in\mathbb{N}^*}$ be a φ-mixing strictly stationary sequence of X-valued r.v.s on Ω. Let F be a real r.v. on $X^{\mathbb{N}^*}$, and set $f_n = F(\xi_n, \xi_{n+1}, \ldots), n\in\mathbb{N}^*$. Clearly, $(f_n)_{n\in\mathbb{N}^*}$ is a strictly stationary sequence of real r.v.s on Ω, which, in general, is no longer φ-mixing. Assume, without any loss of generality, that $\mathsf{E}(f_1)=0$, and set $S_0=0, S_n=\sum_{i=1}^n f_i, n\in\mathbb{N}^*$. For any $n\in\mathbb{N}^*$ let us define the stochastic processes

$$\boldsymbol{\eta}_n^C : \boldsymbol{\eta}_n^C(t) = \frac{1}{\sigma\sqrt{n}}(S_{[nt]} + (nt-[nt])f_{[nt]+1}), \quad t\in[0,1],$$

and

$$\boldsymbol{\eta}_n^D : \boldsymbol{\eta}_n^D(t) = \frac{1}{\sigma\sqrt{n}} S_{[nt]}, \quad t\in[0,1],$$

where σ is a positive number, which will be precisely defined later. Clearly, $\boldsymbol{\eta}_n^C(t)$ and $\boldsymbol{\eta}_n^D(t)$ are real r.v.s on Ω for any $t\in[0,1]$ and, consequently, $\boldsymbol{\eta}_n^C$ and $\boldsymbol{\eta}_n^D$ are r.v.s on Ω which take values in C and D, respectively (see Sections A1.18 and A1.19).

Proposition A3.6. (Popescu (1978)) *Assuming only the strict stationarity of $(\xi_n)_{n\in\mathbb{N}^*}$ and that $\mathsf{E}(f_1^2)<\infty$, the sequences $(\boldsymbol{\eta}_n^C)_{n\in\mathbb{N}^*}$ and $(\boldsymbol{\eta}_n^D)_{n\in\mathbb{N}^*}$ have the same asymptotic behaviour. This means that if $\boldsymbol{\eta}_n^C \Rightarrow \mathsf{Q}$ (in C), then $\boldsymbol{\eta}_n^D \Rightarrow \mathsf{Q}_{\text{ext}}$ (in D) where Q_{ext} is the extension of Q to $\mathscr{B}_D$ defined by $\mathsf{Q}_{\text{ext}}(A)=\mathsf{Q}(C\cap A), A\in\mathscr{B}_D$. Conversely, if $\boldsymbol{\eta}_n^D \Rightarrow \mathsf{Q}$ (in D), then the support of Q is C and $\boldsymbol{\eta}_n^C \Rightarrow \mathsf{Q}$ (in C).*

Theorem A3.7. (Billingsley (1968, Theorem 21.1); Popescu (1978)) *If $\mathsf{E}(|f_1|^{2+\delta})<\infty$ for some $\delta\geqslant 0$ (or $|f_1|\leqslant c<\infty$), $\sum_{n\in\mathbb{N}^*}\varphi^{(1+\delta)/(2+\delta)}(n)<\infty$ (or $\sum_{n\in\mathbb{N}^*}\varphi(n)<\infty$), and*

$$\sum_{n\in\mathbb{N}^*} \mathsf{E}^{1/2}([f_1 - \mathsf{E}(f_1|(\xi_i)_{1\leqslant i\leqslant n})]^2) < \infty, \tag{A3.1}$$

then the series $\sigma^2 = \mathsf{E}(f_1^2) + 2\sum_{n\in\mathbb{N}^*}\mathsf{E}(f_1 f_{n+1})$ converges absolutely, $\sigma^2\geqslant 0$ and $\mathsf{E}(S_n^2/n)\to\sigma^2$ as $n\to\infty$. If $\sigma>0$, then $\mathsf{P}\boldsymbol{\eta}_n^{-1}\Rightarrow\mathsf{B}$, where $\boldsymbol{\eta}_n$ stands for either $\boldsymbol{\eta}_n^C$ or $\boldsymbol{\eta}_n^D$. The last conclusion still holds when P is replaced by any probability $\mathsf{Q}\ll\mathsf{P}$.

An important special case of Theorem A3.7 is obtained when the function F only depends on finitely many coordinates of a current point of $X^{\mathbb{N}^*}$, i.e. when F is a real r.v. on X^k, for a given $k\in\mathbb{N}^*$. In this case $f_n = F(\xi_n, \ldots, \xi_{n+k-1}), n\in\mathbb{N}^*$, and assumption (A3.1) is automatically satisfied. (Note that in this case the sequence $(f_n)_{n\in\mathbb{N}^*}$ is itself φ-mixing.)

For this special case it is possible to prove a weak convergence theorem for the empirical distribution functions. (Note that, under stronger assumptions, we can get the same result for the general case where the function F depends on all the coordinates of the current point of $X^{\mathbb{N}^*}$). More precisely, assume that $0 \leqslant f_n \leqslant 1, n \in \mathbb{N}^*$, and let $F_n^*(t)$ denote the relative frequency of the values f_i, $1 \leqslant i \leqslant n$, which do not exceed t, $0 \leqslant t \leqslant 1$. For any $n \in \mathbb{N}^*$ consider the process

$$\mathbf{Y}_n : \mathbf{Y}_n(t) = \sqrt{n}(F_n^*(t) - \bar{F}(t)), \quad 0 \leqslant t \leqslant 1,$$

where $\bar{F}(t) = \mathsf{P}(f_1 \leqslant t)$. Clearly, $\mathbf{Y}_n$ is a D-valued r.v. on Ω.

Theorem A3.8. (Billingsley (1968, Theorem 22.1, Yokoyama (1980)) *If $\sum_{n \in \mathbb{N}^*} \varphi(n) < \infty$ and $\bar{F}$ is a continuous function, then $\mathsf{P}\mathbf{Y}_n^{-1} \Rightarrow \mathsf{G}$ (in D), where G is a probability on $\mathscr{B}_D$, the support of which is C. Under G the stochastic process $(\mathrm{pr}_t)_{0 \leqslant t \leqslant 1}$ is a Gaussian process, whose trajectories are a.s. continuous, of mean 0 and covariance $r(s,t) = \mathsf{E}(g_s(f_1) g_t(f_1)) + \sum_{n \in \mathbb{N}^*} \mathsf{E}(g_s(f_1) g_t(f_{n+1})) + \sum_{n \in \mathbb{N}^*} \mathsf{E}(g_t(f_1) g_s(f_{n+1}))$, $0 \leqslant s, t \leqslant 1$, both series being absolutely convergent, where*

$$g_t(u) = \begin{cases} 1 - \bar{F}(t), & \text{if } 0 \leqslant u \leqslant t \\ -\bar{F}(t), & \text{if } t < u \leqslant 1. \end{cases}$$

The conclusion still holds when P is replaced by any probability $\mathsf{Q} \ll \mathsf{P}$.

Remark. The hypothesis of continuity of $\bar{F}$ can be ignored, with the loss of the G-a.s. continuity of the trajectories of the limit process.

In the same framework we shall state a functional variant of the law of the iterated logarithm.

Let $K \subset C$ be the collection of all absolutely continuous functions $\mathbf{x} \in C$ for which $\mathbf{x}(0) = 0$ and $\int_0^1 [\mathbf{x}'(t)]^2 \, dt \leqslant 1$. Here $\mathbf{x}'$ stands for the derivative of $\mathbf{x}$ which exists almost everywhere w.r.t. Lebesgue measure. Under the assumptions of Theorem A3.7 (clearly, equation (A3.1) is trivially satisfied in the present case), the series $\sigma^2 = \mathsf{E}(f_1^2) + 2\sum_{n \in \mathbb{N}^*} \mathsf{E}(f_1 f_{n+1})$ converges absolutely and $\sigma^2 \geqslant 0$. If $\sigma > 0$, for any $n \geqslant 3$ put

$$\boldsymbol{\theta}_n(t) = \frac{1}{\sigma\sqrt{2n \log\log n}} (S_{[nt]} + (nt - [nt]) f_{[nt]+1})$$
$$= \frac{1}{\sqrt{2 \log\log n}} \boldsymbol{\eta}_n^C(t), \quad 0 \leqslant t \leqslant 1.$$

Theorem A3.9. (Strassen's law of the iterated logarithm: Heyde & Scott (1973); Iosifescu (1972); Oodaira & Yoshihara (1971a, b)) *Under the assumptions of Theorem A3.7, the sequence $(\boldsymbol{\theta}_n)_{n \geqslant 3}$, viewed as a subset of C, is a relatively compact set, whose derived set coincides P-a.s. with K.*

Corollary A3.10. (Classical law of iterated logarithm) *Under the assumptions of Theorem* A3.9 *we have*

$$\mathbf{P}\left(\limsup_{n\to\infty}\frac{S_n}{\sigma\sqrt{2n\log\log n}}=1\right)=1.$$

Note that under stronger assumptions on the moments of the r.v. f_1 and on the φ-mixing coefficients it is also possible to prove (Reznik (1968)) a classical law of the iterated logarithm for the case where F is a r.v. on $X^{\mathbb{N}^*}$.

Note also that it can be deduced from Theorem A3.9 that the derived set of the sequence $(S_n/\sigma\sqrt{2n\log\log n})_{n\geq 3}$ coincides $\mathbf{P}$-a.s. with the segment $[-1,1]$.

A3.2

Let $(X,\mathscr{X})$ be a measurable space and P a t.p.f. on $(X,\mathscr{X})$. For any $x\in X$, consider the probability space $(\Omega,\mathscr{K},\mathbf{P}_x)$, where $\Omega=X^{\mathbb{N}^*}$ and $\mathscr{K}=\mathscr{X}^{\mathbb{N}^*}$, while $\mathbf{P}_x$ is defined (uniquely, by Theorem A1.3) by

$$\mathbf{P}_x(\mathrm{pr}^{-1}_{\{1,\ldots,n\}}(A_1\times\cdots\times A_n))$$
$$=\begin{cases}P(x,A_1), & \text{if } n=1\\ \int_{A_1}P(x,\mathrm{d}x_1)\int_{A_2}P(x_1,\mathrm{d}x_2)\cdots\int_{A_n}P(x_{n-1},\mathrm{d}x_n), & \text{if } n>1\end{cases}$$

for all $A_i\in\mathscr{X}$, $1\leqslant i\leqslant n$, $n\in\mathbb{N}^*$. Define the X-valued r.v.s ζ_n on Ω by the equations $\zeta_n(\omega)=\mathrm{pr}_n\,\omega$, $n\in\mathbb{N}^*$, $\omega\in\Omega$. It follows immediately that

$$\mathbf{P}_x(\zeta_1\in A)=P(x,A),\quad A\in\mathscr{X},$$
$$\mathbf{P}_x(\zeta_{n+1}\in A\,|\,\zeta_1,\ldots,\zeta_n)=P(\zeta_n,A),\quad \mathbf{P}_x\text{-a.s.}\tag{A3.2}$$

for all $n\in\mathbb{N}^*$ and $A\in\mathscr{X}$. The sequence $(\zeta_n)_{n\in\mathbb{N}}$ with $\zeta_0=x$ is called a homogeneous Markov chain (for short MC), with state space $(X,\mathscr{X})$, t.p.f. P, and initial distribution concentrated at state $x\in X$ (or initial state $x\in X$). Sometimes we also consider a probability $\mathbf{p}$ on $\mathscr{X}$ (which plays the role of the distribution of ζ_0) and the probability $\mathbf{P}_\mathbf{p}$ defined by the equation

$$\mathbf{P}_\mathbf{p}(E)=\int_X\mathbf{p}(\mathrm{d}x)\mathbf{P}_x(E),\quad E\in\mathscr{K}.$$

Clearly, the Markov property (A3.2) also holds under the probability $\mathbf{P}_\mathbf{p}$.

In what follows, the mean value operator w.r.t. the probability $\mathbf{P}_x(\mathbf{P}_\mathbf{p})$ will be denoted $\mathbf{E}_x(\mathbf{E}_\mathbf{p})$.

In the special case where X is a denumerable set, the t.p.f. P is defined by means of a stochastic matrix $\mathbf{P}=(p_{ij})_{i,j\in X}$ setting

$$P(i,A)=\sum_{j\in A}p_{ij},\quad i\in X,\quad A\subset X.$$

Completely similar definitions can be introduced in the non-homogeneous case, where we are given a sequence $({}^nP)_{n\in\mathbb{N}}$ of t.p.f.s on $(X,\mathscr{X})$. In this case $\mathbf{P}_x$ is defined by the equations

$$\mathbf{P}_x(\mathrm{pr}^{-1}_{\{1,\ldots,n\}}(A_1\times\cdots\times A_n))$$
$$=\begin{cases} {}^0P(x,A_1), & \text{if } n=1\\ \displaystyle\int_{A_1}{}^0P(x,\mathrm{d}x_1)\int_{A_2}{}^1P(x_1,\mathrm{d}x_2)\cdots\int_{A_n}{}^{n-1}P(x_{n-1},\mathrm{d}x_n), & \text{if } n>1\end{cases}$$

for all $A_i\in\mathscr{X}$, $1\leqslant i\leqslant n$, $n\in\mathbb{N}^*$. With the r.v.s defined as before, we have

$$\mathbf{P}_x(\zeta_1\in A)={}^0P(x,A),\quad A\in\mathscr{X},$$
$$\mathbf{P}_x(\zeta_{n+1}\in A\,|\,\zeta_1,\ldots,\zeta_n)={}^nP(\zeta_n,A),\quad \mathbf{P}_x\text{-a.s.}$$

for all $n\in\mathbb{N}^*$ and $A\in\mathscr{X}$. The sequence $(\zeta_n)_{n\in\mathbb{N}}$ with $\zeta_0=x$ is called a (non-homogeneous) MC with state space $(X,\mathscr{X})$, t.p.f.s ${}^nP, n\in\mathbb{N}$, and initial distribution concentrated at state $x\in X$ (or initial state $x\in X$).

An equivalent method of defining the concept of an MC, which is quite convenient in stochastic control theory, is as follows. A non-homogeneous MC $(\zeta_n)_{n\in\mathbb{N}}, \zeta_0=x$, with state space $(X,\mathscr{X})$, is a sequence of X-valued r.v.s on a probability space $(\Omega,\mathscr{K},\mathbf{P})$, which is defined recursively by $\zeta_{n+1}=u_n(\zeta_n,\xi_{n+1}), n\in\mathbb{N}$. Here $(\xi_n)_{n\in\mathbb{N}^*}$ is a sequence of independent r.v.s on $(\Omega,\mathscr{K},\mathbf{P})$ which takes values in a measurable space $(Y,\mathscr{Y})$, while u_n is an $(\mathscr{X}\otimes\mathscr{Y},\mathscr{Y})$-measurable map from $X\times Y$ into Y, $n\in\mathbb{N}$. For a proof of the equivalence of the two definitions see Bergmann & Stoyan (1978) and O'Brien (1975). See also Kifer (1986a, Ch. 1).

The transition operator of a homogeneous MC $(\zeta_n)_{n\in\mathbb{N}}$ (or, equivalently, the Markov operator associated with the t.p.f. P of the chain) is the operator U on $B(X,\mathscr{X})$ defined by the equation

$$Uf=\int_X P(\cdot,\mathrm{d}y)f(y).$$

The operator V on $\mathrm{ca}(X,\mathscr{X})$ defined by the equation

$$V\mu=\int_X \mu(\mathrm{d}x)P(x,\cdot)$$

is the adjoint of U, i.e. $(\mu,Uf)=(V\mu,f)$, where

$$(\mu,f)=\int_X f(x)\mu(\mathrm{d}x),\quad \mu\in\mathrm{ca}(X,\mathscr{X}),\quad f\in B(X,\mathscr{X}).$$

Both U and V are positive operators ($Uf\geqslant 0$ if $f\geqslant 0$ and $V\mu\geqslant 0$ if $\mu\geqslant 0$). Since $Uf=f$ if f is a constant, and $V\mu\in\mathrm{pr}(X,\mathscr{X})$ if $\mu\in\mathrm{pr}(X,\mathscr{X})$, we have $|U|=\|V\|=1$.

The iterates of U are given by the equation

$$U^n f = \int_X P^n(\cdot, dy) f(y), \quad n \in \mathbb{N}^*,$$

where P^n is the n-step t.p.f., which is defined recursively as $P^1 = P$ and, for $n > 1$,

$$P^n(x, A) = \int_X P(x, dy) P^{n-1}(y, A), \quad x \in X, \quad A \in \mathscr{X}.$$

Hence

$$P^{m+n}(x, A) = \int_X P^m(x, dy) P^n(y, A), \quad x \in X, \quad A \in \mathscr{X}, \tag{A3.3}$$

for all $m, n \in \mathbb{N}$, with the convention that

$$P^0(x, A) = \chi_A(x) = \begin{cases} 1, & \text{if } x \in A \\ 0, & \text{if } x \notin A. \end{cases}$$

Equation (A3.3) is known as the Chapman–Kolmogorov equation.

The iterates of V are given by the equation

$$V^n \mu = \int_X \mu(dx) P^n(x, \cdot), \quad n \in \mathbb{N}^*.$$

The probabilistic meaning of P^n, U^n and V^n follows from the equations

$$\mathsf{P}_x(\zeta_{m+n} \in A \mid \zeta_1, \ldots, \zeta_m) = P^n(\zeta_m, A), \quad \mathsf{P}_x\text{-a.s.},$$

$$\mathsf{E}_x(f(\zeta_{m+n}) \mid \zeta_1, \ldots, \zeta_m) = U^n f(\zeta_m), \quad \mathsf{P}_x\text{-a.s.},$$

$$\mathsf{P}_{\mathsf{p}}(\zeta_m \in A) = V^m \mathsf{p}(A),$$

valid for all $m, n \in \mathbb{N}^*$, $A \in \mathscr{X}$ and $f \in B(X, \mathscr{X})$.

In the non-homogeneous case we have a family $({}^mU)_{m \in \mathbb{N}}$ of transition operators defined as

$${}^mUf = \int_X {}^mP(\cdot, dy) f(y), \quad m \in \mathbb{N}, \quad f \in B(X, \mathscr{X}).$$

If we put

$${}^mU^n = {}^mU \circ \ldots \circ {}^{m+n-1}U, \quad m \in \mathbb{N}, \quad n \in \mathbb{N}^*,$$

we have

$${}^mU^n f = \int_X {}^mP^n(\cdot, dy) f(y),$$

where ${}^mP^n$ is a t.p.f. on $(X, \mathscr{X})$, which is defined recursively as ${}^mP^1 = {}^mP$, $m \in \mathbb{N}$, and for $n > 1$,

$${}^mP^n(x, A) = \int_X {}^mP(x, dy)\, {}^{m+1}P^{n-1}(y, A), \quad x \in X, \quad A \in \mathscr{X}.$$

Hence

$${}^mP^{l+n}(x, A) = \int_X {}^mP^l(x, dy)\, {}^{m+l}P^n(y, A), \quad x \in X, \quad A \in \mathscr{X},$$

for all $l, m, n \in \mathbb{N}$, with the convention that ${}^mP^0(x, A) = \chi_A(x)$. The above equation is known as the non-homogeneous Chapman–Kolmogorov equation.

The probabilistic meaning of ${}^mP^n$ and ${}^mU^n$ follows from the equations

$$\mathsf{P}_x(\zeta_n \in A) = {}^0P^n(x, A),$$
$$\mathsf{P}_x(\zeta_{m+n} \in A \mid \zeta_1, \ldots, \zeta_m) = {}^mP^n(\zeta_m, A), \quad \mathsf{P}_x\text{-a.s.},$$
$$\mathsf{E}_x(f(\zeta_{m+n}) \mid \zeta_1, \ldots, \zeta_m) = {}^mU^n f(\zeta_m), \quad \mathsf{P}_x\text{-a.s.},$$

valid for all $x \in X$, $A \in \mathscr{X}$, $m, n \in \mathbb{N}^*$ and $f \in B(X, \mathscr{X})$.

A homogeneous MC is φ-mixing iff there exist $n_0 \in \mathbb{N}^*$ and $\delta > 0$ such that

$$|P^{n_0}(x', A) - P^{n_0}(x'', A)| \leqslant 1 - \delta \tag{A3.4}$$

for all $x', x'' \in X$ and $A \in \mathscr{X}$. For a φ-mixing homogeneous MC we have $\varphi_{\mathbf{p}}(n) \leqslant ab^n$, $n \in \mathbb{N}^*$, $\mathbf{p} \in \mathrm{pr}(X, \mathscr{X})$, where $a > 0$ and $0 < b < 1$ are suitably chosen constants. (The subscript $\mathbf{p}$ shows that the φ-mixing coefficients are calculated w.r.t. the probability $\mathsf{P}_{\mathbf{p}}$.) Consequently, a homogeneous MC can only be exponentially φ-mixing. Condition (A3.4) also ensures the existence of a stationary probability, i.e. a probability π on $\mathscr{X}$ such that $V\pi = \pi$; the sequence $(\zeta_n)_{n \in \mathbb{N}}$ is strictly stationary under P_π. Therefore, the possibility of applying limit theorems valid for φ-mixing sequences to MCs is completely clear.

Now, we state two limit theorems for certain homogeneous MCs which are not φ-mixing.

Theorem A3.11. (Breiman (1960)) *Let X be a compact metric space. Consider an MC $(\zeta_n)_{n \in \mathbb{N}}$ with initial state $\zeta_0 = x$, state space X, and t.p.f. P, and assume that the transition operator U takes $C(X)$ into itself, i.e. $Uf \in C(X)$ for any $f \in C(X)$. Assume also that there exists a unique probability π on $\mathscr{B}_X$ such that $V\pi = \pi$. Then, for any $x \in X$ and $f \in C(X)$ we have*

$$\lim_{n \to \infty} \frac{\sum_{k=1}^n f(\zeta_k)}{n} = \int_X f(x)\pi(\mathrm{d}x), \quad \mathsf{P}_x\text{-}a.s.$$

The second theorem also refers to the MCs considered in Theorem A3.11. Given a $\mathscr{B}_X$-measurable real-valued function on X for which $\int_X f^2(x)\pi(\mathrm{d}x) < \infty$, put $f_n = f(\zeta_n), n \in \mathbb{N}^*$. Under P_π, the sequence $(f_n)_{n \in \mathbb{N}^*}$ is strictly stationary. Assume that $\mathsf{E}_\pi(f_1) = 0$ and put $S_0 = 0$, $S_n = \sum_{i=1}^n f_i$, $n \in \mathbb{N}^*$. As in Section A3.1, for any $n \in \mathbb{N}^*$ define the stochastic processes

$$\boldsymbol{\eta}_n^C : \boldsymbol{\eta}_n^C(t) = \frac{1}{\sigma\sqrt{n}}(S_{[nt]} + (nt - [nt]) f_{[nt]+1}), \quad t \in [0, 1],$$

and

$$\boldsymbol{\eta}_n^D : \boldsymbol{\eta}_n^D(t) = \frac{1}{\sigma\sqrt{n}} S_{[nt]}, \quad t \in [0, 1],$$

where σ is a positive number, which will be precisely defined later. Clearly, $\boldsymbol{\eta}_n^C(t)$ and $\boldsymbol{\eta}_n^D(t)$ are real r.v.s on Ω for any $t \in [0, 1]$, and, consequently, $\boldsymbol{\eta}_n^C$ and $\boldsymbol{\eta}_n^D$ are r.v.s on Ω which take values in C and D, respectively. By Proposition A3.6, the sequences $(\boldsymbol{\eta}_n^C)_{n \in \mathbb{N}^*}$ and $(\boldsymbol{\eta}_n^D)_{n \in \mathbb{N}^*}$ have the same asymptotic behaviour under $\mathbf{P}_\pi$. In what follows $\boldsymbol{\eta}_n$ stands for either $\boldsymbol{\eta}_n^C$ or $\boldsymbol{\eta}_n^D$.

The result below is a special instance of more general results proved by Heyde & Scott (1973) and Scott (1973).

Theorem A3.12. *Consider an MC $(\zeta_n)_{n \in \mathbb{N}}$ satisfying the assumptions of Theorem* A3.11. *Assume that*

$$\sum_{n \in \mathbb{N}^*} \mathbf{E}_\pi^{1/2}(\mathbf{E}_\pi^2(f_{n+1} | \zeta_1)) < \infty.$$

Then the limit $\lim_{n \to \infty} \mathbf{E}_\pi(S_n^2/n) = \sigma^2 \geqslant 0$ *exists. If $\sigma > 0$, then $\mathbf{P}_\pi \boldsymbol{\eta}_n^{-1} \Rightarrow \mathbf{B}$, and the sequence $(\boldsymbol{\eta}_n^C/\sqrt{2 \log \log n})_{n \geqslant 3}$, viewed as a subset of C, is a relatively compact set, whose derived set coincides $\mathbf{P}_\pi$-a.s. with K defined in Section* A3.1.

General references

Billingsley (1968), Bradley (1986), Chung (1967), Hall & Heyde (1980), Ibragimov & Linnik (1971), Iosifescu (1980a), Iosifescu & Theodorescu (1969), Peligrad (1986).

Notes and comments

1.1

The first explicit formal definition of the concept of dependence with complete connections was given by Onicescu & Mihoc (1935a), who considered what is now called an OM chain (see Section 5.1). In many urn models, which were also used as basic examples by Onicescu & Mihoc (1936a, b, 1937, 1943) the concept of dependence with complete connections is implicitly contained. (For information on urn models see Johnson & Kotz (1977).) However, just as special cases of Markov chains had been known long before the formal definition of the concept of Markovian dependence, special cases of dependence with complete connections were noted before 1935. For instance, as noticed by Kruskal & Mosteller (1980, p. 185), Tschuprow (1923)

> had an unusually clear idea of probabilistic structure and the growing importance of stochastic dependence among the observations, dependence of the kind that appears in sampling without replacement from a finite population as in a social survey.

The concept of a random system with complete connections (RSCC), as given in Subsections 1.1.1–1.1.3, was defined by Iosifescu (1963a). The 1970s saw attempts to generalize this concept. The concept of a generalized random system with complete connections (GRSCC) was introduced by Le Calvé & Theodorescu (1971), while another generalization of the RSCCs is to be found in Usacev (1971).

Comment to page 15, lines 10-12: In the last 20 years or so, this case has been much investigated under the nomenclature "iterated function system" and viewed in most instances as a genuinely new concept while no reference was made to its special place within dependence-with-complete-connections theory. On the same disregarding line, a general RSCC is renamed an iterated function system with place dependent probabilities.

1.2

Many other examples of RSCCs and GRSCCs are to be found in Arkin & Evstigneev (1987), Bežaeva (1971, 1974, 1977), Böttcher & Nawrotzki (1976), Boudiba (1986), Cohen (1976, 1977), Cohen, Kesten & Newman (Eds.) (1986), Doukhan & Ghindès (1980a, b), Eremin & Mazurov (1979, Ch. II, §5), Gihman & Skorohod (1979), Grey (1980), Grincevičius

(1974, 1978), Herkenrath (1979, 1980a, b), Herkenrath & Theodorescu (1978a, b, c, 1980, 1981), Högnäs (1978), de Hoog, Brown, Saunders & Westcott (1986), Iordache (1987), Kaijser (1972, 1973, 1978c, 1981a, 1983, 1986), Kalin & Theodorescu (1982, 1984), Kantel & Nawrotzki (1975), Kukush (1986), Lakshmivarahan (1981), Letac (1986), Macchi (1976), Madras (1986), Maibaum & Mühlmann (1975), McCabe (1969), Mokkadem (1987), Nawrotzki (1981/82), Norman (1975), Onicescu & Botez (1985), Onicescu & Guiaşu (1965), Popescu (1981), Rhenius (1971, 1974), Slater (1967), Solomon (1975), Sragovič (1976, 1981), Takahashi (1969), Tomlinson (1971) (see Halfin (1975)), Tsetlin (1973), Wierenga (1974), and Wolff (1976).

1.3

The classification problems were approached by Doeblin & Fortet (1937), Kerbrat & Le Calvé (1973a, b), Mihoc & Ciucu (1971), Pruscha & Theodorescu (1978, 1981), and Theodorescu & Tweedie (1983). The RSCCs of type (B) were introduced by Doeblin & Fortet (1937) in the special context of the strict-sense infinite-order-chains (see Section 5.5).

2.1

Theorem 2.1.3 was proved by Ionescu Tulcea (1959). His work is fundamental to the study of the ergodicity of RSCCs. The existence of the probability $\mathbf{P}_\infty$ in Theorem 2.1.5 was stated by Theiler (1967), whose proof, together with the proof in Iosifescu & Theodorescu (1969, p. 135), is incomplete. The proof given here is basically due to Popescu (1978) (see also Iosifescu (1977/78)). Assertions (ii) and (iii) in Theorem 2.1.5 are due to Cohn (1964, 1965a).

2.2–2.4

The exposition in these sections is based on Iosifescu (1963b, 1965b, 1966b, c). Condition FLS(A_0, ν) was used in special contexts by Fortet (1938) and Lamperti & Suppes (1959).

3.1

The study of the asymptotic behaviour of MCs goes back to Doeblin (1937, 1938), who used measure-theoretical tools in deriving his results. A detailed account of Doeblin's work may also be found in the celebrated book by Doob (1953, pp. 192–215). The operatorial method of investigation of the asymptotic behaviour of MCs was initiated by Krylov & Bogoljubov (1937) and developed by Yosida & Kakutani (1941). The applications of quasi-compactness in probability theory and especially in the study of MCs, have been widely explored (see Brunel & Revuz (1974), Foguel (1973), and Lin (1974, 1978)). The interdependence between Doeblin's condition and the quasi-compactness of different operators associated with an MC has also received special attention. Basic references for this topic are Doob (1953, pp. 192–215), Fortet (1978), Herkenrath (1977), Jacobs (1960, Ch. 2), and Neveu (1964, Ch. 5). The concept of a Doeblin–Fortet chain is implicit in Doeblin & Fortet (1937); however, it was formally defined and developed by Norman (1968, 1972). Many results in Subsection 3.1.3, among them the important Theorem 3.1.24, are due to Norman (1972, Ch. I, 2). The study of Markov operators acting on classes of continuous real-valued functions defined on a compact metric space was initiated by Bebutov (1942) and Krylov & Bogoljubov (1937) and developed by Jacobs (1962, Ch. 11) and Jamison (1964, 1965). The results in Subsection 3.1.4 together with Propositions 3.1.8 and 3.1.9 in Subsection 3.1.1, are essentially due to Herkenrath (1977).

3.2–3.3

The interest in the theory of Markov chains with an arbitrary state space was revitalised by Chung (1964), who cites Doeblin (1937) as his forerunner and inspiration. Basic references for the subsequent development of this theory are Cogburn (1975), Nummelin (1984), Orey (1971), Revuz (1975), and Šidák (1967, 1977).

The concept of a compact MC is due to Norman (1968). The exposition in Section 3.2 closely follows Norman (1972, Ch. I, 3). The concept of a continuous MC, as a natural concept in studying the MC associated with a GRSCC, was introduced by Grigorescu (1976a) and Herkenrath (1977) independently of each other. Section 3.3 is essentially based on Grigorescu (1976a, b, 1979).

3.4

The results in this section are essentially due to Norman (1968, 1972) and include as special cases previous results by Blackwell (1957b), Doeblin & Fortet (1937), Fortet (1938), Ionescu Tulcea (1959), Kennedy (1957), and Onicescu & Mihoc (1935a, b). For another approach to the study of the ergodicity of RSCCs see Kaijser (1978a, b, c, 1981a, b), who uses coupling techniques.

4.1

The pertinent references for the law of large numbers for RSCCs are Cohn (1965a), Iosifescu (1964, 1965a), and Norman (1968). Subsection 4.1.3 is essentially based on Popescu (1977, 1978), except for some simplifications and improvements. Corollary 4.1.6 and the two last statements in Corollary 4.1.7 were proved in a direct manner by Cohn (1966, 1968), while Theorem 4.1.12 was proved by Iosifescu in 1966 (see Iosifescu & Theodorescu (1969, p. 147)). The proof of the latter theorem uses the well-known blocking technique of Bernstein (1926). The study of the asymptotic behaviour of the infinite order chain associated with an RSCC was also undertaken – in a different manner – by Norman (1972), who proved the central limit theorem, although under stronger assumptions.

Classical and functional limit theorems for the infinite order chain associated with both an RSCC and a GRSCC were proved – in a unitary manner, but under somewhat restrictive assumptions – by Grigorescu & Oprişan (1976), who used the concept of a $J-X$ process, introduced (in a different setting) by Janssen (1971) and subsequently developed by O'Brien (1974). Theorem 4.1.16 was proved by Iosifescu (1965a), while Theorem 4.1.17 and Corollary 4.1.18 were proved by Popescu (1977, 1978). It would be interesting to compare the results in the last-mentioned corollary with those due to Botez (1965).

4.2

Theorem 4.2.1 was proved by Norman (1968). A version of Theorem 4.2.7 without remainder was proved – under slightly different assumptions – by Grigorescu & Popescu (1973), Kaijser (1972), and Norman (1972, pp. 75–80); a version with remainder giving a non-optimal convergence rate was subsequently obtained by Popescu (1978). Another approach to the central limit theorem with remainder for the associated MC, also giving a non-optimal convergence rate, is to be found in Kaijser (1979). The derivation of Theorem 4.2.7 here is based on Iosifescu (1987c), who adapted a long-established method to the present context (see the references given in the proof of Theorem 4.2.4) whose initiators should be considered to be Onicescu & Mihoc (1939, 1940a).

For recent work on the central limit theorem with remainder for different classes of MCs see Gudinas (1982a, b).

5.1

As has already been pointed out, the OM chains were, historically speaking, the first RSCCs studied. The basic references are Onicescu & Mihoc (1935a, b, 1936a, b, 1937, 1943) and Fortet (1938), who also introduced the nomenclature OM chain. The associated MC was defined by Onicescu & Mihoc (1936b). The first central limit theorem was proved by Onicescu & Mihoc (1940b) for alternate OM chains, while its extension to the case $m > 1$ is due to Iosifescu (1961). See also Mihoc & Ciucu (1961, 1973). It was also Iosifescu (1963a) who defined the OM chains with an arbitrary state space.

5.2

Whole sections or chapters on the metric theory of continued fractions can be found in the books by Billingsley (1965), Ibragimov & Linnik (1971), Khinchin (1964), and Lévy (1954). The exposition in this section follows Iosifescu (1974, 1978), who clarified Doeblin's idea (see Doeblin (1940)) of using dependence with complete connections to study the metric properties of continued fractions. Subsection 5.2.2 is a transcription into the language of dependence with complete connections of the *ad hoc* method used by Lévy (1929) to solve Gauss' problem. The same applies to Subsection 5.2.3 which is based on work by Kuzmin (1928).

For recent results concerning classical and functional limit theorems for quantities related to the continued fraction expansion see Heinrich (1987) and Samur (1987).

For recent work on different types of continued fraction expansions using a dependence-with-complete-connections approach see Kalpazidou (1986d, e, 1987a, b).

Comment to pages 174-190: The topic treated here has been greatly expanded in M. Iosifescu and C. Kraaikamp, *Metric Theory of Continued Fractions*, Kluwer Academic Publishers, Dordrecht-Boston-London, 2002.

5.3–5.4

The first pw.m.t.s studied were the f-expansions, which, as was noted, were introduced by Kakeya (1924). Another early paper, in which a pw.m.t. is called a zig-zag function, is Holladay (1957), where, under stringent assumptions, the existence of an invariant probability is already proved. Rényi (1957) initiated the recent developments. In particular, he proved Theorem 5.3.5 (though only for f-expansions).

The Frobenius–Perron operator of a non-singular transformation of an abstract measure space was introduced by Rechard (1956). The results in Subsection 5.3.3 are based essentially on the work by Rychlik (1983), except for some simplifications and clarifications mainly due to the use of the Ionescu Tulcea–Marinescu Theorem A2.4 (unlike Rychlik, who uses the uniform ergodic theorem). It should be noted that the idea of using the technique of bounded variation in the study of pw.m.t.s is due to Lasota & Yorke (1973). The relevance of the Ionescu Tulcea–Marinescu Theorem A2.4 in this context was first pointed out by Keller (1979) (see also Keller (1982)). Corollary 5.3.13 (as well as Theorems 5.3.22, 5.3.23, and 5.3.24) supersedes many results previously obtained under stronger assumptions.

The study of the differential properties of the invariant probability was initiated by Halfant (1977). The results in Subsection 5.3.4 come from Iosifescu (1988), and are stronger and more general than Halfant's results, due to the use of the functional–theoretic method. As to the remark concluding Subsection 5.3.4, it should be mentioned that the similarity of the pw.m.t. and RSCC settings was first noticed by Sacksteder (1972).

Theorems 5.3.22 and 5.3.23 are due to Hofbauer & Keller (1982), and Theorem 5.3.24 to Rousseau-Egèle (1983). The elegant proof of Theorem 5.3.22, reproduced here, is due to Rychlik (1983).

Important references for the metric theory of f-expansions are Aaronson (1986) (see also Spătaru (1987)), Gordin (1968), Parry (1964), Tran Vinh-Hien (1963), and Walters (1978a). For other aspects of the metric problems occurring in different representations of real numbers, the reader may consult Galambos (1976), Philipp (1971), Rényi (1958), Schweiger (1981), and Waterman (1975).

The results in Subsection 5.4.2 are due to Iosifescu (1974, 1987a).

Comment to page 222, lines 19-31: A solution to this conjecture has been recently sketched by Adriana Berechet, Solving a conjecture about certain f-expansions, *Proc. Romanian Acad. Ser. A Math. Phys. Tech. Sci. Inf. Sci.* **5**(2004), 231-35.

5.5

The basic references for strict-sense infinite-order chains are Doeblin & Fortet (1937), Harris (1955), and Iosifescu & Spătaru (1973). (Note that the publication years of these papers form an arithmetic series!) Theorem 5.5.1 was proved by Iosifescu (1966d) as a generalization of a theorem by Ionescu Tulcea (1959) stated under Conditions FLS$(X, 1)$ and $K_1(X, 0)$. See, however, Problem 13 at the end of this chapter.

Subsection 5.5.3 follows Iosifescu & Spătaru (1973), while Subsection 5.5.4 presents (corrected) results due to Harris (1955) (see Spătaru (1974)).

For other existence theorems for strict-sense infinite-order chains we refer the reader to Kalpazidou (1985b, c, 1986a).

For other aspects of the infinite-order-chain theory we refer the reader to Berbee (1987), Blackwell (1957a, b), Gudinas (1982b), Keane (1972), Lalley (1986) and Ledrappier (1976).

Finally, we wish to highlight several subjects which have not been included in this book.

The most important among them is the statistical inference for RSCCs; this is a field in which there is still a lot to be done, though the recent work of Pruscha (1986) is an important breakthrough. The interested reader may also consult Andrieu & Flavigny (1974), Botez (1966a, b), Bui Trong Lieu & Flavigny (1974), Bui Trong Lieu & Theodorescu (1974), Gorunescu (1981), Mihoc & Craiu (1972), Pruscha (1975, 1976), and Pruscha & Maurus (1979).

Another subject we have not looked at is the theory of controlled RSCCs. This is because there is already a monograph on it by Zidăroiu (1975). See also Zidăroiu (1976, 1978, 1986, 1987).

The dependence with complete connections in the case of a continuous parameter is not yet a consistent theory; which is why we have not referred to it. However, let us mention some pertinent references: Ciucu (1962), Iosifescu (1967, 1968a), Le Calvé & Theodorescu (1973), Onicescu (1954), Onicescu, Oprişan & Popescu (1983), and Pruscha (1983). Again, the last reference might be a genuine breakthrough.

Last but not least, few of the results by T. Kaijser are presented in this book. This is because we feel that his papers on dependence with complete connections deserve to be expanded into a monograph on their own.

Appendix 2

Comment to page 262, lines 27-31: As was shown by H. Hennion, Sur un théorème spectral et son application aux noyaux lipschitziens, *Proc. Amer. Math. Soc.* **118** (1993), 627-34, Condition (ITM_1) is superfluous. Thus, line 28 should stop after a linear subspace of X, and the remaining lines should be ignored.

References

Aaronson, J. (1986) Random f-expansions. *Ann. Probab.* **14**, 1037–57.

Adams, W. W. (1979) On a relationship between the convergents of the nearest integer and regular continued fractions. *Math. Comp.* **33**, 1321–31.

Adler, R. L. (1975) Continued fractions and Bernoulli trials. *Ergodic Theory*, pp. 111–20. Courant Inst. Math. Sci., New York Univ., New York.

Adler, R. L. (1979) Afterword. *Comm. Math. Phys.* **69**, 15–7.

Andrieu, Colette, Bui Trong Lieu, Flavigny, R. & Langrand, C. (1973) Sur le comportement asymptotique de systèmes aléatoires généralisés à liaisons complètes non homogènes. *Studia Math.* **49**, 51–67.

Andrieu, Colette & Flavigny, R. (1973) Certains aspects du comportement asymptotique de systèmes aléatoires généralisés à liaisons complètes. *C.R. Acad. Sci. Paris Sér. A–B*, **276**, A1233–5.

Andrieu, Colette & Flavigny, R. (1974) Une méthode de moindres carrés pour des systèmes aléatoires généralisés à liaisons complètes. *C.R. Acad. Sci. Paris Sér. A–B*, **278**, A965–7.

Araujo, A. & Giné, E. (1980) *The Central Limit Theorem for Real and Banach Valued Random Variables.* Wiley, New York.

Arkin, V. I. & Evstigneev, I. V. (1987) *Stochastic Models of Control and Economic Dynamics.* Academic Press, London.

Babenko, K. I. (1978) On a problem of Gauss. *Soviet Math. Dokl.* **19**, 136–40.

Babenko, K. I. & Jur'ev, S. P. (1978) On the discretization of a problem of Gauss. *Soviet Math. Dokl.* **19**, 731–5.

Barnsley, M. F. & Demko, S. G. (Eds.) (1986) *Chaotic Dynamics and Fractals.* Academic Press, Orlando, Florida.

Baumgärtel, H. (1985) Analytic Perturbation Theory for Matrices and Operators. *Operator Theory: Advances and Applications*, Vol. 15. Birkhäuser, Basel.

Bebutov, M. V. (1942) Markov chains with a compact state space. *Mat. Sb.* **10(52)**, 213–38.

Berbee, H. (1987) Chains with infinite connections: uniqueness and Markov representation. *Probab. Theory Relat. Fields*, **76**, 243–53.

Bergmann, R. & Stoyan, D. (1978) Monotonicity properties of second order characteristics of stochastically monotone Markov chains. *Math. Nachr.* **82**, 99–102.

Bergström, H. (1982) *Weak Convergence of Measures.* Academic Press, New York.

Bernstein, S. N. (1926) Sur l'extension du théorème limite du calcul des probabilités aux sommes de quantités dépendantes. *Math. Ann.* **97**, 1–59.

Bert, Marie-Claude (1968) Convergence d'un processus à liaisons complètes, et application à un processus d'apprentissagee linéaire. *Ann. Inst. H. Poincaré Sect. B* (*N.S.*), **4**, 1–24.

Bert, Marie-Claude, Boudreau, J.-R. & Theodorescu, R. (1985) Sur la stationnarité de certains processus d'apprentissage. *Bull. Sci. Math.* (*2*), **109**, 431–45.

Bert, Marie-Claude, Herkenrath, U. & Theodorescu, R. (1984) Mesures de probabilités stationnaires pour une classe de modèles non-markoviens. *Bull. Sci. Math.* (*2*), **108** 113–27.

Bežaeva, Z. I. (1971) Limit theorems for conditional Markov chains. *Theory Probab. Appl.* **16**, 429–37.

Bežaeva, Z. I. (1974) Ergodic properties of conditional Markov chains. *Theory Probab. Appl.* **19**, 522–31.

Bežaeva, Z. I. (1977) Conditional Markov chains with countable state space. *Theory Probab. Appl.* **22**, 543–53.

Billingsley, P. (1965) *Ergodic Theory and Information.* Wiley, New York.

Billingsley, P. (1968) *Convergence of Probability Measures.* Wiley, New York.

Bissinger, B. H. (1944) A generalization of continued fractions. *Bull. Amer. Math. Soc.* **50**, 868–76.

Blackwell, D. (1957a) On discrete variables whose sum is absolutely continuous. *Ann. Math. Statist.* **28**, 520–1.

Blackwell, D. (1957b) The entropy of functions of finite-state Markov chains. *Trans. First Prague Conf. Information Theory, Statist. Decision Functions, Random Processes,* (*Liblice, 1956*), pp. 13–20. Czechoslovak Acad. Sci., Prague.

Blackwell, D. (1965) Discounted dynamic programming. *Ann. Math. Statist.* **36**, 226–35.

Borel, E. (1909) Les probabilités dénombrables et leurs applications arithmétiques. *Rend. Circolo Mat. Palermo,* **27**, 247–71.

Botez, M. (1965) Sur quelques problèmes non paramétriques pour les systèmes aléatoires à liaisons complètes. *Rev. Roumaine Math. Pures Appl.* **10**, 1437–45.

Botez, M. (1966a) On a method of estimation for systems with complete connections. *Stud. Cerc. Mat.* **18**, 75–86. (Romanian)

Botez, M. (1966b) Remarques sur quelques types de lois limites. *Bull. Math. Soc. Sci. Math. R.S. Roumanie* (*N.S.*), **10(58)**, 203–14.

Böttcher, G. & Nawrotzki, K. (1976) Stationäre Anfangsverteilungen stochastischer Automaten II. *Elektron. Informationsverarb. Kybernet.* **12**, 459–70.

Boudiba, M. A. (1986) La chaîne de Feller $X_{n+1} = |X_n - Y_{n+1}|$ et les chaines associées. *Ann. Sci. Univ. Clermont-Ferrand II Probab. Appl.* No. 5, 91–132.

Bougerol, P. & Lacroix, J. (1985) Products of Random Matrices with Applications to Schrödinger Operators. *Progress in Probability and Statistics,* Vol. 8. Birkhäuser, Basel.

Bowen, R. (1979) Invariant measures for Markov maps of the interval. *Comm. Math. Phys.* **69**, 1–14.

Bradley, R. C. (1986) Basic properties of strong mixing conditions. In: Eberlein & Taqqu (Eds.) (1986), pp. 165–92.

Brandt, A. (1986) The stochastic equation $Y_{n+1} = A_n Y_n + B_n$ with stationary coefficients. *Adv. in Appl. Probab.* **18**, 211–20.

Breiman, L. (1960) A strong law of large numbers for a class of Markov chains. *Ann. Math. Statist.* **31**, 801–3.

Brodén, T. (1900) Wahrscheinlichkeitsbestimmungen bei der gewöhnlichen Kettenbruchentwicklung reeller Zahlen. *Öfversigt af Kongl. Ventenskaps-Akademiens Förhandlingar,* **57**, 239–66.

Broughton, A. & Huff, B. W. (1977) A comment on unions of sigma-fields. *Amer. Math. Monthly,* **84**, 553–4.

Bruneau, M. (1974) Variation totale d'une fonction. *Lecture Notes in Math.,* Vol. 413. Springer, Berlin.

Brunel, A. & Revuz, D. (1974) Quelques applications probabilistes de la quasi-compacité. *Ann. Inst. H. Poincaré Sect. B* (*N.S.*), **10**, 301–37.

Bugiel, P. (1985) A note on invariant measures for Markov maps of an interval. *Z. Wahrsch. Verw. Gebiete,* **70**, 345–9.

Bugiel, P. (1987) Correction and addendum to Bugiel (1985). *Probab. Theory Relat. Fields,* **76**, 255–6.

Bui Trong Lieu & Flavigny, R. (1974) Fonctions aléatoires attachées à un système aléatoire généralisé à liaisons complètes et application à un problème d'estimation. *C.R. Acad. Sci. Paris Sér. A–B*, **278**, A33–5.
Bui Trong Lieu & Theodorescu, R. (1974) Estimateurs du quasi-maximum de vraisemblance pour des systèmes aléatoires généralisés à liaisons complètes. *C.R. Acad. Sci. Paris Sér. A–B*, **278**, A901–3.
Bush, R. R. & Mosteller, F. (1951) A mathematical model for simple learning. *Psychol. Rev.* **58**, 313–23.
Bush, R. R. & Mosteller, F. (1955) *Stochastic Models for Learning.* Wiley, New York.
Chevrolet, D. & Le Calvé, G. (1973) Essai de formalisation de la dynamique de la discussion dans les groupes restreints: de quelques conditions d'apparition d'un comportement limite. *Math. Sci. Humaines*, **11**, no. 1, 13–26.
Chung, K. L. (1964) The general theory of Markov processes according to Doeblin. *Z. Wahrsch. Verw. Gebiete*, **2**, 230–54.
Chung, K. L. (1967) *Markov Chains with Stationary Transition Probabilities*, 2nd edn. Springer, Berlin.
Chung, K. L. (1982) A cluster of great formulas. *Acta Math. Acad. Sci. Hungar.* **39**, 65–7.
Ciucu, G. (1962) L'ergodicità die processi stocastici a vincoli completi. *Atti Accad. Naz. Lincei Rend. Cl. Sci. Fis. Mat. Natur.* (8), **33**, 119–21.
Ciucu, G. & Theodorescu, R. (1960) *Processes with Complete Connections.* Ed. Academiei R. P. Romîne, Bucharest. (Romanian).
Cogburn, R. (1975) A uniform theory for sums of Markov chain transition probabilities. *Ann. Probab.* **3**, 191–214.
Cohen, J. E. (1976) Ergodicity of age structure in populations with Markovian vital rates. I. Countable states. *J. Amer. Statist. Assoc.* **71**, 335–9.
Cohen, J. E. (1977) Ergodicity of age structure in populations with Markovian vital rates. II. General states; III. Finite-state moments and growth rate: an illustration. *Adv. in Appl. Probab.* **9**, 18–37; 462–75.
Cohen, J. E., Kesten, H. & Newman, C. M. (Eds.) (1986) Random Matrices and Their Applications (Brunswick, Maine, 1984). *Contemp. Math.* 50. Amer. Math. Soc., Providence, R. I.
Cohn, H. (1964) Sur les propriétés asymptotiques des'systèmes à liaisons complètes. *Bull. Math. Soc. Sci. Math. Phys. R.P. Roumaine (N.S.)*, **8(56)**, 7–21.
Cohn, H. (1965a) Limit theorems for systems with complete connections. *Stud. Cerc. Mat.* **17**, 757–64. (Romanian)
Cohn, H. (1965b) On a class of dependent random variables. *Rev. Roumaine Math. Pures Appl.* **10**, 1593–606.
Cohn, H. (1966) Limit theorems for stochastic processes. *Stud. Cerc. Mat.* **18**, 993–1027. (Romanian)
Cohn, H. (1968) On the maximum of sums of dependent random variables. *Rev. Roumaine Math. Pures Appl.* **13**, 295–301.
Collet, P. & Eckmann, J.-P. (1980) Iterated Maps on the Interval as Dynamical Systems. *Progress in Physics*, Vol. 1. Birkhäuser, Basel.
Cornfeld, I. P., Fomin, S. V. & Sinai, Ya. G. (1982) *Ergodic Theory.* Springer, Berlin.
van Danzig, D. & Sheffer, C. (1954) On arbitrary hereditary time-discrete stochastic processes, considered as Markov chains, and the corresponding general form of Wald's fundamental identity. *Indag. Math.* **16**, 377–88.
Denker, M. & Keller, G. (1986) Rigorous statistical procedures from data from dynamical systems. *J. Statist. Phys.* **44**, 67–93.
Doeblin, W. (1937) Sur les propriétés asymptotiques de mouvements régis par certains types de chaînes simples. *Bull. Math. Soc. Roumaine Sci.* **39**, no. 1, 57–115; no. 2, 3–61.
Doeblin, W. (1938) Exposé de la théorie des chaînes simples constantes de Markoff à un nombre fini d'états. *Rev. Math. de l'Union Interbalkanique*, **2**, 77–105.

Doeblin, W. (1940) Remarques sur la théorie métrique des fractions continues. *Compositio Math.* **7**, 353–71.

Doeblin, W. & Fortet, R. (1937) Sur des chaînes à liaisons complètes. *Bull. Soc. Math. France*, **65**, 132–48.

Doob, J. L. (1938) Stochastic processes with an integral valued parameter. *Trans. Amer. Math. Soc.* **44**, 87–150.

Doob, J. L. (1953) *Stochastic Processes.* Wiley, New York.

Doos, R. (1980) The Hahn decomposition theorem. *Proc. Amer. Math. Soc.* **80**, 377.

Doukhan, P. & Ghindès, M. (1980a) Etudes du processus: '$X_{n+1} = f(X_n)\varepsilon_n$'. *C.R. Acad. Sci. Paris Sér. A–B*, **290**, A921–3.

Doukhan, P. & Ghindès, M. (1980b) Estimations dans le processus '$X_{n+1} = f(X_n) + \varepsilon_n$'. *C.R. Acad. Sci. Paris Sér. A–B*, **291**, A61–4.

Dubins, L. E. & Freedman, D. A. (1966) Invariant probabilities for certain Markov processes. *Ann. Math. Statist.* **37**, 837–48.

Dubins, L. E. & Freedman, D. A. (1967) Random distribution functions. *Proc. Fifth Berkeley Sympos. Math. Statist. Probab.*, Vol. II, Part 1, pp. 183–214. Univ. California Press, Berkeley, California.

Dudley, R. M. (1966) Convergence of Baire measures. *Studia Math.* **27**, 251–68.

Dudley, R. M. (1976) Convergence of laws on metric spaces, with a view to statistical testing. *Lecture Notes Series*, No. 45. Matematisk Institut, Aarhus Univ., Aarhus.

Dunford, N. & Schwartz, J. T. (1958) *Linear Operators, Part I: General Theory.* Interscience, New York.

Dynkin, E. B. (1965) Controlled random sequences. *Theory Probab. Appl.* **10**, 1–14.

Eberlein, E. & Taqqu, M. S. (Eds.) (1986) Dependence in Probability and Statistics: A Survey of Recent Results (Oberwolfach, 1985). *Progress in Probability and Statistics*, Vol. 11. Birkhäuser, Boston.

Eckmann, J.-P. (1981) Roads to turbulence in dissipative dynamical systems. *Rev. Mod. Phys.* **53**, 643–54.

Eremin, I. I. & Mazurov, V. D. (1979) *The Nonstationary Processes of Mathematical Programming.* Nauka, Moscow. (Russian)

Ermakov, S. M. (1975) *The Monte Carlo Method and Related Questions*, 2nd edn. Nauka, Moscow. (Russian)

Estes, W. K. (1950) Toward a statistical theory of learning. *Psychol. Rev.* **57**, 94–107.

Everett, C. J., Jr. (1946) Representations for real numbers. *Bull. Amer. Math. Soc.* **52**, 861–69.

Feller, W. (1966) *An Introduction to Probability Theory and Its Applications*, Vol. 2. Wiley, New York.

Findeisen, P. (1980) Asymptotic properties of a certain class of Bush–Mosteller learning models. *Adv. in Appl. Probab.* **12**, 922–41.

Flerov, Ju. A. (1976) The limit behaviour and asymptotic optimality of stochastic automata. *Studies in the Theory of Adaptive Systems*, pp. 25–46. Vyčisl. Centr, Akad. Nauk SSSR, Moscow. (Russian)

Foguel, S. R. (1973) The ergodic theory of positive operators on continuous functions. *Ann. Scuola Norm. Sup. Pisa Cl. Sci. (4)*, **27**, 19–51.

Fortet, R. (1938) Sur l'itération des substitutions algébriques linéaires à une infinité de variables et ses applications à la théorie des probabilités en chaîne. *Rev. Ci. (Lima)*, **40**, 185–261; 337–447; 481–528.

Fortet, R. (1940) Sur une suite également répartie. *Studia Math.* **9**, 54–70.

Fortet, R. (1978) Condition de Doeblin et quasi-compacité. *Ann. Inst. H. Poincaré Sect. B (N.S.)*, **14**, 379–90.

Fréchet, M. (1934a) Sur l'allure asymptotique des densités itérées dans le problème des probabilités 'en chaîne'. *Bull. Soc. Math. France*, **62**, 68–83.

Fréchet, M. (1934b) Sur l'allure asymptotique de la suite des itérés d'un noyau de Frédholm. *Quart. J. Math.* **5**, 106–44.

Galambos, J. (1976) Representations of Real Numbers by Infinite Series. *Lecture Notes in Math.*, Vol. 502. Springer, Berlin.

Gihman, I. I. & Skorohod, A. V. (1979) *Controlled Stochastic Processes.* Springer, New York.

Góra, P. (1984) On small stochastic perturbations of mappings of the unit interval. *Colloq. Math.* **49**, 73–85.

Góra, P. (1985) Random composing of mappings, small stochastic perturbations and attractors. *Z. Wahrsch. Verw. Gebiete*, **69**, 137–60.

Gordin, M. I. (1968) On random processes generated by number-theoretic endomorphisms. *Soviet Math. Dokl.* **9**, 1234–7.

Gordin, M. I. (1971a) The behaviour of the dispersion of sums of random variables that generate a stationary process. *Theory Probab. Appl.* **16**, 474–84.

Gordin, M. I. (1971b) Exponentially fast mixing. *Soviet Math. Dokl.* **12**, 331–5.

Gordin, M. I. & Reznik, M. H. (1970) The law of the iterated logarithm for the denominators of continued fractions. *Vestnik Leningrad. Univ.* **25**, no. 13, 28–33. (Russian)

Gorunescu, F. (1981) The method of the maximum likelihood for dependent variables connected in a random system with complete connections. *Rev. Roumaine Math. Pures Appl.* **26**, 417–20.

Grauert, H. & Lieb, I. (1968) *Differential- und Integralrechnung*, III: *Integrationstheorie. Kurven- und Flächenintegrale.* Springer, Berlin.

Gray, J. J. (1984) A commentary on Gauss's mathematical diary, 1796–1814, with an Engligh translation. *Exposition Math.* **2**, 97–130.

Grey, D. R. (1980) The fair charge on a car ferry. *J. Appl. Probab.* **17**, 645–61.

Grigorescu, S. (1975) Notes on the theory of random systems with complete connections. *Lecture Notes.* Dept. of Math., Wales University, Univ. College of Swansea, Swansea.

Grigorescu, S. (1976a) Ergodic decomposition for continuous Markov chains. *Rev. Roumaine Math. Pures Appl.* **21**, 683–98.

Grigorescu, S. (1976b) On the tail σ-algebra for a class of Markov chains. *Stud. Cerc. Mat.* **28**, 439–44. (Romanian)

Grigorescu, S. (1977a) On a class of generalized OM chains. *Rev. Roumaine Math. Pures Appl.* **22**, 1411–4.

Grigorescu, S. (1977b) A mathematical model of the diplomatic talks. *Mathematical Approaches to International Relations*, pp. 371–7. Romanian Academy of Social and Political Sciences, Bucharest.

Grigorescu, S. (1979) Contributions to the theory of generalized random systems with complete connections. I, II. *Stud. Cerc. Mat.* **31**, 277–310; 399–426. (Romanian)

Grigorescu, S. & Oprişan, G. (1976) Limit theorems for J-X processes with a general state space. *Z. Wahrsch. Verw. Gebiete*, **35**, 65–73.

Grigorescu, S. & Popescu, G. (1973) A central limit theorem for a class of Markov chains. *Proc. Fourth Conf. Probability Theory* (*Braşov, 1971*), pp. 153–66. Ed. Academiei R. S. România, Bucharest.

Grigorescu, S. & Popescu, G. (1987) Limit theorems for renewal generalized processes with complete connections. *Rev. Roumaine Math. Pures Appl.* **32**, 687–91.

Grincevičius, A. (1974) A central limit theorem for the group of linear transformations of the real axis. *Soviet Math. Dokl.* **15**, 1512–5.

Grincevičius, A. (1978) An approximation in variation of the distributions of the products of random linear transformations of the line. *Lithuanian Math. J.* **18**, 183–90.

Grincevičius, A. (1981) On a random difference equation. *Lithuanian Math. J.* **21**, 302–5.

Gudinas, P. P. (1982a) Refinements of the central limit theorem for a homogeneous Markov chain. *Lithuanian Math. J.* **22**, 36–45.

Gudinas, P. P. (1982b) B-regularity of a homogeneous Markov chain. *Lithuanian Math. J.* **22**, 265–75.

Guivarc'h, Y. (1980) Exposants de Liapounoff des marches aléatoires à pas markovien. *Séminaire de Probabilités, Rennes, 1980*, Exp. No. 1, 17 pp. Univ. de Rennes, Rennes.

Halfant, M. (1977) Analytic properties of Rényi's invariant density. *Israel J. Math.* **27**, 1–20.
Halfin, S. (1975) Explicit construction of invariant measures for a class of continuous state Markov processes. *Ann. Probab.* **3**, 859–64.
Hall, P. & Heyde, C. C. (1980) *Martingale Limit Theory and Its Application.* Academic Press, New York.
Halmos, P. R. (1950) *Measure Theory.* Van Nostrand, New York. (Reprinted 1974 by Springer, New York.)
Harris, T. E. (1955) On chains of infinite order. *Pacific J. Math.* **5**, 707–24.
Heinrich, L. (1987) Rates of convergence in stable limit theorems for sums of exponentially ψ-mixing random variables with an application to metric theory of continued fractions. *Math. Nachr.* **131**, 149–65.
Herkenrath, U. (1977) Markov processes under continuity assumptions. *Rev. Roumaine Math. Pures Appl.* **22**, 1419–31.
Herkenrath, U. (1979) The associated Markov process of generalized random systems with complete connections. *Rev. Roumaine Math. Pures Appl.* **24**, 243–54.
Herkenrath, U. (1980a) *Eine Klasse von Lernmodellen–Selektive Verfahren und assozierte stochastische Prozesse.* Habilitationsschrift, Universität Bonn, Bonn.
Herkenrath, U. (1980b) *On market-pricing models.* Preprint No. 358, SFB72 (Approximation und Optimierung), Universität Bonn, Bonn.
Herkenrath, U., Kalin, D. & Lakshmivarahan, S. (1981) On a general class of absorbing-barrier learning algorithms. *Inform. Sci.* **24**, 225–63.
Herkenrath, U. & Theodorescu, R. (1978a) On certain aspects of the two-armed bandit problem. *Elektron Informationsverarb. Kybernet.* **14**, 527–35.
Herkenrath, U. & Theodorescu, R. (1978b) General control systems. *Inform. Sci.* **14**, 57–73.
Herkenrath, U. & Theodorescu, R. (1978c) Expediency and optimality for general control systems. In: 'Théorie de l' information. Développements récents et applications (Cachan, 1977)'. *Colloques internationaux du CNRS*, No. 276, pp. 447–55. CNRS, Paris.
Herkenrath, U. & Theodorescu, R. (1980) Sur l'utilisation des procédures d'apprentissage dans un problème de formation des prix sur le marché. *Publ. Económétriques*, **13**, no. 2, 21–9.
Herkenrath, U. & Theodorescu, R. (1981) A mathematical framework for learning and adaption: (generalized) random systems with complete connections. *Trabajos Estadist. Investigación Oper.* **32**, 116–29.
Heyde, C. C. & Scott, D. J. (1973) Invariance principles for the law of the iterated logarithm for martingales and processes with stationary increments. *Ann. Probab.* **1**, 428–36.
Hill, B., Lane, D. & Sudderth, W. (1980) A strong law for some generalized urn processes. *Ann. Probab.* **8**, 214–26.
Hinderer, K. (1970) Foundations of Non-Stationary Dynamic Programming with Discrete Time Parameter. *Lecture Notes in Oper. Research and Math. Systems*, Vol. 33. Springer, Berlin.
Hinderer, K. (1973) On the stability of a dynamic stochastic production and inventory system controlled by an optimal policy. *Z. Operations Res. Ser. A–B*, **17**, 217–37.
Hofbauer, F. (1986a) Piecewise invertible dynamical systems. *Probab. Theory Relat. Fields*, **72**, 359–86.
Hofbauer, F. (1986b) *Generic properties of invariant measures for simple piecewise monotonic transformations.* Preprint, Inst. f. Math., Universität Wien, Wien.
Hofbauer, F. & Keller, G. (1982) Ergodic properties of invariant measures for piecewise monotonic transformations. *Math. Z.* **180**, 119–40.
Högnäs, G. (1978) A note on the product of random elements of a semi-group. *Monatsh. Math.* **85**, 317–21.
Holden, A. V. (Ed.) (1986) *Chaos.* Manchester Univ. Press, Manchester.

Holladay, J. C. (1957) On the existence of a mixing measure. *Proc. Amer. Math. Soc.* **8**, 887–93.
de Hoog, F. R., Brown, A. H. D., Saunders, I. W. & Westcott, M. (1986) Numerical calculation of the stationary distribution of a Markov chain in genetics. *J. Math. Anal. Appl.* **115**, 181–91.
Hostinsky, B. (1931) Méthodes générales du calcul des probabilités. *Mém. Sci. Math.* 52. Gauthier-Villars, Paris.
Ibragimov, I. A. & Linnik, Yu. V. (1971) *Independent and Stationary Sequences of Random Variables.* Wolters-Noordhoff, Groningen.
Ikegami, G. (Ed.) (1986) *Dynamical Systems and Nonlinear Oscillations.* World Scientific Publishing Co., Singapore.
Ionescu Tulcea, C. T. (1959) On a class of operators occurring in the theory of chains of infinite order. *Canad. J. Math.* **11**, 112–21.
Ionescu Tulcea, C. T. & Marinescu, G. (1948) Sur certaines chaînes à liaisons complètes. *C.R. Acad. Sci. Paris,* **227**, 667–9.
Ionescu Tulcea, C. T. & Marinescu, G. (1950) Théorie érgodique pour des classes d'opérations non complètement continues. *Ann. of Math.* (2), **52**, 140–7.
Iordache, O. (1987) *Polystochastic Models in Chemical Engineering.* VNU Science Press, Utrecht & Ed. Academiei, Bucharest.
Iordache, O. & Iosifescu, M. (1981) A discrete stochastic model with applications to mixing. *Proc. Sixth Conf. Probability Theory* (*Braşov, 1979*), pp. 307–12. Ed. Academiei R. S. România, Bucharest.
Iosifescu, M. (1961) On the asymptotic behaviour of chains with complete connections. *Com. Acad. R. P. Romîne,* **11**, 619–24. (Romanian)
Iosifescu, M. (1963a) Random systems with complete connections with an arbitrary set of states. *Rev. Math. Pures Appl.* **8**, 611–45; Addenda *ibid.* **9** (1964), 91–2. (Russian; English translation obtainable from Addis Translations, Menlo Park, California)
Iosifescu, M. (1963b) Sur l'érgodicité uniforme des systèmes aléatoires homogènes à liaisons complètes à un ensemble quelconque d'états. *Bull. Math. Soc. Sci. Math. Phys. R.P. Roumaine* (*N.S.*), **7(55)**, 177–88.
Iosifescu, M. (1964) Sur la loi forte des grands nombres pour les systèmes aléatoires homogènes à liaisons complètes à un ensemble quelconque d'états. *C.R. Acad. Sci. Paris,* **258**, 4421–3.
Iosifescu, M. (1965a) Quelques propriétés asymptotiques des systèmes aléatoires homogènes à liaisons complètes. *Rev. Roumaine Math. Pures Appl.* **10**, 635–41.
Iosifescu, M. (1965b) Systèmes aléatoires non-homogènes à liaisons complètes à un ensemble quelconque d'états. *Bull. Math. Soc. Sci. Math. R.S. Roumanie* (*N.S.*), **9(57)**, 13–29.
Iosifescu, M. (1966a) Conditions nécessaires et suffisantes pour l'ergodicité uniforme des chaînes de Markoff variables et multiples. *Rev. Roumaine Math. Pures Appl.* **11**, 325–30.
Iosifescu, M. (1966b) On the uniform ergodicity of a class of non-homogeneous random systems with complete connections. *Rev. Roumaine Math. Pures Appl.* **11**, 763–72.
Iosifescu, M. (1966c) Some asymptotic properties of the associated system to a random system with complete connections. *Rev. Roumaine Math. Pures Appl.* **11**, 973–8.
Iosifescu, M. (1966d) Un théorème d'existence pour les chaînes d'ordre infini. *Atti Accad. Naz. Lincei Rend. Cl. Sci. Fis. Mat. Natur.*(8), **40**, 211–4.
Iosifescu, M. (1967) Systèmes aléatoires à liaisons complètes à paramètre continu. *Rev. Roumaine Math. Pures Appl.* **12**, 1289–92.
Iosifescu, M. (1968a) Processus aléatoires à liaisons complètes purement discontinus. *C.R. Acad. Sci. Paris Sér. A–B,* **266**, A1159–61.
Iosifescu, M. (1968b) The law of the iterated logarithm for a class of dependent random variables. *Theory Probab. Appl.* **13**, 304–13; Addendum *ibid.* **15**(1970), 160.
Iosifescu, M. (1972) On Strassen's version of the loglog law for some classes of dependent random variables. *Z. Wahrsch. Verw. Gebiete,* **24**, 155–8.
Iosifescu, M. (1974) On the application of random systems with complete connections to

the theory of f-expansions. In: 'Progress in Statistics (European Meeting of Statisticians, Budapest, 1972)'. *Colloq. Math. Soc. János Bolyai*, Vol. 9, pp. 335–65. North–Holland, Amsterdam.

Iosifescu, M. (1976) On two remarkable examples of random systems with complete connections whose mappings are not distance diminishing. *Rev. Roumaine Math. Pures Appl.* **21**, 707–13.

Iosifescu, M. (1977) A Poisson law for ψ-mixing sequences establishing the truth of a Doeblin statement. *Rev. Roumaine Math. Pures Appl.* **22**, 1441–7.

Iosifescu, M. (1977/8) Notes on dependence with complete connections. *Lecture Notes*, Fachbereich Mathematik, Johann Gutenberg Universität, Mainz. (Revised version reprinted in: Sequential Methods in Statistics (Papers XVIIIth Semester, Stefan Banach Internat. Math. Center, Warsaw, 1981). *Banach Center Publ.* 16, pp. 245–62. PWN, Warsaw, 1985.)

Iosifescu, M. (1978) Recent advances in the metric theory of continued fractions. *Trans. Eighth Prague Conf. Information Theory, Statist. Decision Functions, Random Processes (Prague, 1978)*, Vol. A, pp. 27–40. Reidel, Dordrecht.

Iosifescu, M. (1979) Notes on the general stochastic model for learning. *Technical Report*, Division of Systems Studies. University of Bucharest, Bucharest. (Revised version reprinted in: *Studies in Probability and Related Topics: Papers in Honour of Octav Onicescu on His 90th Birthday*, pp. 287–300. Editrice Nagard, Roma, 1983.)

Iosifescu, M. (1980a) Recent advances in mixing sequences of random variables. *Third Internat. Summer School Probab. Theory, Math. Statist. (Varna, 1978)*, pp. 111–38. Bulgarian Acad. Sci., Sofia.

Iosifescu, M. (1980b) *Finite Markov Processes and Their Applications*. Wiley, Chichester & Ed. Tehnică, Bucharest.

Iosifescu, M. (1983) Asymptotic properties of learning models. In 'Mathematical Learning Models–Theory and Algorithms'. *Lecture Notes in Statistics*, Vol. 20, pp. 86-92. Springer, New York.

Iosifescu, M. (1985a) Ergodic piecewise monotonic transformations and dependence with complete connections. In 'Probability and Statistical Decision Theory'. *Proc. 4th Pannonian Sympos. Math. Statist. (Bad Tatzmannsdorf, 1983)*, Vol. A, pp. 167–73. Reidel, Dordrecht.

Iosifescu, M. (1985b) f-expansions: a result and a query. *Rev. Roumaine Math. Pures Appl.* **30**, 749–50.

Iosifescu, M. (1985c) Review 60087. *Zbl.* **546**, 287–8.

Iosifescu, M. (1987a) Mixing properties for f-expansions. In 'Probability Theory and Mathematical Statistics'. *Proc. Fourth Vilnius Conf. (Vilnius, 1985)*, Vol. II, pp. 1–8. VNU Science Press, Utrecht.

Iosifescu, M. (1987b) Mixing properties for f-expansions: the bounded p-variation case. In: 'Mathematical Statistics and Probability Theory'. *Proc. 6th Pannonian Sympos. Math. Statist. (Bad Tatzmannsdorf, 1986)*, Vol. A, pp. 195–9. Reidel, Dordrecht.

Iosifescu, M. (1987c) A class of Markov chains with optimal convergence rate in the central limit theorem. *Proc. 4th Colloq. Probab. Theory and Operations Research (Craiova, 1987)*, pp. 89–94. Craiova Univ., Dept. of Natural Sciences, Craiova.

Iosifescu, M. (1988) On invariant probability densities of piecewise monotonic transformations. *Trans. Tenth Prague Conf. Information Theory, Statist. Decision Functions, Random Processes (Prague, 1986)*, Vol. A, pp. 41–54. Reidel, Dordrecht.

Iosifescu, M. (1989) On mixing coefficients for the continued fraction expansion. *Stud. Cerc. Mat.* **41**, 491–9.

Iosifescu, M. & Mandl, P. (1966) Applications des systèmes à liaisons complètes à un problème de réglage. *Rev. Roumaine Math. Pures Appl.* **11**, 533–9.

Iosifescu, M. & Spătaru, A. (1973) On denumerable chains of infinite order. *Z. Wahrsch. Verw. Gebiete*, **27**, 195–214.

Iosifescu, M. & Theodorescu, R. (1969) *Random Processes and Learning*. Springer, Berlin.

Isaac, R. (1962) Markov processes and unique stationary probability measures. *Pacific J. Math.* **12**, 273–86.

Ishitani, H. (1986) A central limit theorem of mixed type for a class of 1-dimensional transformations. *Hiroshima Math. J.* **16**, 161–88.

Ismail, M. E. H. & Theodorescu, R. (1980) Asymptotic and explicit formulae for a non-Markovian model with linear transition rule. *Biom. J.* **22**, 67–71.

Jabłoński, M. & Malczak, J. (1985) The central limit theorem for transformations on the real line. *Univ. Iagel. Acta Math.* No. 25, 195–201.

Jacobs, K. (1960) *Neuere Methoden und Ergebnisse der Ergodentheorie*. Springer, Berlin.

Jacobs, K. (1962) Lecture Notes on Ergodic Theory, I–II. *Lecture Notes Series*, No. 1. Matematisk Institut, Aarhus Univ., Aarhus.

Jager, H. (1985) Metrical results for the nearest integer continued fraction. *Indag. Math.* **47**, 417–27.

Jamison, B. (1964) Asymptotic behaviour of successive iterates of continuous functions under a Markov operator. *J. Math. Anal. Appl.* **9**, 203–14.

Jamison, B. (1965) Ergodic decompositions induced by certain Markov operators. *Trans. Amer. Math. Soc.* **117**, 451–68.

Jamison, B. & Sine, R. (1969) Irreducible almost periodic Markov operators. *J. Math. Mech.* **18**, 1043–57.

Janssen, J. (1971) Les processus (*J-X*). *Cahiers Centre Études Rech. Opér.* **11**, 181–214.

Johnson, N. L. & Kotz, S. (1977) *Urn Models and Their Application: An Approach to Modern Discrete Probability Theory*. Wiley, New York.

Junge, M. & Theodorescu, R. (1983) Stationarität einiger nicht-Markovscher Ketten. *Studies in Probability and Related Topics: Papers in Honour of Octav Onicescu on His 90th Birthday*, pp. 307–20. Editrice Nagard, Roma.

Kac, M. (1938) Sur les fonctions indépendantes. V. *Studia Math.* **7**, 96–100.

Kaijser, T. (1972) *Some limit theorems for Markov chains with applications to learning models and products of random matrices*. Ph.D. Thesis, Institute Mittag-Leffler, Djursholm (Sweden).

Kaijser, T. (1973) A limit theorem for the conditional distributions of Markov chains with incomplete state information. *Report LiTH-MAT-R-73-7*, Dept. of Math., Linköping Univ., Linköping.

Kaijser, T. (1975) A limit theorem for partially observed Markov chains. *Ann. Probab.* **3**, 677–96.

Kaijser, T. (1978a) On weakly distance diminishing random systems with complete connections. *Report LiTH-MAT-R-78-15*, Dept. of Math., Linköping Univ., Linköping.

Kaijser, T. (1978b) A limit theorem for Markov chains in compact metric spaces with applications to products of random matrices. *Duke Math. J.* **45**, 311–49.

Kaijser, T. (1978c) On distance diminishing random chains. *Report LiTH-MAT-R-78-17*, Dept. of Math., Linköping Univ., Linköping.

Kaijser, T. (1979) Another central limit theorem for random systems with complete connections. *Rev. Roumaine Math. Pures Appl.* **24**, 383–412.

Kaijser, T. (1981a) On couplings, Markov chains and random systems with complete connections. *Proc. Sixth Conf. Probability Theory* (*Braşov, 1979*), pp. 139–59. Ed. Academiei R. S. România, Bucharest.

Kaijser, T. (1981b) On a new contraction condition for random systems with complete connections. *Rev. Roumaine Math. Pures Appl.* **26**, 1075–1117.

Kaijser, T. (1983) A note on random continued fractions. *Probability and Mathematical Statistics: Essays in Honour of Carl-Gustav Esseen*, pp. 74–84. Uppsala Univ., Dept. of Math., Uppsala.

Kaijser, T. (1984) On iterations of random mappings of the circle. *Report No. 18*, Institute Mittag-Leffler, Djursholm (Sweden).

Kaijser, T. (1986) A note on random systems with complete connections and their applications to products of random matrices. In: Cohen, Kesten & Newman (Eds.) (1986), pp. 243–54.
Kakeya, S. (1924) On a generalized scale of notations. *Japan J. Math.* **1**, 95–108.
Kalin, D. & Theodorescu, R. (1982) A learning approach to sequential design: finite horizon. *Elektron. Informationsverarb. Kybernet.* **18**, 471–87.
Kalin, D. & Theodorescu, R. (1984) A learning approach to sequential design: infinite horizon. *Statist. Decisions*, **2**, 23–37.
Kalpazidou, Sofia (1985a) On a random system with complete connections associated with the continued fraction to the nearer integer expansion. *Rev. Roumaine Math. Pures Appl.* **30**, 527–37.
Kalpazidou, Sofia (1985b) On some bidimensional denumerable chains of infinite order. *Stochastic Process. Appl.* **19**, 341–57.
Kalpazidou, Sofia (1985c) Denumerable chains of infinite order and Hurwitz expansion. In: 'Selected Papers Presented at the 16th European Meeting of Statisticians (Marburg, 1984)'. *Statist. Decisions*, Suppl. Issue No. 2, 83–7.
Kalpazidou, Sofia (1986a) A class of Markov chains arising in the metrical theory of the continued fraction to the nearer integer expansion. *Rev. Roumaine Math. Pures Appl.* **31**, 877–90.
Kalpazidou, Sofia (1986b) Some asymptotic results on digits of the nearest integer continued fraction. *J. Number Theory*, **22**, 271–9.
Kalpazidou, Sofia (1986c) On nearest continued fractions with stochastically independent and indentically distributed digits. *J. Number Theory*, **24**, 114–25.
Kalpazidou, Sofia (1986d) On a problem of Gauss–Kuzmin type for continued fractions with odd partial quotients. *Pacific J. Math.* **123**, 103–14.
Kalpazidou, Sofia (1986e) A Gaussian measure for certain continued fractions. *Proc. Amer. Math. Soc.* **96**, 629–35.
Kalpazidou, Sofia (1987a) On the entropy of the expansions with odd partial quotients. In 'Probability Theory and Mathematical Statistics'. *Proc. Fourth Vilnius Conf.* (*Vilnius, 1985*), Vol. II, pp. 55–62. VNU Science Press, Utrecht.
Kalpazidou, Sofia (1987b) On the application of dependence with complete connections to the metrical theory of G-continued fractions. *Lithuanian Math. J.* **27**, no. 1, 32–40.
Kantel, I. & Nawrotzki, K. (1975) Stationäre Anfangsverteilungen stochastischer Automaten. I. *Elektron. Informationsverarb. Kybernet.* **11**, 151–64.
Karlin, S. (1953) Some random walks arising in learning models. I. *Pacific J. Math.* **3**, 725–56.
Kato, T. (1976) *Perturbation Theory for Linear Operators*, 2nd edn. Springer, Berlin.
Katok, A. & Kifer, Y. (1986) Random perturbations of transformations of an interval. *J. Analyse Math.* **47**, 193–237.
Keane, M. (1972) Strongly mixing g-measures. *Inventiones Math.* **16**, 309–24.
Keller, G. (1978) Piecewise monotonic transformations and exactness. *Séminaire de Probabilités, Rennes* 1978, Exp. No. 6, pp. 32 Univ. de Rennes, Rennes.
Keller, G. (1979) Ergodicité et mesures invariantes pour les transformations dilatantes par morceaux d'une région bornée du plan. *C.R. Acad. Sci. Paris Sér. A–B*, **289**, A625–7.
Keller, G. (1980) Un théorème de la limite centrale pour une classe de transformations monotones par morceaux. *C.R. Acad. Sci. Paris Sér. A–B*, **291**, A155–8.
Keller, G. (1982) Stochastic stability in some chaotic dynamical systems. *Monatsh. Math.* **94**, 313–33.
Keller, G. (1984) On the rate of convergence to equilibrium in one-dimensional systems. *Comm. Math. Phys.* **96**, 181–93.
Keller, G. (1985) Generalized bounded variation and applications to piecewise monotonic transformations. *Z. Wahrsch. Verw. Gebiete*, **69**, 461–78.
Keller, G. (1988) Markov extensions, zeta functions, and Fredholm theory for piecewise invertible dynamical systems. *Trans. Amer. Math. Soc.* (To appear).

Kelley, J. L. (1955) *General Topology*. Van Nostrand, Princeton, N.J.
Kelly, F. P. (1981) How a group reaches agreement: a stochastic model. *Math. Soc. Sci.* **2**, 1–8.
Kennedy, M. (1957) A convergence theorem for a certain class of Markov processes. *Pacific J. Math.* **7**, 1107–24.
Kerbrat, Y. & Le Calvé, G. (1973a) Classification des états pour un système aléatoire à liaisons complètes. *Rev. Roumaine Math. Pures Appl.* **18**, 1381–91.
Kerbrat, Y. & Le Calvé, G. (1973b) Classification de états pour les systèmes aléatoires généralisés à liaisons complètes. *C.R. Acad. Sci. Paris Sér. A–B*, **277**, A753–4.
Khinchin, A. Ya. (1964) *Continued Fractions*. Univ. of Chicago Press, Chicago.
Kifer, Yu. (1986a) Ergodic Theory of Random Transformations. *Progress in Probability and Statistics*, Vol. 10. Birkhäuser, Boston.
Kifer, Y. (1986b) General random perturbations of hyperbolic and expanding transformations. *J. Analyse Math.* **47**, 111–50.
Kolmogorov, A. N. (1931) Ueber die analytischen Methoden in der Wahrscheinlichkeitsrechnung. *Math. Ann.* **104**, 415–58.
Kosjakin, A. A. & Sandler, E. A. (1972) Ergodic properties of a certain class of piecewise smooth transformations of a segment. *Izv. Vysš. Učebn. Zaved. Matematika*, No. 3(118), 32–40. (Russian).
Kowalski, Z. S. (1979a) Invariant measure for piecewise monotonic transformation has a positive lower bound on its support. *Bull. Acad. Polon. Sci. Sér. Sci. Math.* **27**, 53–7.
Kowalski, Z. S. (1979b) Piecewise monotonic transformations and their invariant measures. *Bull. Acad. Polon. Sci. Sér. Sci. Math.* **27**, 63–9.
Kraaikamp, C. (1987) The distributions of some sequences connected with the nearest integer continued fraction. *Indag. Math.* **49**, 177–91.
Kruskal, W. & Mosteller, F. (1980) Representative sampling, IV: the history of the concept in statistics, 1895–1939. *Internat. Statist. Rev.* **48**, 169–95.
Krylov, N. & Bogoljubov, N. (1937) La théorie générale de la mesure dans son application à l'étude des systèmes dynamiques de la mécanique non linéaire. *Ann. of Math.* (2), **38**, 65–113.
Kukush, A. G. (1986) Stability theorems for the sequences $\eta_{n+1} = f(\eta_n, \xi_{n+1})$ in Banach and metric spaces. *Theory Probab. Math. Statist.* No. 33, 47–57.
Kuzmin, R. O. (1928) On a problem of Gauss. *Dokl. Akad. Nauk SSSR Ser. A*, 375–80. (Russian; French version in *Atti Congr. Internaz. Mat.* (*Bologna, 1928*), Tomo VI, pp. 83–9. Zanichelli, Bologna, 1932.)
Ladohin, V. I. & Moskvin, D. A. (1971) The estimation of the remainder term in the central limit theorem for sums of functions of independent variables and for sums of the form $\sum f(t2^k)$. *Theory Probab. Appl.* **16**, 116–25.
Lakshmivarahan, S. (1981) *Learning Algorithms–Theory and Applications*. Springer, New York.
Lalley, S. P. (1986) Regenerative representation for one-dimensional Gibbs states. *Ann. Probab.* **14**, 1262–71.
Lalley, S. P. (1987) Distribution of periodic orbits of symbolic and Axiom *A* flows. *Adv. in Appl. Math.* **8**, 154–93.
Lamperti, J. & Suppes, P. (1959) Chains of infinite order and their application to learning theory. *Pacific J. Math.* **9**, 739–54; Correction *ibid.* **15** (1965), 1471–2.
Lange, K. (1979) On Cohen's stochastic generalization of the strong ergodic theorem of demography. *J. Appl. Probab.* **16**, 496–504.
Lasota, A. & Mackey, M. C. (1985) *Probabilistic Properties of Deterministic Systems*. Cambridge Univ. Press, Cambridge.
Lasota, A. & Yorke, J. A. (1973) On the existence of invariant measures for piecewise monotonic transformations. *Trans. Amer. Math. Soc.* **186**, 481–8.
Lasota, A. & Yorke, J. A. (1982) Exact dynamical systems and the Frobenius–Perron operator. *Trans. Amer. Math. Soc.* **273**, 375–84.

Le Calvé, G. (1967) *Propriétés asymptotiques des processus stochastiques d'adaptation fonctionelle*. Thèse de 3e cycle, Univ. de Rennes, Faculté des Sciences, Rennes.
Le Calvé, G. (1969) Systèmes aléatoires à liaisons complètes et processus d'adaptation. *Rev. Roumaine Math. Pures Appl.* **14**, 1483–1509.
Le Calvé, G. & Theodorescu, R. (1967) Systèmes aléatoires à liaisons complètes totalement non homogènes. *C.R. Acad. Sci. Paris Sér. A–B*, **265**, A347–9.
Le Calvé, G. & Theodorescu, R. (1971) Systèmes aléatoires généralisés à liaisons complètes. *Z. Wahrsch, Verw. Gebiete*, **19**, 19–28.
Le Calvé, G. & Theodorescu, R. (1973) Systèmes aléatoires généralisés à liaisons complètes à temps continu. I–III. *Atti. Acad. Naz. Lincei Rend. Cl. Sci. Fis. Mat. Natur.*, (8) **54**, 434–40, 577–83, 898–901.
Ledrappier, F. (1976) Sur la condition de Bernoulli faible et ses applications. In: 'Théorie ergodique'. *Lecture Notes in Math.*, Vol. 532, pp. 152–9. Springer, Berlin.
Ledrappier, F. (1981) Some properties of absolutely continuous invariant measures on an interval. *Ergodic Theory Dynamical Systems*, **1**, 77–93.
Le Page, E. (1982) Théorèmes limites pour les produits de matrices aléatoires. In: 'Probability Measures on Groups (Proceedings, Oberwolfach, 1981)'. *Lecture Notes in Math.*, Vol. 928, pp. 258–303. Springer, Berlin.
Lesigne, E. (1979) Existence et unicité d'une probabilité invariante pour des translations aléatoires de R^n. *Séminaire de Probabilités, Rennes* 1979, Exp. No. 6, 10 pp. Univ. de Rennes, Rennes.
Letac, G. (1986) A contraction principle for certain Markov chains and its applications. In: Cohen, Kesten & Newman (Eds.) (1986), pp. 263–74.
Lévy, P. (1929) Sur les lois de probabilités dont dépendent les quotients complets et incomplets d'une fraction continue. *Bull. Soc. Math. France*, **57**, 178–94.
Lévy, P. (1954) *Théorie de l'addition des variables aléatoires*, 2ème éd. Gauthier-Villars, Paris.
Li, T.-Y. (1976) Finite approximation for the Frobenius–Perron operator. A solution to Ulam's conjecture. *J. Approx. Theory*, **17**, 177–86.
Li, T.-Y. & Yorke, J. A. (1978) Ergodic transformations from an interval into itself. *Trans. Amer. Math. Soc.* **235**, 183–92.
Liedl, R., Reich, L. & Targonski, Gy. (Eds.) (1985) Iteration Theory and Its Functional Equations (Proceedings, Schloss Hofen, 1984). *Lecture Notes in Math.*, Vol. 1163. Springer, Berlin.
Lin, M. (1971) Mixing for Markov operators. *Z. Wahrsch. Verw. Gebiete*, **19**, 231–42.
Lin, M. (1974) On quasi-compact Markov operators. *Ann. Probab.* **2**, 464–75.
Lin, M. (1978) Quasi-compactness and uniform ergodicity of positive operators. *Israel J. Math.* **29**, 309–11.
Lisek, B. (1982) A method for solving a class of recursive stochastic equations. *Z. Wahrsch. Verw. Gebiete*, **60**, 151–61.
Loève, M. (1977) *Probability Theory*, I & II, 4th edn. Springer, New York.
Macchi, C. (1976) Modèles d'apprentissage. *Computer Oriented Learning Processes*, pp. 25–67. Noordhoff, Leyden.
Madras, N. (1986) A process in a randomly fluctuating environment. *Ann. Probab.* **14**, 119–35.
Maibaum, G. & Mühlmann, P. (1975) Ueber mehrdimensionale stochastische Ketten. *Math. Nachr.* **65**, 247–57.
Marliss, G. S. & McGregor, J. R. (1968) Symmetric limiting distributions of response probabilities. *Psychometrika*, **33**, 383–5.
Marliss, G. S. & McGregor, J. R. (1971) The construction of limiting distributions of response probabilities. *J. Appl. Probab.* **8**, 757–66.
May, R. M. (1976) Simple mathematical models with very complicated dynamics. *Nature*, **261**, 459–67.

McCabe, B. J. (1969) The asymptotic behaviour of a certain Markov chain. *Ann. Math. Statist.* **40**, 665–6.
McGregor, J. R. & Hui, Y. Y. (1962) Limiting distributions associated with certain stochastic learning models. *Ann. Math. Statist.* **33**, 1281–5.
McGregor, J. R. & Zidek, J. V. (1965a) Limiting distributions of response probabilities. *Ann. Math. Statist.* **36**, 706–7.
McGregor, J. R. & Zidek, J. V. (1965b) A sequence of limiting distributions of response probabilities. *Psychometrika*, **30**, 491–7.
Mehta, M. L. (1979) On f-representation of real numbers. *Acta Math. Acad. Sci. Hungar.* **34**, 91–100.
Mihoc, G. & Ciucu, G. (1961) Some limit properties in the theory of chains with complete connections. *An. Univ. Bucureşti Ser. Şti. Natur. Mat. Fiz.* **10**, no. 29, 225–31. (Romanian).
Mihoc, G. & Ciucu, G. (1971) Sur l'ergodicité des chaînes à liaisons complètes. *Math. Balkanica*, **1**, 164–70.
Mihoc, G. & Ciucu, G. (1973) Sur la loi normale pour les chaînes à liaisons complètes. *Proc. Fourth Conf. Probability Theory (Braşov, 1971)*, pp. 169–71. Ed. Academiei R. S. România, Bucharest.
Mihoc, G. & Craiu, Mariana (1972) *Statistical Inference for Dependent Variables.* Ed. Academiei R. S. România, Bucharest. (Romanian)
Misevičius, G. (1972) Estimate of the remainder term in the central limit theorem for sums of the form $\sum \varphi(2^j t)$. *Litovsk. Mat. Sb.* **12**, no. 2, 403–7. (Russian)
Misevičius, G. (1981) Estimate of the remainder term in the limit theorem for the denominators of continued fractions. *Lithuanian Math. J.* **21**, 245–53.
Mokkadem, A. (1987) Sur un modèle autorégressif non linéaire, ergodicité et ergodicité géométrique. *J. Time. Ser. Anal.* **8**, 195–204.
Moskvin, D. A. & Postnikov, A. G. (1978) A local limit theorem for the distribution of fractional parts of an exponential function. *Theory Probab. Appl.* **23**, 521–8.
Moy, S.-T. C. (1965) λ-continuous Markov chains. I, II. *Trans. Amer. Math. Soc.* **117**, 68–91; **120**, 83–107.
Mukherjea, A. & Pothoven, K. (1984) *Real and Functional Analysis. Part A: Real Analysis*, 2nd edn. Plenum Press, New York.
Mukherjea, A. & Pothoven, K. (1986) *Real and Functional Analysis. Part B: Functional Analysis*, 2nd edn. Plenum Press, New York.
Nagaev, S. V. (1957) Some limit theorems for stationary Markov chains. *Theory Probab. Appl.* **2**, 378–406.
Nagaev, S. V. (1961) More exact statements of limit theorems for homogeneous Markov chains. *Theory Probab. Appl.* **6**, 62–81.
Nakada, H. (1981) Metrical theory for a class of continued fraction transformations and their natural extensions. *Tokyo J. Math.* **4**, 399–426.
Natanson, I. P. (1961) *Theorie der Funktionen einer reellen Veränderlichen*, 2. Aufl. Akademie-Verlag, Berlin.
Nawrotzki, K. (1981/82) Discrete open systems or Markov chains in a random environment. I–II. *Elektron. Informationsverarb. Kybernet.* **17**, 569–99; **18**, 83–98.
Neveu, J. (1964) *Bases mathématiques du calcul des probabilités.* Masson, Paris.
Norman, M. F. (1968) Some convergence theorems for stochastic learning models with distance diminishing operators. *J. Math. Psych.* **5**, 61–101.
Norman, M. F. (1972) *Markov Processes and Learning Models.* Academic Press, New York.
Norman, M. F. (1974) Markovian learning processes. *SIAM Rev.* **16**, 143–62.
Norman, M. F. (1975) An ergodic theorem for evolution in a random environment. *J. Appl. Probab.* **12**, 661–72.
Nowakowska, Maria (1981) Influence of others on one's decision: a model of group choice. *Math. Soc. Sci.* **2**, 65–73.

Nummelin, E. (1984) *General Irreducible Markov Chains and Non-negative Operators.* Cambridge Univ. Press, Cambridge.

O'Brien, G. L. (1974) Limit theorems for sums of chain-dependent processes. *J. Appl. Probab.* **11**, 582–7.

O'Brien, G. L. (1975) The comparison method for stochastic processes. *Ann. Probab.* **3**, 80–8.

Onicescu, O. (1954) Continuous-time chains with complete connections. *Rev. Univ. 'C. I. Parhon' Politehn. Bucureşti Ser. Şti. Natur.* **3**, no. 4–5, 73–85. (Romanian)

Onicescu, O. & Botez, M. C. (1985) *Uncertainty and Economic Modelling: Information Econometrics.* Ed. Stiinţifică şi Enciclopedică, Bucharest. (Romanian)

Onicescu, O. & Guiaşu, S. (1965) Finite abstract random automata. *Z. Wahrsch. Verw. Gebiete*, **3**, 279–85.

Onicescu, O. & Mihoc, G. (1935a) Sur les chaînes de variables statistiques. *C.R. Acad. Sci. Paris*, **200**, 511–2.

Onicescu, O. & Mihoc, G. (1935b) Sur les chaînes de variables statistiques. *Bull. Sci. Math.* (2), **59**, 174–92.

Onicescu, O. & Mihoc, G. (1936a) Sopra le leggi-limite delle probabilità. *Giorn. Ist. Ital. Attuari*, **7**, 54–69.

Onicescu, O. & Mihoc, G. (1936b) Sur les chaînes statistiques. *C.R. Acad. Sci. Paris*, **202**, 2031–3.

Onicescu, O. & Mihoc, G. (1937) La dépendance statistique. Chaînes et familles de chaînes discontinues. *Act. Sci. Ind.* 503. Hermann, Paris.

Onicescu, O. & Mihoc, G. (1939) Sur les sommes de variables enchaînées. Second mémoire. *Bull. Math. Soc. Roumaine Sci.* **41**, no. 1, 99–116.

Onicescu, O. & Mihoc, G. (1940a) Propriétés asymptotiques des chaînes de Markoff étudiées à l'aide de la fonction caractéristique. *Mathematica (Cluj)*, **16**, 13–43.

Onicescu, O. & Mihoc, G. (1940b) Comportement asymptotique des chaînes à liaisons complètes. *Disquisit. Math. Phys.* **1**, 61–2.

Onicescu, O. & Mihoc, G. (1943) Les chaînes de variables aléatoires. Problèmes asymptotiques. *Etudes et recherches*, XIV. Acad. Roumaine, Bucarest.

Onicescu, O., Oprişan, G. & Popescu, G. (1983) Renewal processes with complete connections. *Rev. Roumaine Math. Pures Appl.* **28**, 985–98.

Oodaira, H. & Yoshihara, K. (1971a) The law of the iterated logarithm for stationary processes satisfying mixing conditions. *Kodai Math. Sem. Rep.* **23**, 311–34.

Oodaira, H. & Yoshihara, K. (1971b) Note on the law of the iterated logarithm for stationary processes satisfying mixing conditions. *Kodai Math. Sem. Rep.* **23**, 335–42.

Oprişan, G. & Popescu, G. (1983) Renewal generalized processes with complete connections. *Studies in Probability and Related Topics: Papers in Honour of Octav Onicescu on His 90th Birthday*, pp. 369–74. Editrice Nagard, Roma.

Orey, S. (1971) *Lecture Notes on Limit Theorems for Markov Chain Transition Probabilities.* Van Nostrand, New York.

Parry, W. (1964) Representations for real numbers. *Acta Math. Acad. Sci. Hungar.* **15**, 95–105.

Parthasarathy, K. R. (1967) *Probability Measures on Metric Spaces.* Academic Press, New York.

Peitgen, H.-O. & Richter, P. H. (1986) *The Beauty of Fractals: Images of Complex Dynamical Systems.* Springer, Berlin.

Peligrad, Magda (1983) A note on two measures of dependence and mixing sequences. *Adv. in Appl. Probab.* **15**, 461–4.

Peligrad, Magda (1986) Recent advances in the central limit theorem and its weak invariance principle for mixing sequences of random variables (A survey). In: Eberlein & Taqqu (Eds.) (1986), pp. 193–223.

Pelikan, S. (1984) Invariant densities for random maps of the interval. *Trans. Amer. Math. Soc.* **281**, 813–25.

Perennou, G. (1976) Reconnaissance des formes. Apprentissage et approximation stochastique. *Ann. Sci. Univ. Clermont No.* 58–*Math.*, 12ᵉ fasc., 30–3.
Philipp, W. (1970) Some metrical theorems in number theory II. *Duke Math. J.* **37**, 447–58; Errata *ibid.* 788.
Philipp, W. (1971) Mixing sequences of random variables and probabilistic number theory. *Mem. Amer. Math. Soc.* **114**.
Philipp, W. (1986) Invariance principles for independent and weakly dependent random variables. In: Eberlein & Taqqu (Eds.) (1986), pp. 225–68.
Philipp, W. & Stackelberg, O. P. (1969) Zwei Grenzwertsätze für Kettenbrüche. *Math. Ann.* **181**, 152–6.
Philipp, W. & Stout, W. (1975) Almost sure invariance principles for partial sums of weakly dependent random variables. *Mem. Amer. Math. Soc.* **161**.
Pianigiani, G. (1980) First return map and invariant measures. *Israel J. Math.* **35**, 32–48.
Pianigiani, G. (1981) Existence of invariant measures for piecewise continuous transformations. *Ann. Polon. Math.* **40**, 39–45.
Pincus, S. (1983) A class of Bernoulli random matrices with continuous singular stationary measures. *Ann. Probab.* **11**, 931–8.
Pollicott, M. (1984) A complex Ruelle–Perron–Frobenius theorem and two counterexamples. *Ergodic Theory Dynamical Systems*, **4**, 135–46.
Pólya, G. (1930) Sur quelques points de la théorie des probabilités. *Ann. Inst. H. Poincaré*, **1**, 117–60.
Popescu, G. (1976) A functional central limit theorem for a class of Markov chains. *Rev. Roumaine Math. Pures Appl.* **21**, 737–50.
Popescu, G. (1977) Functional limit theorems for random systems with complete connections. *Proc. Fifth Conf. Probability Theory* (*Braşov, 1974*), pp. 261–75. Ed. Academiei R. S. România, Bucharest.
Popescu, G. (1978) Asymptotic behaviour of random systems with complete connections. I, II. *Stud. Cerc. Mat.* **30**, 37–68, 181–215. (Romanian)
Popescu, Ileana (1978) Computer generation of random systems with complete connections. *Stud. Cerc. Mat.* **30**, 69–74. (Romanian)
Popescu, Ileana (1981) Calcul des intégrales multiples par la méthode de Monte Carlo. *Rev. Roumaine Math. Pures Appl.* **26**, 619–29.
Postelnicu, Viorica (1973) Renewal theory for a class of processes with dependent inter-renewal times. *Proc. Fourth Conf. Probability Theory* (*Braşov, 1971*), pp. 185–93. Ed. Academiei R. S. România. Bucharest.
Preston, C. (1983) Iterates of Maps on an Interval. *Lecture Notes in Math.*, Vol. 999. Springer, Berlin.
Pruscha, H. (1975) *Die statistische Analyse von ergodischen Ketten mit vollständigen Bindungen.* Dissertation, Univ. München, München.
Pruscha, H. (1976) Maximum likelihood estimation in linear learning models. *Biometrika*, **63**, 537–42.
Pruscha, H. (1983) Learning models with continuous time parameter and multivariate point processes. *J. Appl. Probab.* **20**, 884–90.
Pruscha, H. (1986) Statistical analysis of linear OM processes. *Rev. Roumaine Math. Pures Appl.* **31**, 707–22.
Pruscha, H. & Maurus, M. (1979) Analysis of primate communication by means of multiresponse linear learning model. *Rev. Roumaine Math. Pures Appl.* **24**, 1371–83.
Pruscha, H. & Theodorescu, R. (1977) Functions of event variables of a random system with complete connections. *J. Multivariate Anal.* **7**, 336–62.
Pruscha, H. & Theodorescu, R. (1978) Recurrence and transience for discrete parameter stochastic processes. *Trans. Eighth Prague Conf. Information Theory, Statist. Decision Functions, Random Processes* (*Prague, 1978*), Vol. B, pp. 97–105. Reidel, Dordrecht.

Pruscha, H. & Theodorescu, R. (1981) On a non-Markovian model with linear transition rule. *Colloq. Math.* **44**, 165–73.

Pruscha, H. & Theodorescu, R. (1983) Stationary probability distributions for multiresponse linear learning models. *J. Multivariate Anal.* **13**, 109–17.

Pugh, E. L. (1966) A gradient technique of adaptive Monte Carlo. *SIAM Rev.* **8**, 346–55.

Raikov, D. A. (1936) On some arithmetical properties of summable functions. *Mat. Sb.* **1(43)**, 377–84. (Russian)

Răuţu, G. & Zbăganu, G. (1989) Some Banach algebras of functions of bounded variation. *Stud. Cerc. Mat.* **41**, 513–9.

Rechard, O. W. (1956) Invariant measures for many-one transformations. *Duke Math. J.* **23**, 477–88.

Rényi, A. (1957) Representations for real numbers and their ergodic properties. *Acta Math. Acad. Sci. Hungar.* **8**, 477–93.

Rényi, A. (1958) Probabilistic methods in number theory. *Selected Papers of Alfréd Rényi*, Vol. 2, pp. 215–72. Akadémiai Kiadó, Budapest, 1976.

Revuz, D. (1975) *Markov Chains.* North-Holland, Amsterdam.

Reznik, M. K. (1968) The law of the iterated logarithm for some classes of stationary processes. *Theory Probab. Appl.* **13**, 606–21.

Rhenius, D. (1971) *Markoff'sche Entscheidungsmodelle mit unvollständiger Information und Anwendungen in der Lerntheorie.* Dissertation, Univ. Hamburg, Hamburg.

Rhenius, D. (1974) Incomplete information in Markovian decision models. *Ann. Statist.* **2**, 1327–34.

Rickert, U. & Theodorescu, R. (1983) Stationary probabilities and asymptotic behaviour of a non-Markovian model with linear transition rule. *Biom. J.* **25**, 85–91.

Riéder, U. (1975) Bayesian dynamic programming. *Adv. in Appl. Probab.* **7**, 330–48.

Rieger, G. J. (1977) Die metrische Theorie der Kettenbrüche seit Gauss. *Abh. Braunschweig. Wiss. Ges.* **27**, 103–17.

Rieger, G. J. (1979) Mischung und Ergodizität bei Kettenbrüchen nach nächsten Ganzen. *J. Reine Angew. Math.* **310**, 171–81.

Robbins, H. & Monro, S. (1951) A stochastic approximation model. *Ann. Math. Statist.* **22**, 400–7.

Rockett, A. M. (1980) The metrical theory of continued fractions to the nearer integer. *Acta Arith.* **38**, 97–103.

Rosenblatt, M. (1964a) Equicontinuous Markov operators. *Teor. Verojatnost. i Primenen.* **9**, 205–22.

Rosenblatt, M. (1964b) Almost periodic transition operators acting on the continuous functions on a compact space. *J. Math. Mech.* **13**, 837–47.

Rosenblatt, M. (1967) Transition probability operators. *Proc. Fifth Berkeley Sympos. Math. Statist. Probab.*, Vol. II, Part 2, pp. 473–83. Univ. of California Press, Berkeley, California

Rousseau-Egèle, J. (1983) Un théorème de la limite locale pour une classe de transformations dilatantes et monotones par morceaux. *Ann. Probab.* **11**, 772–88.

Royden, H. L. (1968) *Real Analysis*, 2nd edn. Macmillan, New York.

Ruelle, D. (1980) Strange attractors. *Math. Intelligencer*, **2**, 126–137.

Rychlik, M. (1983) Bounded variation and invariant measures. *Studia Math.* **76**, 69–80.

Sacksteder, R. (1972) On convergence to invariant measures. Mimeographed notes.

Samur, J. D. (1987) On some limit theorems for continued fractions. *Notas de Matemática*, No. 47. Univ. Nacional de la Plata, Dep. Mat., La Plata.

Sanghvi, A. P. (1980) Convergence of successive approximations to a stochastic fixed point. *J. Appl. Probab.* **17**, 297–9.

Schmetterer, L. (1979) From stochastic approximation to the stochastic theory of optimization. *Ber. Math.-Statist. Sekt. Forsch. Graz*, No. 124–32, Ber. No. 127, 40pp.

Schweiger, F. (1981) Ergodic properties of fibered systems. *Proc. Sixth Conf. Probability Theory (Braşov, 1979)*, pp. 221–8. Ed. Academiei R. S. România, Bucharest.

Scott, D. J. (1973) Central limit theorems for martingales and for processes with stationary increments, using a Skorokhod representation approach. *Adv. in Appl. Probab.* **5**, 119–37.

Šidák, Z. (1962). Représentations des probabilités de transition dans les chaînes à liaisons complètes. *Časopis Pěst. Mat.* **87**, 389–98.

Šidák, Z. (1967) Classification of Markov chains with a general state space. *Trans. Fourth Prague Conf. Information Theory, Statist. Decision Functions, Random Processes (Prague, 1965)*, pp. 547–71. Academia, Prague.

Šidák, Z. (1977) Miscellaneous topics in Markov chains with a general state space. *Trans. Seventh Prague Conf. Information Theory, Statist. Decision Functions, Random Processes & Eighth European Meeting of Statisticians (Tech. Univ. Prague, Prague, 1974)*, Vol. A, pp. 531–44. Reidel, Dordrecht.

Siegrist, K. (1981) Random evolution processes with feedback. *Trans. Amer. Math. Soc.* **265**, 375–92.

Širjaev, A. N. (1964) On the theory of decision functions and control by an observation process with incomplete data. *Trans. Third Prague Conf. Information Theory, Statist. Decision Functions, Random Processes (Liblice, 1962)*, pp. 557–81. Czechoslovak Acad. Sci., Prague. (Russian)

Širjaev, A. N. (1967) Some new results in the theory of controlled random processes. *Trans. Fourth Prague Conf. Information Theory, Statist. Decision Functions, Random Processes (Prague, 1965)*, pp. 131–203. Academia, Prague. (Russian; English translation in: *Selected Translations in Mathematical Statistics and Probability*, Vol. 8, pp. 49–130. Amer. Math. Soc., Providence, R. I., 1970.)

Slater, N. B. (1967) Gaps and steps for the sequence $n\theta$ mod 1. *Proc. Cambridge Philos. Soc.* **63**, 1115–23.

Smith, W. L. & Wilkinson, W. (1969) On branching processes in random environments. *Ann. Math. Statist.* **40**, 814–27.

Solomon, F. (1975) Random walks in a random environment. *Ann. Probab.* **3**, 1–31.

Sparrow, C. (1982) *The Lorenz Equations: Bifurcations, Chaos, and Strange Attractors*. Springer, Berlin.

Spătaru, A. (1974) Determining the measures under which the digits in the continued fraction and D-adic expansions are i.i.d. random variables. *Technical Report*, Centre of Mathematical Statistics, Bucharest. (Romanian)

Spătaru, A. (1977a) Some f-expansions producing stochastically independent digits. *Sankhya Ser. A*, **39**, 160–9.

Spătaru, A. (1977b) (φ, f)-expansions of real numbers. *Rev. Roumaine Math. Pures Appl.* **22**, 1459–69.

Spătaru, A. (1978a) Dichotomies for probability measures induced by $(f_1, f_2, \ldots)$-expansions. *Z. Wahrsch. Verw. Gebiete*, **45**, 337–52.

Spătaru, A. (1978b) Stochastic independence of the digits of an f-expansion. *Stud. Cerc. Mat.* **30**, 75–117. (Romanian)

Spătaru, A. (1987) Equivalent expansions for real numbers. *Stud. Cerc. Mat.* **39**, 448–51.

Sragovič, V. G. (1976) *Theory of Adaptive Systems*. Nauka, Moscow. (Russian)

Sragovič, V. G. (1981) *Adaptive Control*. Nauka, Moscow. (Russian)

Stackelberg, O. P. (1966) On the law of the iterated logarithm for continued fractions. *Duke Math. J.* **33**, 801–19.

Stadje, W. (1985) Bemerkung zu einem Satz von Akcoglu und Krengel. *Studia Math.* **81**, 307–10.

Strassen, V. (1964) An invariance principle for the law of the iterated logarithm. *Z. Wahrsch. Verw. Gebiete*, **3**, 211–26.

Szüsz, P. (1961) Ueber einen Kusminschen Satz. *Acta Math. Acad. Sci. Hungar.* **12**, 447–53.

Szüsz, P. (1971) On the law of iterated logarithm for continued fractions. *Z. Wahrsch. Verw. Gebiete*, **19**, 167–71.
Takahashi, Y. (1969) Markov chains with random transition matrices. *Kodai Math. Sem. Rep.* **21**, 426–47.
Thaler, M. (1980) Estimates of the invariant densities of endomorphisms with indifferent fixed points. *Israel J. Math.* **37**, 303–14.
Thaler, M. (1983) Transformations on [0, 1] with infinite invariant measures. *Israel J. Math.* **46**, 67–96.
Theiler, G. (1967) Uniformly ergodic stochastic processes with discrete parameter. *Rev. Roumaine Math. Pures Appl.* **12**, 191–223.
Theodorescu, R. & Tweedie, R. L. (1983) Solidarity properties and a Doeblin decomposition for a class of non-Markovian stochastic processes. *Metrika*, **30**, 37–47.
Thompson, J. M. T. & Stewart, H. B. (1986) *Nonlinear Dynamics and Chaos: Geometrical Methods for Engineers and Scientists.* Wiley, Chichester.
Tomlinson, M. (1971) New automatic equalizer employing modulo arithmetic. *Electron. Lett.* **7**, 138–9.
Tran Vinh-Hien (1963) A central limit theorem for stationary processes arising from number-theoretic endomorphisms. *Vestnik Moskov. Univ. Ser. I Mat. Meh.* No. 5, 28–34. (Russian)
Tschuprow, A. A. (1923) On the mathematical expectation of the moments of frequency distributions in the case of correlated observations. *Metron*, **2**, 461–93, 646–83.
Tsetlin, M. L. (1973) *Automaton Theory and Modelling of Biological Systems.* Academic Press, New York.
Tuominen, P. & Tweedie, R. L. (1979) Markov chains with continuous components. *Proc. London Math. Soc.* (*3*), **38**, 89–114.
Usačev, E. S. (1964) On limit distributions in stochastic learning models. *Soviet Physics Dokl.* **9**, 1045–6.
Usačev, E. S. (1971) On asymptotic properties of random processes occurring in learning models. *Studies in the Theory of Self-Adaptive Systems*, pp. 153–206. Vyčisl. Centr, Akad. Nauk SSSR, Moscow. (Russian)
Vervaat, W. (1979) On a stochastic difference equation and a representation of non-negative infinitely divisible random variables. *Adv. in Appl. Probab.* **11**, 750–83.
Wagner, G. (1979) The ergodic behaviour of piecewise monotonic transformations. *Z. Wahrsch. Verw. Gebiete*, **46**, 317–24.
Walters, P. (1978a) Equilibrium states for β-transformations and related transformations. *Math. Z.* **159**, 65–88.
Walters, P. (1978b) Invariant measures and equilibrium states for some mappings which expand distances. *Trans. Amer. Math. Soc.* **236**, 121–53.
Walters, P. (1982) *An Introduction to Ergodic Theory.* Springer, New York.
Wasan, M. T. (1969) *Stochastic Approximation.* Cambridge Univ. Press, Cambridge.
Waterman, M. S. (1975) Remarks on invariant measures for number theoretic transformations. *Monatsh. Math.* **79**, 157–63.
Weissner, E. W. (1971) Multitype branching processes in random environments. *J. Appl. Probab.* **8**, 17–31.
Whitley, D. (1982) Discrete dynamical systems in dimensions one and two. *Bull. London Math. Soc.* **15**, 177–217.
Wierenga, B. (1974) *An Investigation of Brand Choice Processes.* Rotterdam Univ. Press, Rotterdam.
Williams, D. (1974) A study of a diffusion process motivated by the Sieve of Erathosthenes. *Bull. London Math. Soc.* **6**, 155–64.
Wirsing, E. (1971) Ueber den Satz von Gauss-Kusmin. In: 'Zahlentheorie (Tagung,

Math. Forschungsinst., Oberwolfach, 1970)'. *Ber. Math. Forschungsinst., Oberwolfach*, Heft 5, pp. 229–31. Bibliographisches Inst., Mannheim.

Wirsing, E. (1974) On the theorem of Gauss–Kusmin–Lévy and a Frobenius-type theorem for function spaces. *Acta Arith.* **24**, 507–28.

Wolff, H. (1973) Grenzwertsätze für einen allgemeinen linearen stochastischen Lernprozess. *Elektron. Informationsverarb. Kybernet.* **9**, 107–13.

Wolff, M. (1976) Ueber Produkte abhängiger zufälliger Veränderlicher mit Werten in einer kompakten Halbgruppe. *Z. Wahrsch. Verw. Gebiete*, **35**, 253–64.

Wolfowitz, J. (1954) Generalization of the theorem of Glivenko-Cantelli. *Ann. Math. Statist.* **25**, 131–8.

Wong, S. (1978) Some metric properties of piecewise monotonic mappings of the unit interval. *Trans. Amer. Math. Soc.* **246**, 493–500.

Wong, S. (1979a) Hölder continuous derivatives and ergodic theory. *J. London Math. Soc.* (2), **22**, 506–20.

Wong, S. (1979b) A central limit theorem for piecewise monotonic mappings of the unit interval. *Ann. Probab.* **7**, 500–14.

Wrench, J. W., Jr. (1960) Further evaluation of Khintchine's constant. *Math. Comp.* **14**, 370–1.

Wrench, J. W., Jr. & Shanks, D. (1966) Questions concerning Khintchine's constant and the efficient computation of regular continued fractions. *Math. Comp.* **20**, 444–8; Corrigenda *ibid.* **21** (1967), 130.

Yokoyama, R. (1980) A note on weak convergence of empirical processes for φ-mixing random variables. *Yokohama Math. J.* **28**, 37–40.

Yosida, K. & Kakutani, S. (1941) Operator-theoretical treatment of Markoff's process and mean ergodic theorem. *Ann. of Math.* (2), **42**, 188–228.

Zhdanok, A. I. (1982) Iterative processes as Markov chains. *Latv. Mat. Ezhegodnik*, No. 26, 153–64. (Russian)

Zidăroiu, C. (1975) *Discrete Dynamical Programming.* Ed. Tehnică, Bucharest. (Romanian)

Zidăroiu, C. (1976) Undiscounted random systems with complete connections. *Rev. Roumaine Math. Pures Appl.* **21**, 487–94.

Zidăroiu, C. (1978) Stability properties for the functional equation occurring in the study of random decision systems with complete connections. *Rev. Roumaine Math. Pures Appl.* **23**, 821–7.

Zidăroiu, C. (1986) New stability results for random decision systems with complete connections. *Rev. Roumaine Math. Pures Appl.* **31**, 415–21.

Zidăroiu, C. (1987) Differential stability for random decision systems with complete connections. *An. Univ. Bucuresti Mat.* **36**, 92–8.

Index